D1154449

Nonlinear Optimization 1981

Nonlinear Optimization 1981

Edited by

M.J.D. POWELL

Department of Applied Mathematics and Theoretical Physics, University of Cambridge, Silver Street, Cambridge, England

1982

ACADEMIC PRESS

A Subsidiary of Harcourt Brace Jovanovich, Publishers

London New York
Paris San Diego San Francisco
São Paulo Sydney Tokyo Toronto

Published in cooperation with NATO Scientific Affairs Division

Proceedings of the NATO Advanced Research Institute
held at Cambridge in July 1981 which was sponsored by
the Special Programme Panel on Systems Science of the NATO
Science Committee and the Mathematical Programming Society

ACADEMIC PRESS INC. (LONDON) LTD.
24/28 Oval Road
London NW1

United States Edition published by
ACADEMIC PRESS INC.
111 Fifth Avenue
New York, New York 10003

Copyright © 1982 by
ACADEMIC PRESS INC. (LONDON) LTD.

All Rights Reserved

No part of this book may be reproduced in any form by photostat, microfilm, or
any other means, without written permission from the publishers

British Library Cataloguing in Publication Data
Nonlinear optimization 1981.—(NATO conference series)
1. Mathematical optimization—Congresses
2. Nonlinear programming—Congresses
I. Powell, M.J.D.
515 QA402.5

ISBN 0-12-563860-4

LCCCN 81-71583

Printed in Great Britain by
Whitstable Litho Ltd., Whitstable, Kent

Names and Addresses of Participants

Dr T.E. Baker, Exxon Corporation, PO Box 153, Florham Park, New Jersey 07932, USA.

Dr M.C. Bartholomew-Biggs, School of Information Sciences, Hatfield Polytechnic, Hatfield AL10 9AB, England.

Professor E.M.L. Beale, Scicon Computer Services Ltd., Brick Close, Kiln Farm, Milton Keynes MK11 3EJ, England.

Professor D.P. Bertsekas, Massachusetts Institute of Technology, Room 35—209, Cambridge, Massachusetts 02139, USA.

Dr P.T. Boggs, U.S. Army Research Office, PO Box 12211, Research Triangle Park, North Carolina 27709, USA.

Dr C.A. Botsaris, 26 Archimides Street, Pagrati, Athens 502, Greece.

Dr A. Buckley, Department of Mathematics, Concordia University, 7141 Sherbrooke St W., Montreal, Canada H4B 1R6.

Dr J.C.P. Bus, Stichting Mathematisch Centrum, PO Box 4079, 1009 AB Amsterdam, The Netherlands.

Dr R.H. Byrd, Department of Computer Science, University of Colorado, Boulder, Colorado 80309, USA.

Dr T.F. Coleman, Applied Mathematics Division, Argonne National Laboratory, Argonne, Illinois 60439, USA.

Dr A.R. Conn, Department of Computer Science, Faculty of Mathematics, University of Waterloo, Waterloo, Ontario N2L 3G1, Canada.

Dr I.D. Coope, Department of Mathematics, University of Canterbury, Christchurch 1, New Zealand.

Professor W.C. Davidon, Department of Physics, Haverford College, Haverford, Pennsylvania 19041, USA.

Professor R.S. Dembo, Yale School of Organisation and Management, Box 1A, New Haven, Connecticut 06520, USA.

Professor J.E. Dennis, Department of Mathematical Sciences, Rice University, Houston, Texas 77001, USA.

Dr L.C.W. Dixon, Numerical Optimization Centre, The Hatfield Polytechnic, Hatfield AL10 9AB, England.

Professor A. Drud, Development Research Center, The World Bank, 1818 H Street NW, Washington, DC 20433, USA.

Dr L.F. Escudero, IBM Scientific Centre, Paseo de la Castellana, 4, Madrid 1, Spain.

Dr J. Ch. Fiorot, Laboratoire d'Informatique, Université des Sciences et Techniques de Lille, 59655 Villeneuve d'Ascq, France.

Dr R. Fletcher, Department of Mathematics, The University, Dundee DD1 4HN, Scotland.

Dr D.M. Gay, Bell Laboratories, 600 Mountain Avenue, Murray Hill, New Jersey 07974, USA.

Dr P.E. Gill, Department of Operations Research, Stanford University, Stanford, California 94305, USA.

Professor D. Goldfarb, Department of Computer Sciences, The City College of New York, New York, New York 10031, USA.

Dr A. Griewank, Department of Applied Mathematics and Theoretical Physics, University of Cambridge, Silver Street, Cambridge CB3 9EW, England.

Mr M.M. Gutterman, Mail Code 1801, Standard Oil (Indiana), PO Box 5910A, Chicago, Illinois 60680, USA.

Professor S-P. Han, Department of Mathematics, University of Illinois, Urbana, Illinois 61801, USA.

Professor D.M. Himmelblau, Room 211, EPS, Department of Chemical Engineering, The University of Texas, Austin, Texas 78712, USA.

Dr R.H.F. Jackson, Center for Applied Mathematics, National Bureau of Standards, Washington, DC 20234, USA.

Dr K. Jittorntrum, Division of Computer Applications, Asian Institute of Technology, PO Box 2754, Bangkok, Thailand.

Professor J.S. Kowalik, Systems and Computing, Washington State University, Pullman, Washington 99164, USA.

Dr D. Kraft, German Aerospace Research Establishment, Muenchenerstr. 20, D-8031 Wessling, West Germany.

Dr J.L. Kreuser, Room N-918, The World Bank, 1818 H Street NW, Washington, DC 20433, USA.

Professor L.S. Lasdon, Department of General Business, School of Business Adminstration, University of Texas, Austin, Texas 78712, USA.

Dr C. Lemaréchal, INRIA, B.P. 105, 78153 Le Chesnay, France.

Professor F.A. Lootsma, Department of Mathematics and Informatics, Delft University of Technology, Julianalaan 132, 2628 BL Delft, The Netherlands.

Professor O.L. Mangasarian, Computer Sciences Department, University of Wisconsin, 1210 West Dayton Street, Madison, Wisconsin 53706, USA.

Professor L. McLinden, Department of Mathematics, University of Illinois at Urbana-Champaign, 1409 West Green Street, Urbana, Illinois 61801, USA.

Professor R.R. Meyer, Computer Sciences Department, University of Wisconsin, 1210 West Dayton Street, Madison, Wisconsin 53706, USA.

Professor R. Mifflin, Department of Mathematics, Washington State University, Pullman, Washington 99164, USA.

Dr J.J. Moré, Applied Mathematics Division, Argonne National Laboratory, Argonne, Illinois 60439, USA.

Dr M.L. Overton, Courant Institute, New York University, 251 Mercer Street, New York, New York 10012, USA.

Professor E. Polak, Department of Electrical Engineering and Computer Sciences, University of California, Berkeley, California 94720, USA.

Professor M.J.D. Powell, Department of Applied Mathematics and

Theoretical Physics, University of Cambridge, Silver Street, Cambridge CB3 9EW, England.

Dr K.M. Ragsdell, School of Mechanical Engineering, Purdue University, West Lafayette, Indiana 47906, USA.

Professor Dr M. Rijckaert, Katholieke Universiteit Leuven, de Croylaan 2, 3030 Heverlee, Belgium.

Professor S.M. Robinson, Department of Industrial Engineering, University of Wisconsin, 1513 University Avenue, Madison, Wisconsin 53706, USA.

Professor J.B. Rosen, Computer Science Department, University of Minnesota, 207 Church Street S.E., Minneapolis, Minnesota 55455, USA.

Professor R.W.H. Sargent, Department of Chemical Engineering and Chemical Technology, Imperial College of Science and Technology, Prince Consort Road, London SW7 2AZ, England.

Dr K. Schittkowski, Institut für Angewandte Mathematik und Statistik, Universität Würzburg, Am Hubland, D-87 Würzburg, West Germany.

Professor R.B. Schnabel, Computer Science Department, University of Colorado, Boulder, Colorado 80309, USA.

Professor D.F. Shanno, Department of Management Information Systems, College of Business and Public Administration, University of Arizona, Tucson, Arizona 85721, USA.

Dr D. Sorensen, Applied Mathematics Division, Argonne National Laboratory, Argonne, Illinois 60439, USA.

Dr T. Steihaug, Department of Mathematical Sciences, Rice University, Houston, Texas 77001, USA.

Dr K. Tanabe, The Institute of Statistical Mathematics, 4-6-7 Minami-azabu, Minato-ku, Tokyo, Japan 106.

Professor R.A. Tapia, Department of Mathematical Sciences, Rice University, Houston, Texas 77001, USA.

Professor M.J. Todd, School of Operations Research, Cornell University, Ithaca, New York 14853, USA.

Professor Ph. Toint, Department of Mathematics, Facultés Universitaires de Namur, 61 Rue de Bruxelles, B-5000 Namur, Belgium.

Dr P. Wolfe, Mathematical Sciences Department, IBM Research Center, PO Box 218, Yorktown Heights, New York 10598, USA.

Dr M.H. Wright, Department of Operations Research, Stanford University, Stanford, California 94305, USA.

Professor Xi Shao-lin, Computing Center, Academia Sinica, PO Box 2719, Beijing, China.

Professor F. Zirilli, Istituto Matematica, Universita di Camerino, 62032 Camerino, Italy.

Preface

Sixty one selected participants attended the NATO Advanced Research Institute on "Nonlinear Optimization" that was held at Trinity Hall, Cambridge, from July 13—24, 1981. The meeting itself was highly interesting and stimulating, but the success of the meeting has to be judged by its value to all researchers in optimization, for the purpose of Advanced Research Institutes is for experts to consider the state of development of a subject, and to publish their findings for the benefit of a wider audience. Therefore these proceedings are intended to give the state of the art in nonlinear optimization, not only by mentioning recent and current work, but also by offering opinions on future directions of research. Because of the experience of the contributors, many of the ideas that are presented should prove to be of major importance, and it is hoped also that the reader will enjoy the given perspective on a highly active area of research.

A usual way of obtaining authoritative opinions during Advanced Research Institutes is for small working groups to give careful consideration to their subjects, and for each working group to write a report on its conclusions. This is an excellent way to proceed if new research requires massive capital investment, but nonlinear optimization is not like that; instead most of the advances have come from the wits of individuals, and in many cases major progress has not been helped by fashionable trends. Therefore these proceedings are composed entirely of views of individuals, in order to convey the excitement, interest, and lively exchange of ideas that are within the subject. Probably it would be detrimental to direct research along particular lines, because one would lose the understanding that can be gained by identifying and comparing the merits of several approaches to a problem.

In order to obtain the given opinions, seven discussion sessions were held that were attended by all participants. Each session was three hours long, about half the time (in theory) being given to four or five invited papers, and about half the time being available for unscheduled contributions from the floor. All 31 invited papers are published in these proceedings, and nearly all the discussion contributions are given also. However, several participants asked me as editor for leave to change their spoken contributions to what they would have said if they had had time for further consideration; in all these cases we preferred the version that seemed to have more value to future research. Editorial licence has also been taken in the ordering of the contributions, in order to co-ordinate the remarks on each of the topics that were discussed.

In addition to the discussion sessions, research seminars were presented by most of the participants on their recent work. Only the titles of these seminars are listed in this volume; further information can be obtained from the authors.

All participants are indebted to NATO for the generous financial support that made the Advanced Research Institute possible. We were also pleased to receive the official sponsorship of the Mathematical Programming Society. We wish to record our thanks to Trinity Hall and to Conference Associates (Cambridge) for their willing assistance and attention to our needs during the meeting. Further, even though its contribution was less obvious, thanks are due to the Department of Applied Mathematics and Theoretical Physics at the University of Cambridge for help received from many of its staff, in particular Hazel Felton, and for the use of facilities for the administration of the meeting and for the preparation of these proceedings. We thank Elisabeth Griewank for coming to our aid when the demand for secretarial assistance was too great for the Department. Also we are most grateful to Academic Press, because their co-operation during the publication of this volume was particularly helpful, and we welcomed their contact with us at the time of the meeting.

Individuals who deserve special thanks include those who served with me on the Organising Committee, namely L.S. Lasdon, C. Lemaréchal, F.A. Lootsma and O.L. Mangasarian, and the chairmen of the discussion sessions, namely R.W.H. Sargent, O.L. Mangasarian, J.B. Rosen, S.M. Robinson, P. Wolfe, L.S. Lasdon and D.M. Himmelblau. Finally, I would like to acknowledge the excellent co-operation that I have received from the contributors to these proceedings.

Pembroke College, M.J.D. Powell
Cambridge,
October, 1981.

Contents

Part 3 Linear Constraints

Part 4 Nonlinear Constraints

Part 5 Large Nonlinear Problems

PART 1

Unconstrained Optimization

1.1

Unconstrained Optimization in 1981

Robert B. Schnabel

Department of Computer Science, University of Colorado, Boulder, Colorado 80309, USA.

1. Introduction

This paper summarizes the author's view of current practice and research in solving the unconstrained minimization problem

$$\underset{x \in \mathbf{R}^n}{\text{minimize}}\ f(x), \qquad f : \mathbf{R}^n \to \mathbf{R}, \tag{1.1}$$

where f is assumed to be twice continuously differentiable. We will be concerned almost exclusively with the task of finding a local minimizer of f(x), that is the solution to (1.1) in some open neighbourhood. Mainly we will consider problems of small or medium size, say n between 2 and 50, where the storage of an $n \times n$ matrix, and the solution of an $n \times n$ system of linear equations at each iteration, is acceptable. Some comments are made on the solution of larger problems.

Because this paper is quite concise, some readers may find it useful to consult publications that give more details and discussion of the algorithms that are mentioned. For general reading the books by Dennis and Schnabel (1982), Fletcher (1980), and Gill, Murray and Wright (1981) are recommended.

The remainder of the paper is divided into two sections. In Section 2 we give our view of the established practice in solving small or medium dimension unconstrained minimization problems.

In Section 3 we try to highlight the important recent developments, and remaining research topics, in all of unconstrained minimization. In separating established practice from recent developments, we have tried to err on the side of caution; several items listed as recent developments may be considered established practice by many people. We try to take a broad view of the field in Section 3. However, we apologize for those topics that are omitted, either due to lack of space or the author's ignorance.

We will denote the solution to (1.1) by x_*. Usually we will be speaking of a single iteration, and will denote the current iterate by x_c and the new iterate by x_+. The first iterate will be denoted by x_0, the sequence of iterates by $\{x_k\}$, the vector of first partial derivatives of f at x by $\nabla f(x)$, and the matrix of second partial derivatives of f at x by $\nabla^2 f(x)$.

In both Sections 2 and 3, we distinguish between local and global methods for solving (1.1). By local methods we mean the quickly convergent methods that are used when x_c is sufficiently close to x_* (e.g., Newton's method). These methods correspond to taking x_+ as the minimizer of the current (quadratic) model of f around x_c. By global methods, we mean the modifications of the local methods that are used to cause convergence to a local minimizer when the local method alone would not be successful (e.g., line search or trust region methods).

We remind the reader that much research in optimization is motivated by the solution of inherently expensive problems, those where evaluation of f(x) is expensive relative to the remaining computational cost of the algorithm. This consideration affects much of the following discussion.

2. Established Practice for Small and Medium Dimension Problems circa 1981

2.1 Local methods The standard local methods for unconstrained minimization all stem from considering the quadratic model of f around x_c,

$$m_c(x_c+d) = f(x_c) + \nabla f(x_c)^T d + \tfrac{1}{2} d^T \nabla^2 f(x_c) d, \tag{2.1}$$

which has a unique minimizer at

$$x_+ = x_c - \nabla^2 f(x_c)^{-1} \nabla f(x_c) \tag{2.2}$$

if and only if $\nabla^2 f(x_c)$ is positive definite. The classical local method for (1.1), Newton's method, consists of using (2.2) as the iteration formula. If n is not too large and $\nabla f(x)$ and $\nabla^2 f(x)$ are analytically available, this is usually the local method of choice. It is locally q-quadratically convergent to any x_* where $\nabla f(x_*) = 0$ and $\nabla^2 f(x_*)$ is positive definite, if $\nabla^2 f(x_*)$ is Lipschitz continuous in an open neighbourhood containing x_*.

If $\nabla f(x)$ or $\nabla^2 f(x)$ are not analytically available, they can be replaced by finite difference approximations, for example

$$\nabla f(x_c)_i \approx [f(x_c+h_i e_i) - f(x_c)]/h_i \tag{2.3}$$

$$\nabla^2 f(x_c)_{ij} \approx [f(x_c+\bar{h}_i e_i+\bar{h}_j e_j) - f(x_c+\bar{h}_i e_i) - f(x_c+\bar{h}_j e_j) + f(x_c)]/\bar{h}_i \bar{h}_j, \tag{2.4}$$

where e_i denotes the i-th unit vector and where the stepsizes h and $\bar{h}$ are appropriately chosen (see e.g., Dennis and Schnabel, 1982). Usually, in practice, the use of finite difference approximations hardly affects the performance of (2.2), unless a high degree of accuracy is required in x_* or f(x) is quite noisy (see e.g., Weiss, 1980). In this case a central difference approximation to $\nabla f(x)$ may be desirable. The gradient approximation requires n (or 2n) additional evaluations of f(x) per iteration, but there is no viable alternative when $\nabla f(x)$ is not available analytically. However, the $(n^2+3n)/2$ additional function evaluations required to approximate $\nabla^2 f(x_c)$ are avoided by the class of secant methods.

Secant methods (also known as quasi-Newton or variable metric methods) use a rougher approximation H_c to $\nabla^2 f(x_c)$, that is constructed using only gradient information from current and previous iterations. H_c replaces $\nabla^2 f(x_c)$ in (2.2), so that the local method becomes

$$x_+ = x_c - H_c^{-1} \nabla f(x_c). \tag{2.5}$$

H_c is then updated to H_+, which is a new approximation to $\nabla^2 f(x_+)$. (The choice of the first approximation H_0 is considered in Section 3.1.) Much research has gone into these update methods in the past twenty years, with the seminal work due to Davidon (1959) and Fletcher and Powell (1963). The acknowledged current choice is the BFGS update (Broyden, 1970; Fletcher, 1970a; Goldfarb, 1970; Shanno, 1970)

$$H_+ = H_c + \frac{y_c y_c^T}{y_c^T s_c} - \frac{H_c s_c s_c^T H_c}{s_c^T H_c s_c}, \tag{2.6}$$

where $s_c = x_+ - x_c$ and $y_c = \nabla f(x_+) - \nabla f(x_c)$. This update and all the others obey the secant equation, $H_+ s_c = y_c$, which enables the quadratic model

$$\hat{m}_+(x_+ + d) = f(x_+) + \nabla f(x_+)^T d + \tfrac{1}{2} d^T H_+ d \tag{2.7}$$

to obey $\nabla \hat{m}_+(x_c) = \nabla f(x_c)$ as well as $\nabla \hat{m}_+(x_+) = \nabla f(x_+)$ and $m_+(x_+) = f(x_+)$. In addition H_+ is positive definite as long as H_c is positive definite and $s_c^T y_c > 0$. (The latter condition is usually enforced by the global portion of the algorithm, e.g., (2.10) below.) Broyden, Dennis and Moré (1973) showed that (2.5–6) is locally q-superlinearly convergent to x_* if the conditions for the q-quadratic convergence of Newton's method are met and $\| H_0 - \nabla^2 f(x_0) \|$ is sufficiently small. Numerical testing has shown that the BFGS method usually but not always requires more iterations to converge than Newton's method, but is almost invariably cheaper in function evaluations than Newton's method with finite difference second derivatives.

The superiority of the BFGS update was established by testing in the 1970's (e.g., Brodlie, 1977). Some research has attempted to explain this, and the reasons are now partially understood. The two main competitors perhaps have been the PSB ("Powell—symmetric—Broyden", Powell, 1970a) and DFP updates, given by the choices $v_c = s_c$ and $v_c = y_c$ respectively in the formula

$$H_+ = H_c + \frac{(y_c - H_c s_c) v_c^T + v_c (y_c - H_c s_c)^T}{v_c^T s_c} - \frac{(y_c - H_c s_c)^T s_c \, v_c v_c^T}{(v_c^T s_c)^2} . \quad (2.8)$$

The methods using these two updates share the local q-superlinear convergence of the BFGS formula. However, the PSB update lacks the hereditary positive definiteness property and also the invariance under linear transformations of the variable space that is shared by the DFP and BFGS updates; these disadvantages give some problems in practice. Why the DFP update is computationally inferior to the BFGS update is less clear; perhaps the fact that the global convergence result of Powell (1976) for the BFGS method has not been replicated for the DFP method is a reason. The DFP formula also has a tendency to produce nearly singular Hessian approximations more frequently than the BFGS formula.

In production codes, the BFGS update is not implemented by (2.6), but instead one works with the lower triangular matrix $L_c \epsilon \mathbf{R}^{n\times n}$ for which $L_c L_c^T = H_c$. This matrix is updated directly to the lower triangular matrix $L_+ \epsilon \mathbf{R}^{n\times n}$ such that $L_+ L_+^T = H_+$, where H_+ is given by (2.6). The update can be done in $O(n^2)$ arithmetic operations, using the techniques pioneered by Gill, Golub, Murray and Saunders (1974). Because L_c is lower triangular, the solution of (2.5) requires no matrix factorizations, so that the number of arithmetic operations per iteration is $O(n^2)$, as opposed to $O(n^3)$ for a second derivative method.

2.2 Global methods Of course, the local methods of Section 2.1 do not always converge to a local minimizer if x_0 is too far from x_*. The modifications of (2.2) and (2.5), that are used to achieve global convergence, can be motivated by the well-known theorem (Wolfe, 1969, 1971) that, if f(x) is bounded below and the angles between $\nabla f(x_c)$ and s_c are uniformly bounded below 90°, then the conditions

$$f(x_+) \le f(x_c) + \alpha \nabla f(x_c)^T (x_+ - x_c), \quad \alpha \in (0, \tfrac{1}{2}) \tag{2.9}$$

and

$$\nabla f(x_+)^T (x_+ - x_c) \ge \beta \nabla f(x_c)^T (x_+ - x_c), \quad \beta \in (\alpha, 1) \tag{2.10}$$

are sufficient to guarantee $\nabla f(x_k) \to 0$ as $k \to \infty$; furthermore, they will both be satisfied by (2.2) or (2.5) for $\| x_c - x_* \|$ sufficiently small when the conditions for quadratic or superlinear convergence are met. Therefore what is usually done is to try (2.2) or (2.5) first, and, if this x_+ does not satisfy (2.9—10), then the step is modified so that the conditions hold. This section surveys these modifications, in the case when $\nabla^2 f(x_c)$ or H_c is positive definite; the case when $\nabla^2 f(x_c)$ is not positive definite is considered in Section 3.2. For convenience of notation, H_c will denote the analytic, finite difference or secant Hessian.

The oldest and perhaps simplest modification of (2.2) or (2.5) is the line search

$$x_+ = x_c - \lambda_c H_c^{-1} \nabla f(x_c), \tag{2.11}$$

where $\lambda_c > 0$ is chosen so that x_+ satisfies (2.9—10). It may be shown that, if H_c is positive definite and if f is bounded below, then there exists an interval of satisfactory values of λ_c. In practice, α is usually set quite small (e.g., $\alpha = 10^{-4}$) and β rather close to 1 (e.g., $\beta = 0.9$) so that the line search conditions are easily met. There is a multitude of line search algorithms; they usually try $\lambda_c = 1$ first, and, if adjustment of λ_c is necessary, they model $\hat{f}(\lambda) = f(x_c - \lambda H_c^{-1} \nabla f(x_c))$ by a one-dimensional quadratic or cubic that interpolates function and derivative values of $\hat{f}(\lambda)$, and they try next the value of λ that minimizes the one-dimensional model.

The other major class of global modification methods are the trust region algorithms. An early one is given by Powell (1970b), who reasoned that, when steps much shorter than the Newton step are required, they should be in the steepest descent direction $-\nabla f(x_c)$ instead of the Newton direction. His algorithm keeps a trust region parameter $\tau_c \epsilon \mathbf{R}$ that is adjusted during each iteration. Then x_+ is chosen to be the unique point on the dogleg curve (Figure 2.1) such that $\|x_+ - x_c\| = \tau_c$. (CP, the Cauchy point, is the minimizer of the quadratic model (2.1) in the steepest descent diection.) A double dogleg modification by Dennis and Mei (1979) is now often used in practice; it biases the step more towards the Newton direction, by including a step from CP back to the Newton direction as shown in the figure. The distance $\|CP-x_c\|_2$ is always less than $\|N-x_c\|_2$, and is sometimes very much less in practice.

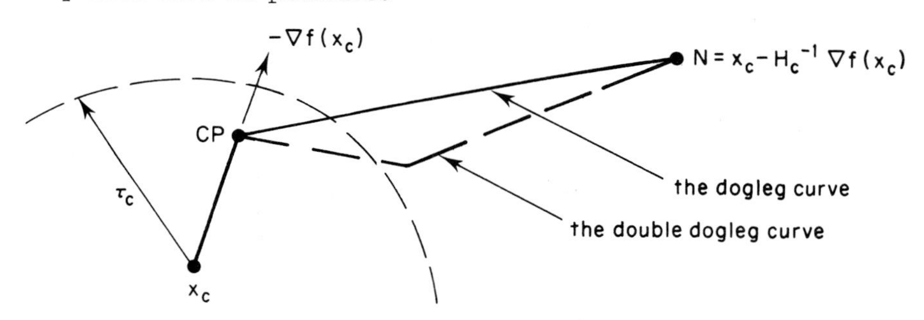

Figure 2.1 The dogleg and double dogleg curves

The dogleg step can be viewed as an approximation to the vector $d(\tau)$ that solves

$$\left.\begin{array}{ll} \underset{d \epsilon \mathbf{R}^n}{\text{minimize}} & m_c(x_c+d) = f(x_c) + g_c^T d + \tfrac{1}{2} d^T H_c d \\ \text{subject to } d^T d \le \tau^2 & \end{array}\right\} . \qquad (2.12)$$

This point of view is appropriate because $d(\tau)$ obeys $d(\tau) = -H_c^{-1}g_c$ when $\tau \geq \|H_c^{-1}g_c\|_2$, and because $\lim d(\tau)/\|d(\tau)\|_2 = -g_c/\|g_c\|_2$ as $\tau \to 0$. (Here g_c denotes the analytic or finite difference gradient.) We will refer to $d(\tau)$ as the constrained local model step. When $\tau < \|H_c^{-1}g_c\|_2$, the solution to (2.12) is $d = -(H_c+\alpha I)^{-1}g_c$ for the $\alpha > 0$ such that $d^Td = \tau^2$.

Minimization algorithms whose iteration consists of solving (2.12) exactly for some τ close to the current trust region τ_c have recently been developed by Moré (1978), Gay (1981) and many others; earlier work along these lines was by Goldfeld, Quandt and Trotter (1966), Hebden (1973) and others. These algorithms may require the solution of more than one linear system per iteration, which the other global methods do not need. All the trust region algorithms require heuristic rules for updating τ_c; they increase or decrease τ_c when the quadratic model predicts $f(x)$ well or poorly, respectively, and aim to allow full Newton steps near the minimizer. For more details, see Sorensen's paper in this volume.

Numerical testing has not revealed any consistently large differences between these three global strategies. Our experience (Weiss, 1980), based on the test problems of Moré, Garbow and Hillstrom (1981a), is that a line search strategy is slightly superior to the trust region strategies when H_c is approximated by the BFGS formula, and that the reverse is true when H_c is the analytic or finite difference Hessian. The tests of Gay (1980a) and Weiss exhibit little difference between the two trust region strategies, and so the dogleg may be preferable to solving (2.12) since it is simpler and less expensive.

2.3 Software Major sources of codes for unconstrained minimization at this time include the NAG library (Oxford), the Harwell library (Harwell), MINPACK (Argonne) and the IMSL library (Houston). Various researchers also distribute codes that are widely used.

3. Recent Developments and Research Topics

3.1 Secant updates One major development in secant updates in recent years was Davidon's (1975) projected and optimally conditioned update. This update is also a symmetric rank two change to the Hessian approximation that preserves positive definiteness in many cases. Projections enable the algorithm to exactly minimize a positive definite quadratic in n+1 or fewer iterations, without the use of exact line searches. Optimal conditioning causes the ℓ_2 condition number of $H_+^{-1}H_c$ to be minimized at each iteration, among a one parameter family of such projected updates. However, numerical testing has not yet revealed a superiority of this update over the BFGS (see e.g., Shanno and Phua, 1978b). In our research, a slightly different projected BFGS update (Dennis and Schnabel, 1981) has performed a little better than the BFGS, but probably not well enough to justify the additional algorithmic complexity.

An important practical development has been the development of scaling strategies for the initial Hessian approximation H_0. It has been traditional to use $H_0 = I$. Shanno and Phua (1978a) demonstrated that modifying H_0 to $\bar{H}_0 = (s_0^T y_0 / s_0^T H_0 s_0) H_0$ prior to the first update (so that $s_0^T \bar{H}_0 s_0 = s_0^T y_0$) can sometimes improve performance considerably, and this has now become standard practice. Other initial scaling strategies are possible (see e.g., Frank and Schnabel, 1981); what is clear is that some initial scaling should be performed. Our experience is that setting H_0 to a finite difference approximation to $\nabla^2 f(x_0)$ is not usually cost effective. Shanno's strategy is a modification of a more general suggestion by Oren (1974).

3.2 Global methods and negative curvature When an analytic or finite difference Hessian is used, a strategy is required for the case when it is not positive definite. This is an active research area, where we believe there is still room for important contributions. In this section we denote the current analytic or finite difference gradient and Hessian by g_c and H_c, respectively.

In line search algorithms, there is an established practice due to Gill and Murray (see e.g., Gill, Murray and Wright, 1981). It is to perform a perturbed Cholesky factorization

$$H_c + E = LDL^T \tag{3.1}$$

of H_c, where L is a lower triangular matrix, D a positive diagonal matrix, and E a non-negative diagonal matrix that is added during the factorization to make $H_c + E$ positive definite and reasonably well-conditioned. Then $H_c + E$ can be used in place of H_c to give the line search direction $-(H_c+E)^{-1}g_c$; when g_c is near zero, a direction of negative curvature that is obtained from the factorization is used instead. Another alternative is to add a multiple of the identity (or a positive diagonal matrix that accounts for the scaling of the variable space) to H_c, in order that $H_c + \alpha I$ is positive definite and reasonably well conditioned. This strategy, which is used in the algorithms of Dennis and Schnabel (1982) and Weiss (1980), is motivated by the algorithm discussed next.

The main recent development in trust region algorithms has been the use of the constrained local model approach (2.12) even when H_c is indefinite. It is easily shown (see e.g., Gay, 1981, or Sorensen, 1980b) that the solution is $d(\tau) = -(H_c+\alpha I)^{-1}g_c$, if there exists $\alpha \geq 0$ such that $H_c + \alpha I$ is positive definite, and such that $\|d(\tau)\|_2 = \tau$. Otherwise, in the indefinite case, g_c is orthogonal to the eigenvector, v say, of the most negative eigenvalue of H_c, $-\alpha$ say, and $d(\tau) = -(H_c+\alpha I)^+g_c+\beta v$ for some $\beta \in \mathbf{R}$, where the superscript "+" denotes the generalized inverse. Work to produce an efficient implementation of this promising approach continues by various researchers — see Sorensen's paper (this volume).

The use of curvilinear line searches has been investigated by many researchers, including recently McCormick (1979), Moré and Sorensen (1979) and Goldfarb (1980a). These methods are not yet commonly used in practice, but seem to be an interesting research area. Comparative testing of all these approaches to negative curvature is still required, along with an appropriate set of test problems.

3.3 Convergence analysis An important recent development is the analysis of Sorensen (1980b), showing that a trust region algorithm based on (2.12) and using analytic first and second derivatives converges globally and quadratically to a strong local minimizer of $f(x)$ in most cases. This result gives a strong theoretical justification for the use of the constrained local model approach, and is also notable for ruling out convergence to saddle points. Another interesting development has been a series of papers by Ritter (1979, 1980, 1981) exhibiting the local and global convergence of a wide range of secant algorithms, under somewhat restrictive assumptions. Two analyses that this author would find interesting are any further explanation of the superiority of the BFGS update, and an analysis of the BFGS algorithm on problems with negative curvature.

3.4 Nonquadratic models A paper by Davidon (1980) has started research on the use of alternatives to the quadratic model (2.1). as the basis of steps in optimization algorithms (see also Davidon, this volume). He suggested replacing the quadratic model by the conic model

$$\bar{m}_c(x_c+d) = f(x_c) + \frac{\nabla f(x_c)^T d}{1+p_c^T d} + \frac{\frac{1}{2} d^T A_c d}{(1+p_c^T d)^2}, \qquad (3.2)$$

where $A_c \in \mathbf{R}^{n\times n}$ and $p_c \in \mathbf{R}^n$. (The level sets of $\bar{m}_c(x)$ are conic sections.) This model can also be viewed as a quadratic in the new variable $s = d/(1+p_c^T d)$; steps in s-space can be converted back to d-space by the collinear scaling $d = s/(1-p_c^T s)$. Davidon's development and Sorensen's (1980a) subsequent analysis are more along this line.

An advantage of the conic model (3.2) is that it can interpolate more information about f(x) than a quadratic model can. Davidon shows how to use (3.2) in a secant method, to interpolate f(x) and $\nabla f(x)$ at x_c and the previous iterate x_-. Schnabel (1981a) proposes the use of (3.2) in a second derivative method, where $\bar{m}_c(x)$ fits $f(x_c)$, $\nabla f(x_c)$, $\nabla^2 f(x_c)$, and f(x) at up to n past iterates. Stordahl (1980) programmed a line search algorithm using this approach, and found that it made noticeable improvements in efficiency over the same algorithm using quadratic models. In our opinion, research into the general field of nonquadratic models for optimization is an exciting new research direction.

3.5 Large dimensional problems Much recent work has gone into developing algorithms for large problems ($n \geq 50$, say) whose Hessian matrix is sparse. Much of this work has involved extending secant methods to this case; however, second derivative based methods or conjugate gradient methods may be the more promising approaches. This is definitely an area where considerable interesting work remains to be done.

Secant updates that preserve symmetry and sparsity and satisfy the secant equation $H_+ s_c = y_c$ were first developed by Marwil (1978), Toint (1977), and Dennis and Schnabel (1979). These updates are generalizations of the PSB formula, and do not necessarily preserve positive definiteness. Shanno (1980) and Toint (1980) developed sparse extensions of the BFGS update, and Toint showed how to sometimes produce sparse positive definite updates in theory. However, recent numerical results by Thapa (1980) and Powell and Toint (1981) report rather discouraging performance of these methods in comparison to methods using second derivative information.

The viability of second derivative methods has been enhanced by the development of efficient finite differencing schemes for sparse Hessians, notably by Curtis, Powell and Reid (1974), Powell and Toint (1979), and Coleman and Moré (1981). These approximations may be used in conjunction with the methods of Section 3.2, but new alternatives are also being considered. An interesting new method of this type by Steihaug (1981) makes use of both the dogleg and conjugate gradient techniques.

The conjugate gradient approach is the other major method for solving large problems. It has been studied by many authors, see e.g., Shanno (1978), Buckley (1978a), and Buckley (this volume). The advantage of this approach is that it does not require any matrix storage. Recent research has focused on using a variable amount of vector storage, trading storage for time efficiency. Currently, this is probably the best developed approach for large problems; it is far too early to know which approach will be best for large sparse problems.

3.6 Software considerations ("Grunge") Several practical issues in solving optimization problems are often overlooked until software is developed. Two important ones are stopping criteria, and the scaling of the variable and function space. Failure of production codes to solve practical problems is often due to insufficient attention to these issues.

While Newton's method and the BFGS formula are invariant under linear transformations of the variable space, trust region algorithms, the choice of H_0 in secant algorithms and stopping criteria usually are not. Therefore many algorithms are adversely affected when the units of the variables differ greatly. Several recent codes e.g., Gay (1980a) and MINPACK, do make adjustments dependent on the scaling of the variable space, but more work on automatic scaling is desirable. The development of stopping criteria that are responsive to the users' requirements in solving the problem is also needed. Another practical topic requiring further attention is the adjustment of optimization codes to accommodate noisy objective functions.

3.7 Global optimization The problem of finding the global minimizer of f(x), the $x_* \in \mathbf{R}^n$ such that $f(x_*) \le f(x)$ for all $x \in \mathbf{R}^n$, is sometimes very important in practice. The reason it has not received sufficient attention may be that it cannot be solved by any elegant mathematical approach in general, but inevitably requires probabilistic or heuristic algorithms. The diversity of approaches that have been taken are exhibited by the probabilistic, clustering and other approaches in the two volumes edited by Dixon and Szegö (1975, 1978), the factorable function approach of Beale and Forrest (1976) and McCormick (1976), the interval arithmetic approach of Mancini and McCormick (1979) and Hansen (1980), and the tunnelling approach of Levy and Montalvo (1979). While we are not sufficiently familiar with this work to comment critically on it, surely it is an important future research area.

Acknowledgements

The author thanks T. Steihaug for pointing out a mistake in the original draft, E.M.L. Beale, L.C.W. Dixon and R. Mifflin for supplying additional references, and J. Moré for teaching him the correct spelling of D. Sorensen's last name.

1.2

Conjugate Gradient Methods

A. Buckley

Department of Mathematics, Concordia University,
Montreal, Canada.

The problem is to minimize a smooth function $f(x)$, $x \in \mathbf{R}^n$. The class of conjugate gradient (CG) methods consists of those for which the iterates $x_0, x_1, x_2, \ldots$ are generated according to the rules

$$x_i = x_{i-1} + \alpha_i d_i, \tag{1a}$$

$$\beta_i = \frac{g_i^T H y_i}{d_i^T y_i}, \tag{1b}$$

$$d_{i+1} = -Hg_i + \beta_i d_i. \tag{1c}$$

Here x_0 is given, $g(x) = \nabla f(x)$, $g_i = g(x_i)$, $d_1 = -Hg_0$ and $y_i = g_i - g_{i-1}$. Also, α_i is a suitably chosen scalar and H is a known positive definite (pd) matrix. We refer to (1) as algorithm CG_H. When $H = I$, (1) becomes CG_I, the original CG algorithm of Fletcher and Reeves (1964). For quadratic f, β_i may alternatively be given by $g_i^T Hg_i/g_{i-1}^T Hg_{i-1}$ or by $g_i^T Hy_i/g_{i-1}^T Hg_{i-1}$ (Fletcher and Reeves, 1964; Polak and Ribière, 1969), but the form (1b) is most suitable for our purposes.

The object here is to present a point of view from which we feel CG algorithms should be approached. It is, roughly, that CG algorithms are actually quasi-Newton (QN) algorithms for which the choice of update should be modified because it has been poorer than need be. Our theme will be the development of a new CG algorithm in a way that emphasizes the new point of view. This algorithm forms part of the thesis of LeNir (1981), and will be fully described in a forthcoming paper.

Many recent papers have examined algorithm (1) with $H \neq I$. This case is often called the "preconditioned" CG algorithm. Because H is pd, the properties of CG_H are obtained immediately from those of CG_I by considering the transformation $\tilde{x} = H^{-\frac{1}{2}}x$, for in the new coordinates CG_H becomes CG_I. Much of this work has centered on the relationship between CG_H and certain QN algorithms (see, for example, Buckley, 1978a; Buckley, 1978b; Nazareth, 1979; Shanno, 1978). In particular, it was first shown by Myers (1968) that, when f is quadratic and $H_0 = I$, CG_I generates the same points as the BFGS QN algorithm, which is the only QN algorithm that we will consider. In the BFGS algorithm, H_0 is a given pd matrix, and, for $i = 1,2,3,\ldots$, we use the formulae

$$d_i = -H_{i-1}g_{i-1} ,$$

$$x_i = x_{i-1} + \alpha_i d_i .$$

The matrix H_{i-1} is then updated according to $H_i = U(H_{i-1},i)$, where $s_i = x_i - x_{i-1} = \alpha_i d_i$ and

$$U(H,i) \equiv H - \frac{s_i y_i^T H + H y_i s_i^T}{s_i^T y_i} + \left(1 + \frac{y_i^T H y_i}{s_i^T y_i}\right) \frac{s_i s_i^T}{s_i^T y_i} . \qquad (2)$$

The principal advantage of CG algorithms is their low storage requirement. For practical purposes, it is important to retain this advantage, but it would be desirable to improve their performance. The means to do this exists in the relationship between CG_H and BFGS. To emphasize the QN-like character of the CG_I algorithm, one may write (1c), with $H = I$, as Perry (1978) did:

$$d_{i+1} = -\left(I - \frac{s_i y_i^T}{s_i^T y_i}\right) g_i \equiv -\hat{Q}_i g_i . \tag{3}$$

Similarly, for CG_H, one can write

$$d_{i+1} = -\left(H - \frac{s_i y_i^T H}{s_i^T y_i}\right) g_i . \tag{4}$$

Shanno (1978) expanded on this approach by modifying (3) to

$$d_{i+1} = -\left[I - \frac{s_i y_i^T + y_i s_i^T}{s_i^T y_i} + \left(1 + \frac{y_i^T y_i}{s_i^T y_i}\right)\frac{s_i s_i^T}{s_i^T y_i}\right] g_i$$

$$\equiv -Q_i g_i . \tag{5}$$

In (3), $\hat{Q}_i$ is singular and fails to satisfy the QN equation (i.e. $\hat{Q}_i y_i \neq s_i$). However, Shanno's Q_i is pd, $Q_i y_i = s_i$, and the directions generated by (3) and by (5) are identical when line searches are exact. Shanno notes that (5) is a BFGS update of the identity, i.e. $Q_i = U(I,i)$. Similarly, (4) can be extended to a BFGS update of the fixed matrix H, i.e.

$$d_{i+1} = -\left[H - \frac{s_i y_i^T H + H y_i s_i^T}{s_i^T y_i} + \left(1 + \frac{y_i^T H y_i}{s_i^T y_i}\right)\frac{s_i s_i^T}{s_i^T y_i}\right] g_i = -H^* g_i , \tag{6}$$

where $H^* = U(H,i)$.

Shanno's "memoryless" QN algorithm (MQN) is based on (5) and is quite successful. It uses a fixed seven vectors (of length n) of storage, which is unfortunate since, as problems become larger, one might expect more than 7n locations to be available, albeit without having the $O(n^2)$ locations needed to use QN. Also, for functions f which are relatively inexpensive to evaluate, the $O(n^2)$ operations per iteration needed by QN methods often dominate the computation. Thus one would like to extend Shanno's approach to obtain an O(n) algorithm in which provision of additional storage can mean improved performance.

The approach that we recommend follows from an extension that Shanno made to his MQN algorithm in order to incorporate Beale's (1972) restart CG algorithm (BCG). The BCG algorithm includes an integer t which is the index of a point x_t commonly termed a restart point. For i = t formulae (1b) and (1c) are used, but for i > t they are replaced by

$$\gamma_i = \frac{g_i^T \bar{H} y_t}{d_t^T y_t}, \qquad \bar{\beta}_i = \frac{g_i^T \bar{H} y_i}{d_i^T y_i}, \tag{7b}$$

$$d_{i+1} = -\bar{H} g_i + \bar{\beta}_i d_i + \gamma_i d_t . \tag{7c}$$

We could also write (7c) as

$$d_{i+1} = -\left(\bar{H} - \frac{s_i y_i^T \bar{H}}{s_i^T y_i} - \frac{s_t y_t^T \bar{H}}{s_t^T y_t} \right) g_i \equiv -\bar{Q}_i g_i,$$

so this CG algorithm is also QN-like in form, but $\bar{Q}_i$ also lacks symmetry and positive definiteness, and $\bar{Q}_i y_i \neq s_i$.

We use the notation $\bar{H}$ and $\bar{\beta}_i$ in (7) instead of H and β_i, in order to show a close analogy to equation (1) by letting H be a modification of $\bar{H}$. Specifically we write (7c) as

$$d_{i+1} = -\left(\bar{H} - \frac{s_t y_t^T \bar{H}}{s_t^T y_t} \right) g_i + \bar{\beta}_i d_i$$

$$\equiv - H g_i + \bar{\beta}_i d_i . \tag{8}$$

This is almost algorithm CG_H, except that $\bar{\beta}_i$ in (8) is given by (7b) instead of (1b). However, for quadratic f and exact searches, the term $y_t^T \bar{H} g_i$ is zero for i > t, which implies $\bar{\beta}_i = \beta_i$. Thus (8) is indeed CG_H, so we take the point of view that Shanno's introduction of Beale restarts just amounts to using (1c) with a different preconditioner H. Of course, H defined by (8) is not pd. Once again this may be rectified, just as in the extension of (4) to give (6); then (8), which is analogous to (1c), becomes $d_{i+1} = -H^*g_i$ with $H^* = U(H,i)$ and $H = U(\bar{H},t)$.

This point of view has the following advantage. Shanno's MQN derivation was tied to Beale's 2-stage recurrence; hence it generated a 2-update algorithm. But in (1) or (6), any preconditioner can be chosen for H. If we let H in (6) be defined at a restart point x_t as $H = U(I,t)$, i.e. $\bar{H} = I$ in (7), then (6) yields Shanno's 2-update MQN, with $H^* = U(H,i)$ for $i > t$.

But suppose storage is available for m rank-2 update terms (as in Buckley, 1978a). Then a QN update of m terms could be used to define H. For example, after a restart at x_t, one could set $H_1 = U(I,t)$, $H_{i+1} = U(H_i, t+i)$ for $i = 1,2,\ldots,m-1$, and $H^* = U(H_m, t+i)$ for $i \geq m$. Then $d_{t+i} = -H_i g_{t+i-1}$ for $i = 1,2,\ldots,m$, and $d_{t+i} = -H^* g_{t+i-1}$ for $i > m$. Note that $m = 1$ gives MQN so we have an extension of Shanno's MQN. It is also a preconditioned BCG algorithm (7) with $\bar{H} = H_{m-1}$. The finite termination properties of this algorithm follow from the results in Buckley (1978b). Indeed, this algorithm is an implementation of that in Buckley (1978a) where the CG steps are done in the form (6).

This emphasizes what we see as the main contribution of Shanno's MQN: that CG steps should be implemented in the "QN-like" form (6) instead of with (1). In other words, CG algorithms should be viewed as QN algorithms in which the approximating matrix ($\hat{Q}_i$, Q_i or $\bar{Q}_i$) is inadequate. One is unlikely to suggest that the QN formula (2) should be modified by discarding the matrices $Hy_i s_i^T$ and $s_i s_i^T$, which make no contribution anyway to d_{i+1} when line searches are exact, because then $s_i^T g_i = 0$. Yet that is precisely what has been taking place implicitly in conjugate gradient methods to date as is clear from equation (4). We recommend that equation (5) or (6) should be used instead. The computations in Shanno (1978), which are based on (5), indicate that for Beale restarts this approach works well. By using (6) in place of (5), we suggest it could work well in other contexts.

Part of the reason for this claim is that (6) relieves some pressure from the line search routine. Because H* is pd provided only that $s_i^T y_i > 0$, it is simple to ensure d_{i+1} is downhill. In the form (1), the value of α_i must be chosen under more stringent conditions. By loosening the strict requirement for highly accurate line searches, which is always viewed as a limitation of CG algorithms, we feel that the CG_H algorithm will become an acceptable alternative in those contexts where it is applicable.

For example, we have suggested above an extension of Shanno's MQN. It can also be viewed as a QN algorithm, where one does QN updates until the available storage is exhausted; then CG_H as in (6) is applicable. It is not clear how frequently such an algorithm should be restarted. Moreover, consider the situation where the Hessian of f is sparse. Toint (1981) considers the strategy of periodically letting H be the sparse Hessian, or a finite difference approximation thereto, and then using sparse QN updates of H. Because of its simplicity, an alternative could be to use the CG_H algorithm (6) in place of the sparse quasi-Newton updates. In Shanno (1981), a similar idea is presented (but again with reference to Beale restarts).

In summary then, equation (6) states the fundamental form in which we feel the CG_H algorithm should be viewed. What is important is to use the freedom in the choice of H to full advantage, for the use of (6) should ensure effectiveness of the CG_H steps.

1.3

Conjugate Directions for Conic Functions

William C. Davidon

Department of Mathematics, Haverford College, Haverford, PA 19041, USA.

Conjugate direction algorithms, such as conjugate gradient algorithms (Hestenes and Stiefel, 1952) using O(n) operations/iteration and variable metric algorithms (Davidon, 1959) using $O(n^2)$ operations/iteration, minimize quadratic objective functions by minimizing sequentially along lines in conjugate directions. While the concept of conjugacy used in these algorithms is invariant under affine transformations $S(w) = x_0 + Jw$, it is a special case of an older concept from projective geometry (Coxeter, 1964) which is invariant under the larger group of collineations $S(w) = x_0 + Jw/(1 + h^T w)$. This paper suggests optimization algorithms using this more general concept of conjugacy.

Definitions A map $S:W \to X$ between open convex sets $W, X \subseteq \mathbf{R}^n$ is a *collineation* iff $S(w) = x_0 + Jw/(1 + h^T w)$ for some invertible $J \in \mathbf{R}^{n \times n}$. A function $f:X \to \mathbf{R}$ is *conic* iff its composition $fS:W \to \mathbf{R}$ with some collineation $S:W \to X$ is quadratic. It is *normal* iff this quadratic composition has a positive definite Hessian. Two lines in the domain X of a normal conic function $f:X \to \mathbf{R}$ are *conjugate* iff they are images of orthogonal ones in W under any collineation $S:W \to X$ which makes $fS:W \to \mathbf{R}$ quadratic with unit Hessian. The *horizon* of a conic function is its singular hyperplane.

The significance of conjugate directions for optimizers is

that minimizing sequentially along conjugate lines locates the minimizer in their affine hull. Figure 1 relates the minimizer, a neighbouring ellipsoidal level set, the polar horizon, and some conjugate lines for a normal conic function. The minimizer of a quadratic function is centered in its level set, its horizon

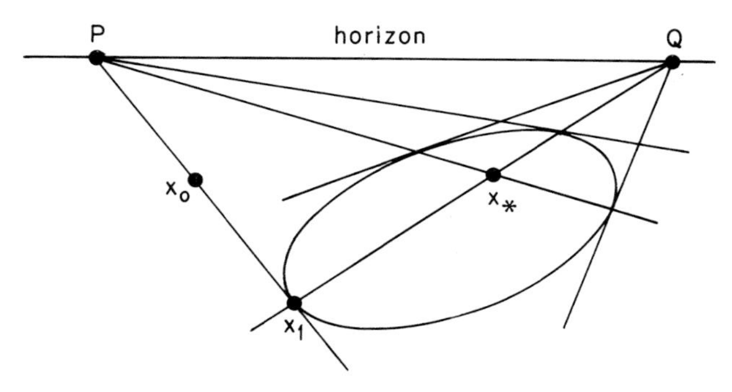

Fig. 1. The ellipse is a level set for a conic function f with minimizer x_* and horizon PQ. Lines through P are conjugate to lines though Q.

is at infinity, and lines meeting at the horizon are parallel.

Algebraically, a function is conic iff

$$f(x_0 + s) = f_0 + \frac{g_0^T s}{1 - a_0^T s} + \tfrac{1}{2} \frac{s^T A_0 s}{(1 - a_0^T s)^2}$$

for all $s \in \mathbf{R}^n$ with $a_0^T s < 1$, where $f_0 \in \mathbf{R}$, $g_0 \in \mathbf{R}^n$, and $A_0 + a_0 g_0^T + g_0 a_0^T \in \mathbf{R}^{n \times n}$ are the value, gradient, and Hessian of f at $x_0 \in \mathbf{R}^n$. The vector $a_0 \in \mathbf{R}^n$ locates the horizon of f relative to x_0. The symmetric matrix A_0 is positive definite iff f is normal, and two lines through x_0 in directions $d_\pm$ are conjugate iff $d_+^T A_0 d_- = 0$. For these reasons, we define a_0 to be the *horizon vector* and A_0 the *conjugacy matrix* for f at x_0. This conjugacy matrix can be viewed as an average of the Hessian over the interval between x_0 and the minimizer. The conjugacy matrix and Hessian of f at x_0 are equal iff f is quadratic or x_0 is a minimizer.

The gradients $g_{\pm}$ of a quadratic function at two points $x_{\pm}$ suffice not only for locating the minimizer on the line through $x_{\pm}$, but also the hyperplane through this minimizer containing all lines conjugate to the line through $x_{\pm}$. However, the gradients of a conic function at just two points determine neither the minimizer on the line through $x_{\pm}$ nor the conjugate hyperplane. Function as well as gradient values of a normal conic at two points $x_{\pm}$ do locate the minimizer on the line through $x_{\pm}$, and these with the gradient at any third point on the line determine both the conjugate hyperplane through the minimizer as well as the horizon of f. Thus the use of conic rather than quadratic approximations makes more use of information in the values and gradients of more general objective functions.

Quantitatively, if $f_{\pm}$ and $g_{\pm}$ are the values and gradients of a normal conic at points $x_{\pm}$, then there is just one ratio γ_+/γ_- satisfying

$$f_+ - f_- - \frac{\gamma_-}{\gamma_+} g_-^T(x_+ - x_-) = f_- - f_+ + \frac{\gamma_+}{\gamma_-} g_+^T(x_+ - x_-) > 0.$$

These $\gamma_{\pm}$ are proportional to the displacements of $x_{\pm}$ from the horizon. The minimizer along the line through $x_{\pm}$ is at

$$\frac{x_- \, g_+^T d \, \gamma_+^3 - x_+ \, g_-^T d \, \gamma_-^3}{g_+^T d \, \gamma_+^3 - g_-^T d \, \gamma_-^3},$$

where d is any nonzero multiple of $x_+ - x_-$. If g_0 is the gradient at any third point x_0 along the line through $x_{\pm} = x_0 + \lambda_{\pm} d$, then the horizon vector a_0 satisfies

$$a_0 = \gamma_+ u_+ - \gamma_- u_-,$$

and

$$a_0^T d = (\gamma_+ - \gamma_-)/(\gamma_+ \lambda_- - \gamma_- \lambda_+),$$

where

$$u_{\pm} = (\gamma_{\pm} g_{\pm} - \gamma_0 g_0)/\lambda_{\pm}(\gamma_+^2 \sigma_+ - \gamma_-^2 \sigma_-),$$

where

$$\gamma_0 = (\gamma_- \lambda_+ - \gamma_+ \lambda_-)/(\lambda_+ - \lambda_-),$$

and where $\sigma_{\pm} = g_{\pm}^T d$ are directional derivatives. Further, the conjugacy matrix A_0 satisfies

$$(\gamma_0^2/\gamma_+\gamma_-) A_0 d = \gamma_+ \sigma_+ u_- - \gamma_- \sigma_- u_+ ,$$

and

$$(\lambda_+ - \lambda_-)(\gamma_0^4/\gamma_+\gamma_-) d^T A_0 d = \gamma_+^2 \sigma_+ - \gamma_-^2 \sigma_- .$$

The conjugate direction algorithms to be presented share the following properties with previous ones:

1) The kth step $x_k - x_{k-1}$ is a linear combination of previous steps and $A_0^{-1} g_{k-1}$, where A_0 is an initial estimate for a conjugacy matrix, or preconditioning matrix; e.g., $A_0 = I$.

2) The line through x_k and x_{k-1} is conjugate to the lines through x_j and x_{j-1} for all $j < k$.

3) The point x_k is the minimizer of the restriction of f to the line through x_k and x_{k-1}. It is also the minimizer of the restriction of f to the affine hull of $\{x_j \in \mathbf{R}^n: j \le k\}$.

One of the two algorithms to be presented generalizes the conjugate gradient algorithm and uses O(n) operations/iteration other than for evaluating the function and its gradient. The other generalizes the variable metric algorithm and uses $O(n^2)$ operations/iteration to update approximations A_k to the conjugacy matrix; A_n is the conjugacy matrix as well as the Hessian of a normal conic objective function at its minimizer x_n. Both algorithms start with an initial estimate A for a conjugacy matrix. In the O(n) algorithm, this estimate is fixed and chosen so that

$Ad = -g$ can be solved for d in $O(n)$ operations. In the $O(n^2)$ algorithm, either A^{-1} or a Cholesky factor of A is updated so that, in each iteration, $Ad = -g$ can be solved for d in $O(n^2)$ operations. Both algorithms begin their k^{th} iteration with f_{k-1} and g_{k-1} at x_{k-1}, and they use these along with f_{-k} and g_{-k} at a trial point x_{-k} to locate the minimizer x_k on the line through x_{k-1} and x_{-k}. They then use the three gradients g_{k-1}, g_{-k}, and g_k at the collinear points x_{k-1}, x_{-k}, and x_k to determine the next step. For conic objective functions and exact arithmetic, both algorithms determine the same sequence of points x_k.

The Algorithms

Since these algorithms are intended only to illustrate the concepts discussed in this paper, they do not include those additional features needed for robust performance with objective functions which are not well approximated by conic ones.

Step 0 (Input) Set $x \in \mathbf{R}^n$, $a \in \mathbf{R}^n$, and $A \in \mathbf{R}^{n \times n}$, which are estimates for the minimizer, horizon vector, and conjugacy matrix of the objective function respectively; e.g. $a = 0$, $A = I$.

Step 1 Call f and g at x. Set $d := -A^{-1}g$. In the $O(n)$ algorithm set $\gamma := 1$ and $x_s := x$.

Step 2 (Beginning of iteration) Set $\sigma_0 := g^T d$. If σ_0 is sufficiently small then exit with x the minimizer.

Step 3 Set $\lambda_- := 1/(1 + a^T d)$, $f_0 := f$, $g_0 := g$, $x_0 := x$,

$x := x_0 + \lambda_- d$, and call f and g at x.

Step 4 Set $\sigma_- := g^T d$,

$\rho := ((f - f_0)^2 - \sigma_0 \sigma_- \lambda_-^2)^{\frac{1}{2}}$,

$$\gamma_- := -\lambda_-\sigma_0/(f_0 - f + \rho),$$
$$\lambda_+ := -\lambda_-\sigma_0/(\gamma_-^3\sigma_- - \sigma_0) \text{ and}$$
$$\gamma_+ := \gamma_-(\gamma_-^2\sigma_- - \sigma_0)/(\gamma_-^3\sigma_- - \sigma_0).$$

Step 5 Set $g_- := g$, $x := x_0 + \lambda_+ d$, and call f and g at x.

Step 6 Set $u_0 := (g_0 - \gamma_+ g)/(-\lambda_+(\sigma_0 - \gamma_-^2\sigma_-))$,
$$u_- := (\gamma_- g_- - \gamma_+ g)/((\lambda_- - \lambda_+)(\sigma_0 - \gamma_-^2\sigma_-)),$$
$$a := u_0 - \gamma_- u_- \text{ and}$$
$$b := \sigma_0 u_- - \gamma_-\sigma_- u_0.$$

Step 7 In the O(n) algorithm, set $\gamma := \gamma\gamma_+$,
$$c := \gamma^2(A^{-1}g - (x - x_s)a^T A^{-1}g\gamma) \text{ and}$$
$$d := -c + d\, b^T c/b^T d, \text{ using } b^T d = 2\rho\gamma_-/(\gamma_+\lambda_-)^2.$$
In the $O(n^2)$ algorithm set $A := (I + \lambda_+ ad^T)A(I + \lambda_+ da^T)/\gamma_+^2$,
$$A := A - \frac{Add^T A}{d^T Ad} + \frac{bb^T}{2\rho}\lambda_-^2 \text{ and } d := -A^{-1}g.$$

Step 8 Return to Step 2.

While other algorithms for minimizing conic functions have been discussed elsewhere (Bjorstad and Nocedal, 1979; Sorensen, 1980a; Davidon, 1980), they have not made explicit use of this general concept of conjugacy, nor have they used gradients at three collinear points to determine the horizon vector a and the subsequent search direction.

1.4

Trust Region Methods for Unconstrained Minimization*

D.C. Sorensen

Applied Mathematics Division, Argonne National Laboratory,
Argonne, Il. 60439, USA.

1. Introduction

Unconstrained minimization of a real valued differentiable function of several real variables is a fundamental problem of mathematical programming. There are many algorithms for obtaining a numerical approximation to a solution of an unconstrained minimization problem. The most effective algorithms are usually iterative and are based upon some variant of Newton's method for finding a zero of the gradient of the objective function. It is well known that Newton's method must be modified in various ways in order to produce a robust optimization algorithm. Such modifications are generally designed to force convergence of the sequence of iterates to a point which satisfies as many necessary conditions for minimization as possible. The most familiar modifications of this type generally compute a search direction followed by a line search to obtain a new iterate from the previous iterate (Gill and Murray, 1974a). Recently, various strategies have been introduced which have been designed to make better use of second order information when it is available

*Work supported in part by the Applied Mathematical Sciences Research Program (KC-04-02) of the Office of Energy Research of the U.S. Department of Energy under Contract W-31-109-Eng-38.

(Goldfarb, 1980a; Moré and Sorensen, 1979). These strategies have resulted in iterations which converge to critical points which satisfy second order necessary conditions for minimization.

In this paper an alternate approach to safeguarding Newton-like methods is discussed. The approach is well known. It is appropriately called a model trust region method in that the step to a new iterate is obtained by minimizing a local model to the objective function over a restricted ellipsoidal region centered about the current iterate. The diameter of this region is expanded and contracted in a controlled way based upon how well the local model predicts behaviour of the objective function. This strategy will force convergence of the iterates to a critical point which satisfies second order necessary conditions for minimization under very reasonable assumptions about the objective function.

This approach has not enjoyed the popularity of the line search strategies. However, recent work of Moré (1978), Fletcher (1980), Gay (1981), and Sorensen (1980b) would indicate that this type of algorithm can provide very efficient, robust and intuitively appealing algorithms which are supported by an elegant convergence theory. Inequalities which are basic to the convergence theory have been shown to hold in the presence of finite precision arithmetic. This has direct implications concerning the robustness of the algorithm.

2. Model Trust Region Methods

Conceptually this class of algorithms is based upon a very simple and intuitively appealing idea. The numerical problem we wish to solve is:

Given $\phi: \mathbf{R}^n \to \mathbf{R}$ produce a sequence $\{x_k\} \subset \mathbf{R}^n$ such that $x_k \to x^*$, where x^* is a local minimizer of ϕ. (2.1)

An algorithm for producing such a sequence is obtained by specifying a rule for producing a new approximation x_+ to a local minimizer based upon local information at a given point x. To accomplish this we may introduce a local scaling of the variables $X(w): \mathbf{R}^n \to \mathbf{R}^n$ with the property $X(0) = x$, and then consider the scaled function

$$\hat{\phi}(w) \equiv \phi(X(w)).$$

A local model $\psi(w)$ to $\hat{\phi}(w)$ is constructed which is considered valid in a trust region

$$\Gamma_\Delta \equiv \{w: \|w\| \le \Delta\}. \tag{2.2}$$

In principle the norm $\|.\|$ may be any convenient norm on $\mathbf{R}^n$. The step to the minimizer of $\hat{\phi}$ is predicted by solving the model problem

$$\min\{\psi(w): w \in \Gamma_\Delta\}. \tag{2.3}$$

If p is a solution to (2.3) then we predict the minimum of $\hat{\phi}$ to be $\hat{\phi}(p)$ and the local minimizer of ϕ to be

$$x_+ = X(p).$$

Now we compute

$$\rho = \underline{\text{ared}}/\underline{\text{pred}} = [\hat{\phi}(0) - \hat{\phi}(p)]/[\psi(0) - \psi(p)],$$

the ratio of actual reduction observed (ared) to predicted reduction (pred). The number ρ is compared to 1 and certain decisions are made based upon this comparison. If ρ is too small then the model has not given a good prediction within the region Γ_Δ so the control parameter Δ is decreased and the model problem (2.3) is solved with this value of Δ to obtain a new prediction. Otherwise, x_+ is accepted. Local information is obtained at this new iterate and the process is repeated unless convergence requirements are satisfied at x_+. If very good agreement between the model and the scaled objective function is observed ($\rho \sim 1$) then the control parameter might be increased.

The conceptual algorithm is given by:

Algorithm (2.4)

1) Let $0 < \mu < \eta < 1$, $0 < \gamma_1 < 1 < \gamma_2$ be specified;
2) Given x_0, Δ_0;
3) *For* $k = 0,1,2,\ldots$ *until* "convergence" execute the following sequence (a) — (g).
   (a) Construct $X_k(w)$ with $X_k(0) = x_k$;
   (b) Construct a local model $\psi_k(w)$ to $\hat{\phi}_k(w) \equiv \phi(X_k(w))$;
   (c) Compute a solution p_k to the model problem $\min\{\psi_k(w) : w \in \Gamma_{\Delta_k}\}$;
   (d) Compute $\rho_k = \underline{ared_k}/\underline{pred_k}$;
   (e) *If* $\rho_k < \mu$ *then*
      (i) $\Delta_k := \gamma_1 \Delta_k$;
      (ii) *go to* c;
   (f) *If* $\mu \le \rho_k$ *then* $x_{k+1} := X_k(p_k)$;
   (g) *If* $\eta < \rho_k$ *then* $\Delta_{k+1} := \gamma_2 \Delta_k$ *else* $\Delta_{k+1} := \Delta_k$.

Of course, many strategies for revising the value of Δ are possible. Some of these can be found in (Fletcher, 1980a). In practice noticeable improvements can be achieved with some of these strategies. The simplest scheme is given here in the interests of brevity and clarity.

3. Properties of Trust Region Algorithms

The most important observation to make about this class of algorithms is that, once a local model is chosen with "desirable" approximation properties, then the burden of the local prediction rests entirely upon the control parameter Δ. The algorithmic advantage of this is great because one need not make *ad hoc* modifications to the local model in order to control the step length. All of this is automatically taken care of within the mechanism for revising Δ based upon the comparison of <u>ared</u> to <u>pred</u>.

The conceptual algorithm has been outlined here in its most general form to emphasize the generality of the approach. The detail is usually dictated by requirements of the structure of the class of optimization problems which the specific algorithm is designed to solve. The choice of the local model is, of course, crucial to the overall performance of the algorithm. Generally, a model is chosen which will provide excellent local convergence properties when the resulting unrestricted iteration is begun at a point that is "close" to a local minimizer. It is also important that <u>ared/pred</u> $\to 1$ as $\Delta \to 0$ so that a new iterate is always accepted after a finite number of trials.

In order to realize an implementable algorithm, the local model problem must be solvable numerically in an efficient and stable manner. This requirement has a great deal of influence upon the choice of norm and the local scaling of variables as well as the choice of a local model. There are certain subtleties to the model problem even in the simplest of cases. In fact, until recently this has probably been the main reason these algorithms have not been so pervasive as line search type algorithms. However, in certain cases the subtleties are now well understood and algorithms based upon these choices of local model for specific problems have been implemented with very satisfying results. Two interesting examples are an algorithm for the nonlinear least squares problem, and an algorithm for the general unconstrained minimization problem when second order information is available.

For the nonlinear least squares problem, the objective function is

$$\phi(x) = \| f(x) \|_2, \tag{3.1}$$

where $f : \mathbf{R}^n \to \mathbf{R}^m$ with $m \geq n$. The local scaling is typically of the form $X_k(w) = x_k + w$, and the local model is

$$\psi_k(w) = \| f_k + J_k w \|_2. \tag{3.2}$$

In (3.2) $f_k = f(x_k)$, and J_k is the Jacobian of f evaluated at x_k. The choice of norm in (2.2) is often $\|w\| \equiv \|D_k w\|_2$, where D_k is a diagonal matrix. It is possible (Moré, 1978) to choose these scaling matrices so that the iteration is invariant with respect to diagonal scaling of the variables, however the full potential of this possibility has not been thoroughly researched. The only requirement on the D_k is that they have uniformly bounded condition numbers.

For the unconstrained minimization problem let us assume that second order information is available in the form of actual calculation or finite difference approximation of the Hessian of the objective function ϕ. The local scaling may take the form

$$X_k(w) = x_k + Z_k w, \tag{3.3}$$

where Z_k is an n×n real nonsingular matrix. The local model is of the form

$$\psi_k(w) = \phi_k + g_k^T w + \tfrac{1}{2} w^T B_k w, \tag{3.4}$$

where $\phi_k = \phi(x_k)$, $g_k = Z_k^T \nabla\phi(x_k)$, and $B_k = Z_k^T \nabla^2 \phi(x_k) Z_k$. The choice of norm in (2.2) has the same possibilities as the least squares problem.

The algorithms resulting from the above choices of model problem are supported by a very strong theory when applied to the problems which satisfy appropriate (mild) assumptions. The resulting algorithm in the nonlinear least squares case is a version of the Levenberg-Marquardt algorithm. It can be shown (Moré, 1978) that the iteration will force

$$\lim_{k\to\infty} \| J_k^T f_k \| = 0, \tag{3.5}$$

under the assumption that the Jacobian of f is uniformly continuous. In the case of unconstrained minimization with second order information the resulting algorithm is a modified Newton's method. In this case it can be shown (Sorensen, 1980b) that every limit point of the sequence produced is a critical point of ϕ which satisfies the second order necessary conditions for minimization. Moreover, if any limit point of the sequence satisfies the second order sufficient conditions, then the entire sequence will converge to this point and the local rate of convergence will ultimately be quadratic. The assumptions required are continuous second derivatives on a level set, which contains only finitely many critical points of ϕ, and uniform boundedness of the scaling matrices $\{\|Z_k\|\}$ and $\{\|Z_k^{-1}\|\}$.

4. Solution of the Model Problem

The numerical solution to the model problem can be interesting in its own right even in the simplest of cases. For example, in the case of unconstrained minimization ((3.4) above with the 2-norm taken in (2.2)) a solution to the model problem (2.3) is found by computing a scalar λ and a vector $p \in \mathbf{R}^n$ such that the matrix $B + \lambda I$ is positive semidefinite and satisfies

$$\begin{array}{l} \text{(i)} \quad (B + \lambda I)\, p = -g \\ \textit{with} \ \text{(ii)} \ \|p\| = \Delta \ \text{and} \ \lambda > 0 \ \textit{or} \ \text{(iii)} \ \|p\| \le \Delta \ \text{and} \ \lambda = 0. \end{array} \quad (4.1)$$

In (4.1) the subscript k has been omitted. The choice of the 2-norm in (2.2) is important here because satisfying (4.1) provides a global solution to the model problem. This will not be true in general when other norms (e.g. ∞-norm) are used. Moreover, these conditions show that, if the Newton step does not lie within the trust region, then the solution to the model problem is a highly structured zero finding problem in λ.

If p_λ denotes a solution to (4.1 i) corresponding to a given value of λ then conditions (4.1) are satisfied if either $\|p_\lambda\| = \Delta$ or $\lambda = 0$ and $\|p_0\| \le \Delta$. Reinsch (1967, 1971) and Hebden (1973) have observed that great advantage could be taken of the fact that the function $\|p_\lambda\|^2$ is a rational function in λ with second order poles a subset of the negatives of the eigenvalues of the symmetric matrix B. Knowledge of this functional form suggests that Newton's zero finding method should be very efficient when applied to the problem

$$\frac{1}{\Delta} - \frac{1}{\|p_\lambda\|} = 0. \tag{4.2}$$

This is because problem (4.2) is almost linear in λ near a solution. Merely knowing the functional form of $\|p_\lambda\|$ makes it possible to apply Newton's iteration to problem (4.2) using the Cholesky factorization $R^TR = B + \lambda I$ to evaluate the necessary derivatives. Explicit computation of the eigensystem of the matrix B, as suggested by Goldfeld, Quandt and Trotter (1966), is not required. Unfortunately, if the gradient g is "nearly" orthogonal to the eigenspace corresponding to the smallest eigenvalue of B, then Hebden's scheme may still require a large number of iterations to produce a sufficiently accurate solution to the model problem. This is not acceptable since a matrix factorization is required for each of these iterations. A numerically stable method is reported in Moré and Sorensen (1981), which overcomes this problem and which requires only 1.5 to 2 factorizations per model problem on the average. The scheme is similar to Moré's (1978) scheme for nonlinear least squares combined with a clever use of the condition estimator routine in LINPACK (Dongarra, Moler, Bunch and Stewart, 1979).

5. Future Developments

Much of the recent interest in trust region methods can be traced to the work of Powell (1975). He demonstrated very powerful convergence properties for a quasi-Newton algorithm for the unconstrained minimization problem. In this algorithm the matrix B in (3.4) was estimated by a certain quasi-Newton updating formula. This technique has been extended to the sparse case by Toint (1979). In this scheme a very relaxed numerical solution to the model problem (2.3) is used. It is known as the "dog-leg step". This strategy has been explored further by Dennis and Mei (1979). All of these strategies are viable in the context of a quasi-Newton method. There has been interest in determining whether more extensive effort to compute the step is justified when second order information is not explicitly available. At the moment the additional work does not seem to be justified since it would increase the arithmetic by an order of magnitude at each step over existing quasi-Newton schemes. Moreover, the trust region calculation described in Section 4 produces a step that is heavily influenced by the eigensystem of the matrix B. For a quasi-Newton approximation the eigensystem of B may bear little relation to the true Hessian matrix. Other difficulties arise when positive definite updates such as BFGS are used.

In the large sparse case it would be of interest to develop a model trust region step calculation which does not require the storage of a matrix factor. An important question here is whether the extensive calculation required to compute a step is justified at an iterate that is "far" from a solution. In the small to medium size problems (those which easily fit in core) the calculations described above are justified because the computation of second order information will generally dominate the calculation. This may not be valid in large scale problems. Dembo, Eisenstat and Steihaug (1980) and Dembo and Steihaug

(1980) have made important contributions in this area.

Extension of these techniques to the constrained optimization problem is also of interest. For linear constraints there is little apparent difficulty. However, there seems to be difficulty in the case of nonlinear constraints. One might try to include the trust region as an additional constraint to local quadratic-linear approximations to the constrained problem. However, the resulting quadratic program will not be straightforward to solve with such a constraint. Fletcher (1981c) has made some progress in this direction. He considers the trust region method applied to an objective function ϕ, which is a composite of the given objective function plus a convex function of the components of the constraints. A local (second order accurate) model is constructed and the scheme described in Section 3 is carried out. He is able to obtain very nice convergence results. However, the implementation is not thoroughly discussed. Additional work along these lines is certainly called for in light of the excellent performance of this class of algorithms in the unconstrained case.

Finally, it might be very profitable to devise algorithms for restricted classes of problems which have special structure. We have only mentioned two specific applications of the trust region approach here. In both of these the local model was chosen to reflect important features of the class of optimization problems under consideration. Richard Byrd (see Paper 2.2 of this volume) has developed a promising class of algorithms for robust regression problems along these lines. He constructs local models which majorize the objective function over the trust region. The majorization is possible due to the structure of the objective function. The resulting algorithms achieve faster overall convergence than competitive algorithms for the same problem class.

1.5

The Use of Ordinary Differential Equations in the Solution of Nonlinear Systems of Equations*

Francesco Zirilli

Istituto Matematico, Università di Camerino,
62032 Camerino (MC), Italy.

1. Introduction

Let $\mathbf{R}^n$ be the n-dimensional real Euclidean space, let

$$x = (x_1, x_2, \ldots, x_n)^T \in \mathbf{R}^n$$

be a generic vector in $\mathbf{R}^n$, and let $f_i(x)$, $i = 1,2,\ldots,n$, be real valued regular functions, so $f(x) = (f_1(x), f_2(x), \ldots, f_n(x))^T \in \mathbf{R}^n$.

We will consider the problem of solving the system of simultaneous equations

$$f(x) = 0. \tag{1.1}$$

In order to solve this system many methods based on ordinary differential equations have been proposed. Our purpose here is to review some of them, and to discuss recent trends of research on this subject.

2. Continuation Methods

The solution of nonlinear equations by a one parameter opera-

*Work partially supported by Ministero della Pubblica Istruzione under contract T.V. 31.8.33 n. 1592/1980 cap. 8551.

tor imbedding has been considered by several authors. The most commonly used imbeddings are:

given $x_0 \in \mathbf{R}^n$

$$H(t,x) = f(x) + (t-1)f(x_0), \qquad t\in[0,1], \tag{2.1}$$

or

$$H(t,x) = (1-t)(x-x_0) + tf(x), \qquad t\in[0,1], \tag{2.2}$$

or

$$H(t,x) = f(x) - e^{-t}f(x_0), \qquad t\in[0,+\infty). \tag{2.3}$$

The solution to the original problem (1.1) is found by following the solution curve $x(t)$ of $H(t,x) = 0$.

The use of differential equations to follow the solution curve $x(t)$ has been considered by several authors (Davidenko, 1953; Jacovlev, 1964; Meyer, 1968).

Along the solution curve $x(t)$ we have

$$\frac{dH(t,x(t))}{dt} \equiv \frac{\partial H(t,x(t))}{\partial t} + \frac{\partial H(t,x(t))}{\partial x}\frac{dx(t)}{dt} = 0. \tag{2.4}$$

Let $J(x) = \partial f/\partial x$ be the Jacobian of f with respect to x. For the choice of $H(t,x)$ given by (2.3), equation (2.4) reduces to

$$\frac{dx}{dt} = -J^{-1}(x(t))\ f(x(t)), \qquad t\in[0,+\infty), \tag{2.5}$$

with the initial condition

$$x(0) = x_0. \tag{2.6}$$

If the solution $x(t)$ of the initial value problem [(2.5), (2.6)] exists in $t\in[0,+\infty)$ and $x^* = \lim_{t\to\infty} x(t)$ exists, then x^* is a solution of (1.1).

Numerical methods for the solution of (1.1) can be obtained by integrating numerically the initial value problem [(2.5), (2.6)] or the ones corresponding to (2.4) for the choices of $H(t,x)$ given by (2.1) or (2.2).

Meyer (1968) considered the problem of integrating numerically the initial value problems that may be derived from (2.1) and from (2.2). Since the independent variable t is in [0,1], and since the approximate value computed for x(1) will be accepted as the solution of the original problem (1.1), the accuracy of the integration method chosen is of primary importance.

In integrating numerically [(2.5), (2.6)], however, where the independent variable t varies on $[0,+\infty)$, good accuracy is less important since, hopefully, further refinement of a numerically computed approximate solution will always be possible by continuing the integration in t.

Let x^* be a solution of (1.1) such that $J(x^*)$ is nonsingular; then $x(t) = x^*$ is an asymptotically stable solution of (2.5). That is, there exists a neighbourhood $U \subseteq \mathbf{R}^n$ of x^* such that, if $x_0 \in U$, then

$$\lim_{t\to\infty} \| x(t,x_0) - x^* \| = 0,$$

where $x(t,x_0)$ is the solution of [(2.5), (2.6)].

The problem of the numerical integration of (2.5) from the point of view of obtaining solutions of (1.1) has been considered by Boggs (1971) and by Abbott and Brent (1975).

Boggs considered the idea of using linear multistep A-stable methods, and chose the trapezoidal rule from the linear multistep A-stable methods. The trapezoidal rule is an implicit scheme so that at each integration step a nonlinear system of equations has to be solved. To avoid this difficulty, a predictor corrector method is used. The use of the predictor corrector technique forces the integration step to be relatively small, so that some of the advantages of using an A-stable method seem to be lost.

Abbott and Brent observed that at a solution x^* of (1.1) with $J(x^*)$ nonsingular, the Frechet derivative of $J(x)^{-1}f(x)$ is always the identity, so that no stiffness occurs at a solution. Instead of implicit A-stable methods, they preferred explicit

Runge–Kutta methods. Because of the special structure of (2.5) for a particular subclass of Runge–Kutta methods, they were able to obtain R-superlinear convergence of the iterates to x^*.

3. "Globally Convergent Methods" for Special Classes of Problems

When $J(x)$ becomes singular on the path defined by [(2.5), (2.6)], the previously described continuation method fails to converge. To circumvent this difficulty, Branin (1972) proposed to give up the property, implicitly contained in (2.3), that each component of f goes to zero monotonically.

Branin considered the differential equation

$$\frac{dx}{dt} = -\sigma(x(t))\ J^{-1}(x(t))\ f(x(t)), \tag{3.1}$$

where $\sigma(x) = \text{sign}(\det J(x))$. Several related choices, where $\sigma(x)$ is a suitable function whose sign is that of $\det J(x)$, are considered by Smale (1976) and by Hirsch and Smale (1979).

Hirsch and Smale have proved that, if $\Omega \subset \mathbf{R}^n$ and if f satisfies certain boundary conditions on $\partial\Omega$, then the trajectories defined by [(3.1), (2.6)] will contain at least one solution of (1.1) for almost every $x_0 \in \partial\Omega$. The "boundary conditions" for f are locally satisfied around a solution x^* of (1.1) with $J(x^*)$ nonsingular, and they are also satisfied on $\partial\Lambda$, where Λ is a compact, convex set of $\mathbf{R}^n$, such that the function $f(x) + x$ maps Λ into itself.

This special class of problems, called Brouwer fixed point problems, is the earliest case where "globally convergent" algorithms are available.

Singular Jacobians $J(x)$ on the path described by [(3.1), (2.6)] do not cause problems in the proof of the results of Hirsch and Smale, but they cause difficulties in the actual computation of the trajectories. In fact, when the Jacobian becomes singular along a trajectory, several cancellations take place in (3.1).

Some of these difficulties have been bypassed by using a slightly different approach due to Keller (1978) and to Kellog, Li and Yorke (1977). Specifically, let us consider

$$H(\lambda,x) = f(x) - (1-\lambda)f(x_0) \tag{3.2}$$

or

$$H(\lambda,x) = \lambda f(x) + (1-\lambda)(x-x_0), \tag{3.3}$$

and try to track the zero curve of (3.2) or (3.3) emanating from $x = x_0$ at $\lambda = 0$. However, we think now of λ as a new independent variable and not as a parameter as we did in Section 2.

The zero curves of (3.2) and (3.3) satisfy the differential equation

$$\frac{\partial H}{\partial \lambda}\frac{d\lambda}{dt} + \frac{\partial H}{\partial x}\frac{dx}{dt} = 0. \tag{3.4}$$

In this system we have (n+1) unknowns and n equations. We add the equation

$$\left(\frac{d\lambda}{dt}\right)^2 + \left\|\frac{dx}{dt}\right\|^2 = 1, \tag{3.5}$$

which implies that t is the arclength on the path $(\lambda(t), x(t)) \subset \mathbf{R}^{n+1}$.

The paths generated by [(3.4), (3.5)] have similar "global convergence" properties to the ones generated by (3.1) but they can be computed more efficiently. Some interesting numerical experience obtained with these methods on problems from different areas of applied physics can be found in Watson (1980). Further, special methods for finding all solutions to polynomial systems of equations have been introduced by Garcia and Zangwill (1980).

4. Classical Mechanics and Systems of Equations

Up to now we have considered only first order systems of differential equations. The use of higher order systems was originally proposed by Poljak (1964).

We define

$$F(x) = f^T(x) \cdot f(x) = \sum_{i=1}^{N} f_i^2(x). \tag{4.1}$$

It is easy to see that x* is an isolated solution of (1.1) if and only if x* is an isolated minimizer of F(x) and F(x*) = 0. Incerti, Parisi and Zirilli (1979) proposed the following second order system of differential equations:

$$\mu(t) \frac{d^2x}{dt^2} = \sigma(t) D \frac{dx}{dt} - \nabla F(x(t)), \tag{4.2}$$

where μ and σ are real valued continuous functions, where $\mu(t) > 0$, where D is an n×n positive matrix, and where ∇F is the gradient of F with respect to x.

The equation (4.2) represents Newton's second law (mass × acceleration = force) for a particle of mass $\mu(t)$ moving in $\mathbf{R}^n$ subject to the force $-\nabla F$ given by the potential F and to the force $-\sigma(t)$ D dx/dt. When $\sigma(t) > 0$, the force $-\sigma(t)$ D dx/dt is dissipative. The methods related to (4.2) are studied in Aluffi, Incerti and Zirilli (1980a, b), Incerti, Parisi and Zirilli (1981) and Zirilli (1981).

In order to use equation (4.2) to solve the system of simultaneous equations (1.1), the following two main approaches seem possible:

(A) Initial value approach If x* is a minimizer of F(x) then $x(t) \equiv x^*$ is a solution of (4.2). Consider the Cauchy data

$$\left.\begin{aligned} x(0) &= x_0 \\ dx(0)/dt &= w_0 \end{aligned}\right\} , \tag{4.3}$$

and let $x(t, x_0, w_0)$ be the solution of the initial value problem [(4.2), (4.3)]. Then, under some assumptions, it can be proved that, if $\| x_0 - x^* \|$ and $\| w_o \|$ are less than certain positive constants, then

$$\lim_{t\to\infty} \| x(t, x_0, w_0) - x^* \| = 0. \tag{4.4}$$

Therefore, in order to solve the system of nonlinear simultaneous equations (1.1), we integrate numerically the Cauchy problem [(4.2), (4.3)].

(B) <u>Boundary value approach</u> Given $x_0, x_1 \in \mathbf{R}^n$, consider the boundary conditions

$$\left.\begin{aligned} x(0) &= x_0 \\ x(\delta) &= x_1 \end{aligned}\right\} , \tag{4.5}$$

where δ is a positive constant. Under some regularity assumptions on F, it can be shown that there exists a value of δ, depending on x_0, x_1 and F, such that the boundary value problem [(4.2), (4.5)] has a solution. In particular, if $x_1 = x^*$, x^* being a solution of the system (1.1), this means that there exists a trajectory of (4.2) connecting x_0 and x^*. Since x^* is unknown, this trajectory has to be computed as a solution of the initial value problem [(4.2), (4.3)] for an appropriate choice of $w_0 = w^*$. The right choice of w^* can be constructed from an initial guess through an iterative procedure.

Some of the advantages of methods that make use of (4.2) over methods using systems of first order differential equations are:

i) from the same initial guess, x_0, different choices of w_0 give rise to different solutions x^* of (1.1);

ii) no restrictions on the rank of $J(x)$ along the trajectory are needed to guarantee convergence. In particular, regions where $J(x) \equiv 0$ may be crossed by the trajectories of (4.2);

iii) because of the inertial term (i.e., the second order term) in equation (4.2), local minima of $F(x)$ that are not solutions of (1.1) may be crossed.

The initial value approach (A) is studied in detail in Aluffi *et al.* (1980a, b) and Zirilli (1981); in particular linearly implicit A-stable methods for the numerical integration of [(4.2), (4.3)] are proposed.

Some interesting numerical experience on test problems is shown in Incerti *et al.* (1981) and Zirilli (1981).

The use of second order differential equations in the global minimization of a scalar function is considered by Griewank (1981).

Discussion on "Unconstrained Optimization"

1.6 Lootsma

How do the optimization methods described in Papers 1.1 and 1.2 behave with numerical approximations to the first derivatives (differences of function values) and possibly with numerical approximations to the second derivatives (differences of first derivatives)? Does numerical differentiation affect their relative performance with respect to each other?

1.7 Schnabel

Virtually all our tests have used finite difference gradients and, in second derivative methods, finite difference Hessians, with the step sizes chosen intelligently by the algorithm. Initially, in some tests made by Weiss (1980), algorithms using analytic gradients and Hessians were compared with the same methods using finite difference derivatives. Because the results were virtually indistinguishable, in terms of the number of iterations required and even the individual iterates that were generated, we have since used only finite difference derivatives, for reasons of programming convenience.

1.8 Buckley

All of the computations which I have made with the new algorithm of Paper 1.2 have used analytic gradients. However, Gill and Murray (1979) suggest that conjugate gradient algorithms should only be used for low accuracy solutions. In this case the use of finite differences (properly done of course!) should not make any significant difference to the algorithm performance. It is hoped to do testing soon to see if this is indeed the case.

1.9 Gay

Because of structure in the optimization problems to which conjugate gradient algorithms are applied, I suspect it is generally much faster to compute the gradients analytically than to approximate them by finite differences. Is this not so?

1.10 Gill

At present there are no efficient algorithms for minimizing smooth large scale problems when first derivatives are unavailable. Straightforward finite difference approximations of the gradient vector are unsuitable in the large scale case. We would hope to solve the problem in the number of function evaluations that would be required just to estimate the gradient vector at the starting point.

1.11 Gutterman

When using central differences for gradient estimation, the forward and backward differences are easily available also. The difference between these latter two is some sort of measure of the noise content of the functions. We have been using it to warn the user of possible troubles. I suggest that some algorithmic use might be made of this technique.

1.12 Beale

I recommend the following correction to the description in Paper 1.2 of Beale's conjugate gradient algorithm. In Buckley's notation, but with $\bar{H} = I$, the algorithm sets

$$x_{i+1} = x_i + \alpha_{i+1} d_{i+1},$$

where

$$d_{i+1} = -g_i + \bar{\beta}_i d_i + \gamma_i d_t.$$

We wish to make d_{i+1} orthogonal to the gradient difference vectors y_i and y_t associated with the search directions d_i and d_t. For a quadratic function this implies that d_{i+1} is conjugate to d_i and d_t. Therefore we solve the equations

$$\bar{\beta}_i d_i^T y_i + \gamma_i d_t^T y_i = g_i^T y_i$$

$$\bar{\beta}_i d_i^T y_t + \gamma_i d_t^T y_t = g_i^T y_t.$$

In the quadratic case $d_i^T y_t = d_t^T y_i = 0$, while in general $d_i^T y_t = 0$ but $d_t^T y_i \neq 0$. Hence it is natural to write (as Buckley does)

$$\gamma_i = g_i^T y_t / d_t^T y_t,$$

but the formula for $\bar{\beta}_i$ should be equivalent to the equation

$$\bar{\beta}_i = (g_i - \gamma_i d_t)^T y_i / d_i^T y_i.$$

1.13 Buckley

It is true that one may modify the formula for $\bar{\beta}_i$ in the Beale restart recurrence by solving a 2×2 triangular set of equations, in order to ensure that the new direction d_{i+1} is orthogonal both to the change in gradients y_i and to the change in gradients y_t (or alternatively y_{i-1}) even for nonquadratic f. This idea is suggested also in Gill and Murray (1979). In one sense the idea would make no difference to the claim that the Beale restart formula is equivalent to a preconditioned CG formula, for that claim was made for the quadratic case. However it may be true that one could rewrite the restart recurrence relation with the new value for $\bar{\beta}_i$ as a preconditioned conjugate gradient algorithm with a different preconditioner for the non-quadratic case, and that might be useful.

1.14 Dixon

Let us assume we are solving a problem in $\mathbf{R}^n$, where n is so large that we only wish to apply the variable metric method m times (m<<n). In these circumstances, instead of forming the variable metric matrix, we store it in product form. If the BFGS method is used to update B, and if B is factorised as JJ^T, then the updated J has the form

$$J^+ = \left(I + \frac{(cy-Bd)\,d^T}{d^T Bd}\right)J = (I+zd^T)J.$$

Thus the matrix J can be viewed as the product of at most m special matrices, that are each stored by holding the 2n elements of z and d. All the matrix-vector products can then be formed efficiently.

In this structure B is positive definite for any combination of the special matrices. In particular it remains positive definite if the oldest of the accumulated corrections are discarded to make way for new updates. Whether the remaining matrix has any other desirable properties is however an open question.

1.15 Davidon

While using just the most recent m steps for an update may be appropriate when these are linearly independent, inclusion of information from earlier steps may be useful when the recent ones are linearly dependent. Therefore the technique of retaining information from only the m most recent iterations may be inefficient.

1.16 Kowalik

Please comment on numerical results for the conic method described in Paper 1.3.

1.17 Davidon

Bjorstad and Nocedal (1979) have analysed and tested algorithms using conic models for one dimensional optimization and found them generally superior to those using cubic models. Sorensen (1980a) has run some tests on an n-dimensional algorithm of this type which have been encouraging though preliminary. Robert Schnabel will be telling us of some of his results later in the conference.

1.18 Overton

I would like to point out the work on conjugate gradient methods using conic models in Bourgeon and Nocedal (1981). One possible advantage of these method is that they help the choice of a suitable initial step in the line search.

1.19 Kowalik

What objective functions are most suitable for conic approximation techniques? Do you expect that these techniques will be useful for general functions or only for some special cases?

1.20 Davidon

In the one-dimensional case, Bjorstad and Nocedal (1979) have found that the use of conic models in a line search algorithm gives better results than a cubic model for many test problems, and they have derived a condition on the first four derivatives of the objective function for which the conic models give more rapid asymptotic convergence. There is less information available on the n-dimensional case, but the following two features of objective functions can be incorporated into conic but not quadratic models. (1) A singularity, or rapid growth in the function near some hyperplane, as with exponential or penalty functions, can be included. (2) The model can provide asymmetric behaviour of the function about its minimizer. Moreover it is useful sometimes that the direction to the minimizer of the model is not the direction of the Newton step.

1.21 Griewank

The first three terms of the Taylor expansion are usually considered as the best quadratic model of a given smooth function in some neighbourhood of a given point and in particular near its minimizer. Is it possible to define correspondingly a unique conic model which yields in some sense an optimal local approximation to the function at a given point?

1.22 Davidon

A conic model can match function values $f_{\pm}$ and gradients $g_{\pm}$ at two points $x_{\pm}$ iff

$$(f_+-f_-)^2 \geq g_+^T(x_+-x_-)\; g_-^T(x_+-x_-) \;.$$

This condition usually holds, and then one-dimensional conic interpolations can be used in line searches. A conic model can also be used to match the gradients at three points along a line to determine not only the one-dimensional minimizer, but also the hyperplane through this minimizer that contains all directions conjugate to the original line. Thus nonlinear dependence of the gradient on position can be used when choosing step directions.

1.23 Griewank

Within the framework of Paper 1.4 the trust region radius Δ depends only on the function value at the new point and is adjusted before the new gradient and possibly the new Hessian are evaluated. This would seem inappropriate since we may confidently take a step from the new point that is greater than the given

Δ if the new gradient is very large, but if the new gradient is rather small, then even a step of size Δ may be much too adventurous.

1.24 Sorensen

There are several sophisticated strategies for adjusting the size of the trust region. I only mentioned the simplest idea in the interests of clarity. Certainly the more sophisticated strategies which take derivative information into account are important in practice and tend to work very well. Your idea of making use of cubic information sounds interesting and should be investigated. However, in my computational experience, on small dense problems, I have not encountered difficulty due to adjusting the size of the trust region.

1.25 Toint

Concerning the possibility of using conjugate gradients to solve the model problem approximately for a trust region, I think it has a certain advantage over a "dog-leg" type method when applied to large dimensional problems. Specifically it usually requires much less work to obtain an acceptable step when the ordinary quasi-Newton step is infeasible. I have been trying different schemes with variable success, and these experiments make me believe that research in this area should be pursued further.

1.26 Gill

It has been stated many times in the literature that a trust region strategy is more "elegant" than a line search strategy. However, these two approaches give fundamentally different methods for determining the *scale* of the search direction in the indefinite case. When the Hessian is indefinite, the quadratic model is unbounded below and can be "minimized" by an infinite step along *any* direction of negative curvature, so it is not possible to determine an *a priori* scaling for the search direction. Both the trust region and line search methods compute x_{k+1} using information from the positive definite part of G_k. With a line search method this is done directly from an appropriate modification of the Hessian and from a line search. With a trust region method, however, the scale is implicitly determined by the size of steps that were taken when G_k was positive definite.

A good implementation of a trust region method should include many features of a line search method and vice versa. For example, a line search method should always be implemented so that $\|x_{k+1}-x_k\|$ is bounded above by a positive scalar. Similarly, a good trust region method will use safeguarded interpolation to adjust the size of the trust region.

1.27 Moré

Another difference between the trust region and line search strategies is that the trust region approach can be made invariant with respect to changes in the scale of the variables, while this is not the case for the line search approach.

1.28 Goldfarb

I would like to point out that the direction of search $-(H_c+E)^{-1}g_c$, computed by Gill and Murray's (1974a) modified

Newton method, tends to be biased towards directions of negative curvature, but the original motivation for the method was to facilitate the numerically stable computation of the Cholesky factorization of the perturbed Hessian matrix.

A second comment that I would like to make is that, when a variable metric algorithm is used to solve a linearly constrained problem, it is important to scale the variable metric matrix appropriately whenever a constraint is dropped from the active set as well as at the start of the computation. In older versions of such algorithms, if this is not done, then in poorly scaled cases the rank of the projected Hessian may become less than it should. In numerically stable versions of these algorithms, even though loss of rank cannot occur, poor scaling can seriously slow down the method.

1.29 Sargent

The snag about "trust region" methods is that they trust the user to formulate his problem in the right frame of reference — in the sense that they assume that the steepest descent direction is a good direction in case of difficulty. If one assumes that the user's frame of reference has no particular merit, and that the current LDU factorization does indeed provide a natural scaling, then the iterative refactorization to determine λ in $(B+\lambda I)$ can be avoided. Instead one may use the Gill and Murray (1974a) idea in the form $L(D+\lambda I)U$, where a suitable choice of λ is immediate. I do not know if this idea has been tried.

1.30 Dembo

One should note that the Gill and Murray (1974a) method is not invariant under a permutation of the variables. This is perhaps unimportant in small scale problems, but in a large scale setting it can be serious since, to prevent fill-in, one will always permute the variables.

1.31 Goldfarb

When the Hessian matrix H is indefinite, its LDU factorization (actually LDL^T since H is symmetric) is not numerically stable, and may not even exist, unless pivoting is used. Therefore, for trust region methods involving possibly indefinite Hessian matrices, I proposed using the numerically stable Bunch and Parlett (1971) symmetric factorization (Goldfarb, 1980b), and the factorizations of Dax and Kaniel (1977) and of Aasen (1971) are also suitable. All of these factorizations can be expressed as $H = P^T LDL^T P$. Consequently, by making the change of variables $y = L^T Px$, one obtains the equation $(D+\alpha I)d_y = \nabla_y f$, which can easily be solved for the step d_y in the y-variables for several values of α.

1.32 Fletcher

I would like to comment that the distortion of the trust region caused by the transformation mentioned in the previous contribution may be substantial and have an adverse effect on the method. Usually pivot strategies allow a certain growth in elements of a factorization, and hence a certain amount of induced ill-conditioning or distortion. Furthermore, the pivot selection and the factors themselves change from one iteration to the next, which can have an adverse effect on the strategy for making suitable adjustments to the radius of the trust region. I am aware that the 'natural' space supplied by the user is arbitrary and may lead to an ill-conditioned Hessian, but at least it is fixed and the ill-conditioning can be controlled to a certain extent by an *a priori* diagonal scaling.

1.33 Sorensen

Using the Bunch and Parlett (1971) factorization is an alternative which I have also considered. It is true that the model problem is easier to solve in this case, but structure of the Hessian is lost due to the pivoting strategy required for the factorization. The solution to the trust region problem proposed by Moré and Sorensen (1981) only requires the Cholesky factorization of a positive (semi)definite matrix. Sparse matrix techniques are well developed for this purpose and could be used. We have observed that this method requires about 1.5 to 2 factorizations on the average per iteration in a minimization subroutine to solve the model problem. Because of this I do not see an advantage in preferring the Bunch–Parlett factorization.

1.34 Dennis

In my opinion, too much has been made of the difference between line searches and trust regions and not enough has been made of their essential similarity. The choice of step-length and trust region radius can, and probably should, be the same algorithm. The difference is in the choice of step direction, but this depends entirely on the scaling of the independent variables used to define the trust region ellipse. In particular, if the scaling is by the approximate Hessian, then the dog-leg, Moré–Hebden, and line search steps are all in the quasi-Newton direction. The essential difference is that the trust-region is generally defined in terms of an ellipse oriented with the user-supplied coordinate system, i.e. a diagonal scaling matrix is used.

1.35 Sorensen

I agree that the trust region and line search methods (for safeguarding Newton's method) are similar when the Hessian matrix is positive definite. In fact they are equivalent for a certain choice of norm. However, when the Hessian matrix is indefinite they are quite different. Two important features of the trust region approach are the automatic relative scaling of positive and negative curvature directions, and that the search path is a space curve in n dimensions rather than a search along a straight line or along a curve restricted to a two dimensional subspace.

1.36 Powell

I find the trust region techniques for forcing convergence far more satisfactory than the crude angle tests, that were often used about ten years ago, for bounding the angle between the search and steepest descent directions away from 90°.

1.37 Rosen

The formulation and numerical solution of nonlinear programming problems as a system of ordinary differential equations was one of the first methods proposed for such problems (Arrow, Hurwicz and Uzawa, 1958). It still has considerable appeal, since such an approach would permit the use of existing ODE software packages for solving the NLP problems. The difficulty is that a great deal of computational effort is likely to be wasted in following and artificially created trajectory, when in fact all that is wanted is the limit point of this trajectory. Thus, while some useful ideas may be obtained by considering different discrete numerical ODE schemes applied to NLP, it does not seem likely that direct computational use of such a scheme will lead to an efficient method.

1.38 Boggs

The use of differential equations (or homotopy or continuation methods) for solving optimization problems will probably not be competitive with well-implemented Newton or quasi-Newton methods on many classes of problems. I think that their strength lies in providing powerful theories and potentially practical techniques for rescuing standard methods when their progress slows or stops due to singularities. In addition, there appear to be particular types of problems for which differential equation methods are directly suitable.

1.39 Conn

I agree that trajectory following techniques have global convergence properties provided one follows the trajectory well, but, in order to avoid slow convergence, in practice I wish to relax the requirement of being close to the trajectory. In order to achieve this, I require some measure of success when the variables are changed: for instance one may require a merit function to decrease monotonically. Therefore my main reservation of the differential equation/homotopy approach is the lack of a suitable criterion for accepting changes to the variables. It is obvious that monotonicity of the objective function does not hold in general, but it is not the only possible measure of success.

1.40 Zirilli

Given the numerical method of integration for the differential equations, the accuracy with which the trajectories are followed depends on the integration step chosen.

Since the non-negative function $F(x) = \| f(x) \|_2^2$ is zero at a solution, monitoring $F(x)$ during the computation can provide various heuristic procedures, which increase the integration step when $F(x)$ is decreasing and which decrease the integration step when $F(x)$ is increasing.

1.41 Todd

I would like to make two points with regard to the need to follow the trajectories of the differential equations very accurately. First, there is a trade-off between accuracy and possibly sacrificing global convergence. In order to guarantee global convergence under the appropriate conditions it is necessary to follow accurately a path of zeros to a fixed homotopy that may depend on the initial point x_0. On the other hand, if a point slightly off this path is generated, it will usually lie on a path of zeros of a slightly perturbed homotopy, possibly corresponding to a perturbed initial point $\tilde{x}_0$. There will usually be little harm in now following this new path. Secondly, if we forgo the differential equation approach, we can make a piecewise-linear approximation to the homotopy and follow a path of zeros of the approximation accurately with no problems. Furthermore, the approximation can be designed to be inaccurate when the homotopy is far from the problem of interest and become progressively more accurate as the true problem is approached. This permits large steps in the early stages of the homotopy.

1.42 Goldfarb

Some people may not be aware that there is a connection between the gradient path (i.e., differential trajectory) methods and trust region methods. If one solves the system of differential equations

$$\dot{x}(t) = -g(x(t)), \quad x(0) = x_0$$

by the backward Euler method

$$x(t) - x_0 = -tg(x(t))$$

after linearizing $g(x(t))$, i.e. $g(x(t)) = g(x_0) + G(x_0)[x(t)-x_0]$, one obtains

$$(I+tG(x_0))(x(t)-x_0) = -tg(x_0)$$

where t above corresponds to $1/\alpha$ in the Goldfeld, Quandt and Trotter (1966) trust region method.

1.43 Botsaris

Consider the differential equation of the steepest descent curve, namely

$$\dot{x}(t) = -g(x), \quad x(0) = x_k,$$

where g(.) is the gradient of the objective function f(.), and where x_k is the starting point of the k-th iteration. Expanding the right-hand side in a Taylor series we obtain the linear system

$$\dot{x}(t) = -g(x_k) - H(x_k)(x-x_k), \quad x(0) = x_k,$$

where H(.) is the Hessian of f(.). The solution to the above system of differential equations is given by

$$x(t) = x_k + [e^{-tH(x_k)} - H]\, H^{-1}(x_k) g(x_k)$$

$$= x_k + \left[\sum_{i=1}^{n} \frac{e^{-t\lambda_i(x_k)} - 1}{\lambda_i(x_k)} u_i(x_k) u_i^T(x_k) \right] g(x_k)$$

where $\{\lambda_i(x_k),\ u_i(x_k);\ i = 1,2,\ldots,n\}$ is the eigensystem of $H(x_k) \in \mathbf{R}^{n\times n}$. The given vector $x(t)$ is always defined and downhill even when the Hessian is singular or nonpositive definite, i.e. $\lambda_i(x_k) \le 0$ for some i. Moreover, for small t we have $x(t) \approx x_k - tg(x_k)$, whereas for large enough t and provided that $H(x_k) > 0$, Newton's point $x_k - H^{-1}(x_k) g(x_k)$ is obtained. Thus $x(t)$ for $0 \le t \le \infty$ describes a continuous changeover from the steepest descent to Newton's method.

By integrating the differential equation numerically using the backward Euler method (which is the simplest A-stable method) with a stepsize t_k, we obtain the implicit equation

$$x_{k+1} = x_k - t_k g(x_{k+1})$$

for x_{k+1}. Applying Newton's method to solve this equation gives

$$y_{\ell+1} = y_\ell - [t_k H(y_\ell) + I]^{-1} [t_k g(y_\ell) + y_\ell - x_k]$$

for $\ell = 0,1,2,\ldots,$ and some approximation y_0 to x_{k+1}. This formula is an extension to the Goldfeld, Quandt and Trotter (1966) method that has just been mentioned by Goldfarb.

Finally, instead of considering the homotopy

$$\mathbf{H}(x,t) \equiv g(x) - e^{-t} g(x_k) = 0, \qquad t \in [0,\infty),$$

yielding the continuous Newton curve

$$\dot{x}(t) = H^{-1}(x) g(x), \qquad x(0) = x_k,$$

let us introduce the homotopy

$$\mathbf{H}(x,t) \equiv e^{-t}[x(t) - x_k] + (1 - e^{-t}) g(x) = 0.$$

By differentiation we obtain

$$\dot{x}(t) = -[e^{-t}I + (1-e^{-t})H(x)]^{-1}g(x), \quad x(0) = x_k,$$

which, as t increases in $0 \leq t \leq \infty$, changes continuously from the continuous steepest descent to the continuous Newton curve.

PART 2

Nonlinear Fitting

2.1

Algorithms for Nonlinear Fitting

J.E. Dennis, Jr.

Mathematical Sciences Department, Rice University, Houston, Texas 77001, USA.

1. Introduction

This paper is to introduce the session which was called "Algorithms for nonlinear fitting, especially least squares calculations" in preliminary announcements of the meeting. However, this paper and the others in the session are more about alternatives to least squares. The reasons for deemphasizing nonlinear least squares are that there is not much left for us to do except some loose ends (admittedly important ones like modifying our codes to handle constraints of the type found in Section 4), and that our more sophisticated clients are beginning to prefer fitting criteria that are more robust or more statistically relevant than least squares criteria. This is a common pattern in numerical analysis: a client asks for help and when we succeed in giving some, then he tells us about the larger or more nonlinear problem he now thinks he really wanted all along to solve. Of course, this ensures us of steady employment.

One valid objection to the least squares criterion for fitting is its sensitivity to outliers or exceptional points in the data. It is ironic that this same complaint, which caused a deemphasis of the minimax criterion a decade ago and put fitting algorithms firmly in the realm of differentiable optimization, has now led to the current interest in ℓ_1 fitting and put fitting algorithms back in both the differentiable and nondifferentiable camps. In this volume, Claude Lemaréchal has a paper on the significant progress in nondifferentiable optimization over the last few years and Michael Overton has one on nonlinear ℓ_1 fitting with constraints. Bartels and Conn (1981) is another timely reference on the same subject.

Robust regression is a less extreme alternative to ℓ_1 for dealing with outliers. The optimization problem usually is at least C^1 and piecewise C^2, but, even with linear fitting, one generally loses convexity. Dennis (1977) gives an introduction to the subject and a local analysis of some Gauss–Newton-type algorithms. Richard Byrd's paper in this volume is on a new global convergence technique (Byrd and Pyne, 1979), developed to deal with the standard algorithms for this problem. The technique applies to the important facility location problem as well.

This paper will begin with a brief discussion of the common structure found in fitting problems and how it can be exploited in our algorithms. It then gives some examples of fitting problems we should consider for algorithm development. Most of this portion will be based on joint work and discussions with David Gay and Roy Welsch (Dennis, Gay and Welsch, 1981; Gay, 1980b — Gay, 1980b is the more complete reference). The details of the nonlinear least squares problem are so familiar to numerical analysts that there seems little point in setting aside a separate section for it, since our understanding has changed little since Dennis (1977). Anyway, it is a special case of the next section.

2. Algorithms for Composite Minimization

The most common form of a fitting problem is

$$\min_{\beta \in \mathbf{R}^p} f(F(\beta)),$$

where β is some parameter vector, $F:\mathbf{R}^p \to \mathbf{R}^n$ and $f:\mathbf{R}^n \to \mathbf{R}$. The ℓ_1, ℓ_2 and ℓ_∞ problems correspond to taking f respectively equal to $\|\cdot\|_1$, $\|\cdot\|_2^2$ and $\|\cdot\|_\infty$ on $\mathbf{R}^n$, and then allowing F full generality. This illustrates another rather common feature: derivatives of f are either cheap or nonexistent and those of F are either expensive, inconvenient, or unavailable in analytic form. Thus, it is very intriguing in the differentiable case that, once we have

$$\nabla_\beta f(F(\beta)) = F'(\beta)^T \nabla f(F(\beta)),$$

we also have a portion of

$$\nabla_\beta^2 f(F(\beta)) = F'(\beta)^T \nabla^2 f(F(\beta)) F'(\beta) + \sum_{i=1}^{n} \partial_i f(F(\beta)) \nabla_\beta^2 F_i(\beta)$$

$$\equiv G(\beta) + S(\beta).$$

There are various ways to argue for various fitting problems that $S(\beta)$ can be neglected for n sufficiently large. This leads to Gauss–Newton-type algorithms, which are based on choosing the next iterate by minimizing a local quadratic model of $f(F(\beta))$, whose Hessian is $G(\beta)$, rather than $G(\beta) + S(\beta)$ as in Newton's method. A more direct way to obtain the Gauss–Newton-type algorithms is to abandon the idea of a local quadratic model of the optimization problem, and instead to choose a trial change to an estimate of a solution of the nonlinear fitting problem by solving a local linear-fitting model problem. In other words, we linearize F at the current iterate β_c and then solve

$$\min_\beta f(F'(\beta_c)(\beta - \beta_c) + F(\beta_c))$$

to obtain the next iterate. This is a useful way to proceed when f is not differentiable.

We all know that the Gauss–Newton algorithm can solve nonlinear least squares problems quite effectively, but it can be quite slow if $S(\beta^*)$ is a significant part of $\nabla^2 f(F(\beta^*))$ at a solution β^* to the fitting calculation. For this reason, biological and social science data are often fitted using general unconstrained minimization algorithms. I have always regarded this is an embarrassment and a challenge. Several attempts have been made to incorporate the attainable portion $G(\beta)$ into an approximate Hessian $H(\beta)$: see Dennis, Gay and Welsch (1981) for a discussion, list of references, and a description of NL2SOL, which is a particular routine of the type considered below.

Dennis and Schnabel (1979) give a general procedure for obtaining $H(\beta) \approx \nabla_\beta^2 f(F(\beta))$ in the form $H(\beta) = G(\beta) + S$, by using least-change secant approximations to update S. The method has worked well in its NL2SOL implementation for nonlinear least squares and Dennis and Walker (1980) show local q-superlinear convergence for the general technique. NL2SOL also uses an interesting adaptive strategy to decide at each iteration whether to use $G(\beta)$ or $H(\beta)$, since q-linear methods are not always slower than q-superlinear methods in achieving convergence numerically.

The idea behind the S_c to S_+ update is a simple one. It gives the secant condition

$$\{G(\beta_+) + S_+\}s_c = y_c \equiv \nabla_\beta f(F(\beta_+)) - \nabla_\beta f(F(\beta_c)),$$

where $s_c = \beta_+ - \beta_c$. In order to make the update we satisfy the secant condition by applying the DFP formula to $\{G(\beta_+) + S_c\}$ to give $\{G(\beta_+) + S_+\}$. Thus S_+ is calculated, and we obtain a Hessian approximation that incorporates our knowledge of $G(\beta_+)$, and that enjoys the good properties of the DFP update with respect to scale invariance and positive definiteness. Further details are given by Dennis, Gay and Welsch (1981).

We can refine this procedure by using a more sophisticated secant condition than $\{G(\beta_+) + S_+\}s_c = y_c$ but the idea and the geometry are the same, although the computational results improve a bit. In particular, we can replace y_c by $\tilde{y}_c = y_s + G(\beta_+)s_c$, where y_s is the vector

$$y_s \equiv F'(\beta_+)^T \nabla_\beta f(F(\beta_+)) - F'(\beta_c)^T \nabla_\beta f(F(\beta_+)).$$

3. The Exponential Family of Maximum Likelihood Estimates

Maximum likelihood estimation is a statistical topic that gives rise to some very interesting composite optimization problems. In the parametric form, there is some data $y \in \mathbf{R}^n$ and a parameterized model $M(\beta) \in \mathbf{R}^n$. There is also a likelihood function $\rho(y_i; M_i(\beta))$, which measures the likelihood of observing y_i from a random variable depending on $M_i(\beta)$ and subject to noise. If the noise at each observation is independent and identically distributed, then the likelihood of observing $y \in \mathbf{R}^n$ is

$$\prod_{i=1}^{n} \rho(y_i; M_i(\beta)).$$

The idea is to choose the parameter $\beta \in \mathbf{R}^p$ to maximize the likelihood of observing the y that was observed. For a rigorous treatment based on Bayes Theorem, see Tapia and Thompson (1978).

Generally, we have the optimization problem:

$$\min_{\beta \in \mathbf{R}^p} L(\beta) = -\sum_{i=1}^{n} \log \rho(y_i; M_i(\beta)),$$

where the objective function is now the negative (to switch from maximization to minimization) of the log (to switch from a product to a sum) of the likelihood function.

The function ρ is often of the form

$$\rho(y_i; M_i(\beta)) = \exp[y_i M_i(\beta) - g(M_i(\beta)) + h(y_i)]$$

except for some nuisance parameters. The minimization problem is then

$$\min_{\beta} L(\beta) = -\sum_{i=1}^{n} \{y_i M_i(\beta) - g(M_i(\beta)) + h(y_i)\},$$

and the derivative of L has the value

$$L'(\beta) = - \sum_{i=1}^{n} \{y_i M_i'(\beta) - g'(M_i(\beta)) M_i'(\beta)\}$$

or

$$\nabla L(\beta) = -M'(\beta)^T \{y - g'(M(\beta))\},$$

where $g'(M(\beta))$ is the n-vector whose i-th component is $g'(M_i(\beta))$. We see that $y - g'(M(\beta))$ is a sort of residual in the sense we are used to, and its expected value is zero under standard large sample assumptions.

The functions $g(t) = t^2/2 = -h(t)$ give the sum of squares $L(\beta) = \frac{1}{2} \Sigma_i (y_i - m_i(\beta))^2$. Here there are effective programs like those of Moré (1978) and Dennis, Gay and Welsch (1981), which are both available through IMSL. There is a need for research on the incorporation of constraints into this problem, especially simple bounds on β and on $y_i - M_i(\beta)$.

Two important problems that need our attention are the logistic problem and the Poisson problem. In order to break out of the ℓ_2 mode, it is instructive to give some detail on one case. Logistic regression has to do with predicting the number of successes from a given number of trials by modelling the probability of a success. Let y_i be the number of successes observed in n_i independent trials. If Π_i is the probability of success in any trial of the i-th set of trials and if there are n sets of trials, then the likelihood of the observed number of successes is

$$\prod_{i=1}^{n} \binom{n_i}{y_i} \Pi_i^{y_i} (1 - \Pi_i)^{n_i - y_i} .$$

Therefore, except for a redundant constant, we have the objective function

$$L(\Pi) = - \sum_{i=1}^{n} \{y_i \log \frac{\Pi_i}{1-\Pi_i} + n_i \log (1-\Pi_i)\} ,$$

where $\Pi = (\Pi_1, \Pi_2, \ldots, \Pi_n)$. However, it is simpler to parameterize according to

$$M_i(\beta) = \log \frac{\Pi_i}{1-\Pi_i} ,$$

which leads to

$$L(\beta) = \sum_{i=1}^{n} \{-y_i M_i(\beta) + n_i \log(1+e^{M_i(\beta)})\} .$$

If $n_i = n_t$ for every i, then $L(\beta)$ has the standard form, where $h(t) = 0$ and $g(t) = -n_t \log(1+e^t)$. The nearest thing to a residual in this case is the vector whose i-th coordinate is

$$y_i = n_t e^{M_i(\beta)} / (1+e^{M_i(\beta)}).$$

In the notation of Section 2, $F(\beta) = M(\beta)$ and $f(v) = \Sigma_i\{-y_i v_i + n_t \log(1+\exp(v_i))\}$.

Poisson regression is another case when the observations are of a nonnegative, integer-valued random variable. The standard form, obtained from $g(t) = e^t$ and $h(t) = -\log t$, is:

$$L(\beta) = \sum_{i=1}^{n} \{-y_i M_i(\beta) + e^{M_i(\beta)} + \log y_i\}.$$

We drop $\Sigma_i \log y_i$, which gives $F(\beta) = M(\beta)$ and $f(v) = \Sigma\{\exp(v_i) - y_i v_i\}$, in the notation of Section 2.

4. The Mixture Density Problem

This section introduces a very important maximum likelihood estimation problem which has received almost no attention from numerical optimizers. The statement of the problem follows Redner and Walker (1981), but, since the wording is mine, I may have introduced misconceptions. Some sources are given in Dempster, Laird and Rubin (1977) and Redner (1980).

This time we assume that $y \epsilon \mathbf{R}^n$ is still n independent observations, but that each y_i comes with unknown probability α_j from the j-th of m random variables, where one can construct a parameterized model for each random variable. Therefore, for parameter vectors $\beta_j \epsilon \mathbf{R}^{p_j}$, $j = 1,2,\ldots,m$, we let $M_{ji}(\beta_j)$ be the term that corresponds to $M_i(\beta)$ in the previous section. Thus the likelihood of observing y_i is a mixture of m likelihoods, and it is assumed to be of the form

$$\rho(y_i;\ \alpha,\beta) = \sum_{j=1}^{m} \alpha_j \rho_j(y_i;\ M_{ji}(\beta_j)).$$

It follows from the independence assumption that the likelihood of observing y is

$$\rho(y;\ \alpha,\beta) = \prod_{i=1}^{n} \sum_{j=1}^{m} \alpha_j \rho_j(y_i; M_{ji}(\beta_j)).$$

Due to the sum, the log-likelihood function no longer has the nice form it had for the exponential family. Before writing the minimization problem down, we restrict ourselves to a very important special case. We let ρ_j have the form

$$\rho_j(y_i;\ M_{ji}(\beta_j)) = \frac{1}{\sqrt{(2\pi)}\,\sigma_j} \exp\left(\frac{\{y_i - M_{ji}(\beta_j)\}^2}{-2\sigma_j^2}\right) \equiv \rho_j(y_i;\ \beta_j,\ \sigma_j),$$

so the parameters of the calculation are $\alpha \epsilon \mathbf{R}^m$, $\sigma \epsilon \mathbf{R}^m$ and $\beta \epsilon \mathbf{R}^p$, where $p = \Sigma_j p_j$ $(j = 1,2,\ldots,m)$, and where we have separated σ_j from β_j. In this case the negative log-likelihood function is

$$L(\alpha,\ \beta,\ \sigma) = -\sum_{i=1}^{n} \log\left[\sum_{j=1}^{m} \frac{\alpha_j}{\sqrt{(2\pi)}\,\sigma_j} \exp\left(\frac{\{y_i - M_{ji}(\beta_j)\}^2}{-2\sigma_j^2}\right)\right]$$

$$= -\sum_{i=1}^{n} \log\left[\sum_{j=1}^{m} \alpha_j \rho_j(y_i;\ \beta_j,\ \sigma_j)\right]$$

$$= -\sum_{i=1}^{n} \log \rho(y_i;\ \alpha,\ \beta,\ \sigma).$$

and our problem is

$$\min_{\alpha,\beta,\sigma} L(\alpha, \beta, \sigma) \quad \text{subject to } \alpha_j \geq 0, \quad \sum_{j=1}^{m} \alpha_j = 1 \text{ and } \sigma_j > 0.$$

The positivity constraints $\sigma_j > 0$ require further discussion. Since the feasible region is not a compact set, it is not clear that a solution exists. Indeed, if $y_i = M_{ji}(\bar{\beta}_j)$ for some choice $\bar{\beta}$ of β, then for any $\bar{\alpha}_j \neq 0$,

$$\lim_{\sigma_j \to 0+} L(\bar{\alpha}, \bar{\beta}, \sigma) = -\infty .$$

Probably the most useful discussion of ways to try and circumvent this difficulty is to be found in Quandt and Ramsey (1978) and the several discussion papers and rejoinder that follow it. It is clear that the situation is far from clear and often in practice one tries a different initial guess if the optimization goes toward $\sigma_j = 0$. The apology offered for this procedure is a proof under strong regularity assumptions, that, if none of the true parameters α_j, σ_j are zero, and if N is a small enough neighbourhood of the true parameters, then with probability 1 as n approaches infinity, there is a unique local minimizer of L in N (Peters and Walker, 1978).

This is not a happy state of affairs; we only know that if there is a sufficient amount of data, then there is a useful local minimizer which we can try to find, but the global minimizer never exists. Redner (1980) has investigated a penalized likelihood statement of the problem that does give a minimization problem whose global minimizer can be expected to be near the true parameters for large n, but it has not been tested and does not seem suited for computation. Quandt and Ramsey (1978) have suggested another formulation, but theirs introduces at least 5 additional parameters. Beale and Thompson in separate conversations with me have suggested restricting the ratios $r_{ij} = \sigma_i/\sigma_j$. The Quandt and Ramsey discussion papers mention work on equality constraints $r_{ij} = c_{ij}$, but a lower bound $\min r_{ij} \geq c$ would blunt most of their objections. The reformulation into a well-posed computationally convenient optimization problem seems worth investigation, but we leave it for now to consider the currently favoured algorithm for finding the right minimizer.

If we take the gradients of L with respect to each parameter type, we obtain

$$\nabla_\alpha L(\alpha,\beta,\sigma)_\ell = -\frac{\partial L}{\partial \alpha_\ell} = -\sum_{i=1}^{n} \frac{\rho_\ell(y_i;\beta_\ell,\sigma_\ell)}{\sum_{j=1}^{m} \alpha_j \rho_j(y_i;\beta_j,\sigma_j)} = -\sum_{i=1}^{n} \frac{\rho_\ell(y_i;\beta_\ell,\sigma_\ell)}{\rho(y_i;\alpha,\beta,\sigma)}$$

$$\nabla_{\beta_\ell} L(\alpha,\beta,\sigma) = -\sum_{i=1}^{n} \frac{\alpha_\ell \nabla_{\beta_\ell} \rho_\ell(y_i;\beta_\ell,\sigma_\ell)}{\rho(y_i;\alpha,\beta,\sigma)}$$

$$\nabla_\sigma L(\alpha,\beta,\sigma)_\ell = -\sum_{i=1}^{n} \alpha_\ell \frac{\partial \rho_\ell(y_i;\beta_\ell,\sigma_\ell)/\partial \sigma_\ell}{\rho(y_i;\alpha,\beta,\sigma)} .$$

Because of the constraints on α, we do not expect $\nabla_\alpha L(\alpha^*, \sigma^*, \beta^*)$ to be zero at the solution. The "EM algorithm" (see Dempster, Laird and Rubin, 1977, or Redner and Walker, 1981) revises α by the formula

$$\alpha_j^+ = \frac{1}{n} \sum_{i=1}^{n} \frac{\alpha_j^c \rho_j(y_i;\ \beta_j^c,\ \sigma_j^c)}{\rho(y_i;\ \alpha^c,\ \beta^c,\ \sigma^c)} , \qquad j = 1,2,\ldots,m,$$

where α^c, $\{\beta_\ell^c;\ \ell = 1,2,\ldots,m\}$ and σ^c are current estimates of the parameters. Thus α_j^+ is an average of *posterior probability estimates* that y_i is an observation of the j-th random variable given that parameters α^c, β^c, σ^c are correct. Since $\rho_j > 0$ for every j, it is clear that α^+ satisfies the constraints.

Now let us consider the equations

$$\nabla_{\beta_\ell} L(\alpha,\ \beta,\ \sigma) = 0.$$

The derivative

$$\nabla_{\beta_\ell} \rho_\ell(y_i;\beta_\ell,\sigma_\ell) = \frac{1}{\sqrt{(2\pi)}\,\sigma_\ell^3} \exp\left(\frac{\{y_i - M_{\ell i}(\beta_\ell)\}^2}{-2\sigma_\ell^2}\right) \{y_i - M_{\ell i}(\beta_\ell)\}\ \nabla_{\beta_\ell} M_{\ell i}(\beta_\ell),$$

implies

$$\nabla_{\beta_\ell} L(\alpha,\beta,\sigma) = -\frac{1}{\sigma_\ell^2} \sum_{i=1}^{n} \frac{\alpha_\ell \rho_\ell(y_i;\beta_\ell,\sigma_\ell)}{\rho(y_i;\alpha,\beta,\sigma)} \{y_i - M_{\ell i}(\beta_\ell)\}\ \nabla_{\beta_\ell} M_{\ell i}(\beta_\ell).$$

Thus the condition

$$\nabla_{\beta_\ell} L(\alpha,\ \beta,\ \sigma) = 0$$

can be written in the form

$$M_\ell'(\beta_\ell)^T W_\ell(\alpha,\ \beta,\ \sigma)\{y - M_\ell(\beta_\ell)\} = 0,$$

where

$$W_\ell(\alpha,\ \beta,\ \sigma) = \text{diag}\left(\frac{\alpha_\ell \rho_\ell(y_i;\ \beta_\ell,\ \sigma_\ell)}{\rho(y_i;\ \alpha,\ \beta,\ \sigma)}\right).$$

Therefore the condition is similar to ones that are obtained by equating gradients of weighted least squares objective functions to zero. The EM algorithm is only written for the case when $M_\ell(\beta_\ell) = X\beta_\ell$, where X is a matrix that is independent of ℓ. Therefore the algorithm calculates β_ℓ^+ from the equation

$$0 = X^T W_\ell^c (y - X\beta_\ell^+) = X^T W_\ell^c - X^T W_\ell^c X\beta_\ell^+,$$

where

$$W_\ell^c = W_\ell(\alpha^c,\ \beta^c,\ \sigma^c).$$

Further, we could generalize to a EM-Gauss–Newton method by linearizing $M_\ell(\beta_\ell)$ about β_ℓ^c to obtain

$$0 = M_\ell'(\beta_\ell^c)^T W_\ell^c \{y - M_\ell(\beta_\ell^c) - M_\ell'(\beta_\ell^c)(\beta_\ell^+ - \beta_\ell^c)\}.$$

Finally, let us consider σ. Since

$$\partial\rho_\ell(y_i;\beta_\ell,\sigma_\ell)/\partial\sigma_\ell = (2\pi)^{-\frac{1}{2}}\left[-\frac{1}{\sigma_\ell^2}\exp\left(\frac{\{y_i - M_{\ell i}(\beta_\ell)\}^2}{-2\sigma_\ell^2}\right)\right.$$

$$\left. + \frac{\{y_i - M_{\ell i}(\beta_\ell)\}^2}{\sigma_\ell^4}\exp\left(\frac{\{y_i - M_{\ell i}(\beta_\ell)\}^2}{-2\sigma_\ell^2}\right)\right]$$

$$= \rho_\ell(y_i;\beta_\ell,\sigma_\ell)\left[\frac{\{y_i - M_{\ell i}(\beta_\ell)\}^2}{\sigma_\ell^3} - \frac{1}{\sigma_\ell}\right],$$

we have

$$\frac{\partial L}{\partial\sigma_\ell} = \sum_{i=1}^{n}\frac{\alpha_\ell\rho_\ell(y_i;\beta_\ell,\sigma_\ell)}{\rho(y_i;\alpha,\beta,\sigma)}\left[\frac{1}{\sigma_\ell} - \frac{\{y_i - M_{\ell i}(\beta_\ell)\}^2}{\sigma_\ell^3}\right].$$

This expression suggests the simple updating formula

$$(\sigma_\ell^+)^2 = \frac{\sum_{i=1}^{n}\alpha_\ell^c\frac{\rho_\ell(y_i;\beta_\ell^c,\sigma_\ell^c)}{\rho(y_i;\alpha^c,\beta^c,\sigma^c)}\{y_i - M_{\ell i}(\beta_\ell^c)\}^2}{\sum_{i=1}^{n}\alpha_\ell^c\frac{\rho_\ell(y_i;\alpha^c,\sigma^c)}{\rho(y_i;\alpha^c,\beta^c,\sigma^c)}}$$

which is used by the EM algorithm.

Our references show that much is known about the global behaviour of the EM algorithm in theory and practice, but the nonlinear case presented here is probably new.

5. Acknowledgements

Because statistics is unfamiliar ground to me, I am especially indebted to Neely Atkinson, Martin Beale, Richard Redner, David Scott, Richard Tapia, George Terrell, Jim Thompson, and Homer Walker for invaluable conversations. Any errors and misconceptions that remain are mine.

2.2

Algorithms for Robust Regression

R.H. Byrd

Computer Science Department, University of Colorado,
Boulder, Colorado 80309, USA.

1. Introduction

The general robust regression problem involves choosing a parameter vector x so as to make a vector of residuals $\{f_i(x);\ i = 1,2,\ldots,m\}$ as close to zero as possible. There is an implicit statistical assumption that, for some x, the values $f_i(x)$ are error terms in observations of a model, and are thus samples of zero-mean, independent, identically distributed, random variables. Estimation of the parameter vector x can be accomplished by minimizing a quantity $\Sigma\rho(f_i(x))$, where ρ is a scalar function that is symmetric about zero and increasing in $|f_i(x)|$. In the ideal case where the error is Gaussian, the maximum likelihood estimate of x is the minimizer of $\Sigma(f_i(x))^2$. In the absence of this assumption the least squares estimate is often too sensitive to components $f_i(x)$ which are very large in absolute value, but some other loss functions are less sensitive or are more "robust", a well known example being the least absolute value case $\rho(r) = |r|$. Statisticians have proposed other loss functions (see Gay, 1980b), all of which are symmetric about zero, are monotone in $|r|$, and increase more slowly than r^2.

The special form of the objective function $\Sigma\rho(f_i(x))$ can often be taken advantage of in an optimization algorithm. In the least squares case, when f_i is nonlinear, the Gauss–Newton and Levenberg–Marquardt algorithms, and their modifications that are obtained from secant conditions (Dennis, 1977), have a number of advantages over general unconstrained minimization algorithms. Analogous algorithms can be constructed for other loss functions (Gay, 1980b).

2. Some Known Algorithms for the Linear Case

In the linear case when f(x) = Ax-b, the function to be minimized takes the form

$$F(x) \equiv \sum_{i=1}^{m} \rho(a_i^T x - b_i) \equiv \sum_{i=1}^{m} \rho(r_i).$$

For this objective function, Newton's method is particularly simple since the Hessian is $A^T \mathrm{diag}[\rho''(r_i)]A$. However, two other algorithms have also been used for this problem, and they possess strong global convergence properties. They are the projected gradient method in residual space as proposed by Huber and Dutter (1974)

$$x^{k+1} = x^k - [\max \rho''(r)]^{-1} (A^T A)^{-1} \nabla F(x^k), \tag{1}$$

and the iteratively reweighted least squares algorithm

$$x^{k+1} = x^k - (A^T \mathrm{diag}[\rho'(r_i)/r_i]A)^{-1} \nabla F(x^k), \tag{2}$$

which has been proposed in a number of contexts (Dutter, 1977; Huber and Dutter, 1974).

These two methods replace the second derivative of ρ with other quantities, and are, at least locally, slower than Newton's method. However, it should be noted that in the projected gradient method (1) the matrix in the linear system to be solved does not change from step to step. This is of course a great saving, which sometimes counterbalances the fact that (1) usually requires the greatest number of iterations of the three algorithms.

To ensure global convergence, Newton's method must be used with a line search, and, for a number of loss functions, there is also the difficulty that $A^T\mathrm{diag}[\rho''(r_i)]A$ may be singular or indefinite. However, almost all proposed loss functions satisfy $\rho'(r)/r > 0$, and then, as long as A has full rank, the matrix of the iteratively reweighted least squares algorithm (2) is always positive definite.

The striking fact about algorithms (1) and (2) is that, for many loss functions, they are globally convergent without a line search (Byrd and Pyne, 1979; Dutter, 1977). This fact can be derived by noting that, as in many unconstrained optimization algorithms, each iterate is the minimizer of a quadratic model of the objective function, which has the form

$$q(x^k+d) = F(x^k) + \nabla F(x^k)^T d + \tfrac{1}{2}d^T(A^T\mathrm{diag}[w_{ii}]A)d.$$

This expression is just the sum of quadratic models of each residual term, where, for $i = 1,2,\ldots,m$, the model of $\rho(r_i^k + a_i^T d)$ is given by

$$q_i(r_i^k+a_i^T d) = \rho(r_i^k) + \rho'(r_i^k)(a_i^T d) + \tfrac{1}{2}w_{ii}(a_i^T d)^2 .$$

The quadratic term w_{ii} is just

$\rho''(r_i)$	for Newton's method,
$\max \rho''(r)$	for Huber's projected gradient (1),
$\rho'(r_i)/r_i$	for iteratively reweighted least squares (2).

Under reasonable conditions on the loss function ρ it can be shown that, for algorithms (1) and (2),

$$q_i(r_i^k + a_i^T d) \geq \rho(r_i^k + a_i^T d) \qquad (3)$$

for all d. Therefore, because $F(x^k) = q(x^k)$, and because the iteration reduces q, the iteration also gives a decrease in F. The conditions on ρ that imply (3) for algorithms (1) and (2) are symmetry about zero, nondecreasing monotonicity of $\rho(r)$ for $r \geq 0$, and nonincreasing monotonicity of $\rho'(r)/r$ for $r \geq 0$. They are satisfied by almost all the standard loss functions. In particular they hold if $\rho(r) = |r|^p$, $p \in [1,2]$, but for $p < 2$ there are theoretical problems with derivative discontinuities.

Some other algorithms also use such majorizing quadratic models, in particular the Weiszfeld iteration for optimal facility location and its generalization (Ostresh, 1978), and an algorithm proposed by Dembo (1979) for the geometric programming dual problem. The E-M algorithm for estimating mixture densities involves a majorizing model which is not quadratic, but the model is sufficiently separable to be minimized easily (Redner and Walker, 1981).

3. Locally Majorizing Quadratic Models

It is clear that there are some severe limitations in the use of globally majorizing quadratic models. Even for objective functions which are the sum of relatively simple functions, the objective function must be bounded above by a quadratic. When such a model is possible, the quadratic model is usually such an extreme overestimate that the steps taken are very short. This is found in practice; iteratively reweighted least squares (2) and Huber's projected gradient method (1) tend to be very slow. They can be shown to be linearly convergent, but convergence factors of about 0.9 are common. In order to achieve faster convergence and still maintain global convergence over a broad class of objective functions, the condition of global majorization must be weakened. We propose to do this by using an analogue of the trust region; we choose a region about the current iterate over which the quadratic model majorizes the objective. In this case the size of the region will determine the model, and we refer to this region as a "prescriptive trust region".

We now consider a step of such an algorithm for the robust regression problem, where σ is the radius of the trust region.

(1) For each residual r_i compute a quadratic model of the form,

$$\hat{q}_i(x^k+d) = \rho(a_i^T x^k - b_i) + \rho'(a_i^T x^k - b_i) a_i^T d + \tfrac{1}{2} w_i (a_i^T d)^2$$

with w_i chosen so that

$$\hat{q}_i(x^k+d) \geq \rho(a_i^T(x^k+d) - b_i) \qquad (4)$$

for all d such that $\|d\| \leq \sigma$.

(2) Minimize the quadratic model

$$\hat{q}(x^k+d) = \Sigma \hat{q}_i(x^k+d),$$

subject to

$$\|d\| \leq \sigma.$$

(3) Let $x^{k+1} = x^k + d$ and determine a new σ.

The choice of the quadratic term w_i could be made in a number of ways, but to get the best majorizing approximation we choose it to be the smallest value such that the majorizing condition (4) is satisfied. For example, in the case of the Huber loss function

$$\rho(r) = \begin{cases} \tfrac{1}{2} r^2 & |r| \leq H \\ H|r| - \tfrac{1}{2} H^2 & |r| \geq H, \end{cases}$$

this condition yields the following formula for w_i:

$$w_i = \begin{cases} 1 & , \quad |r_i^k| \le H \\ H/|r_i^k| & , \quad H \le |r_i^k| \le \tfrac{1}{2}\sigma_i \\ 4H(\sigma_i - |r_i^k|)/\sigma_i^2 & , \quad \tfrac{1}{2}\sigma_i \le |r_i^k| \le \sigma_i - H \\ (H+\sigma_i - |r_i^k|)^2/\sigma_i^2 & , \quad \max[\sigma_i - H, H] \le |r_i^k| \le \sigma_i + H \\ 0 & , \quad \sigma_i + H < |r_i^k|, \end{cases}$$

where $\sigma_i = \sigma/\|a_i\|_2$.

We note that the basis for choosing σ is totally different from that of a usual trust region. To ensure superlinear convergence, σ must tend to zero close to the solution. It is also wasteful if the unconstrained minimizer of the quadratic model falls far outside the trust region. On the other hand, too large a radius results in too large a quadratic term and thus too short a step. Ideally σ should be about the same size as the unconstrained step that results from it. For this reason we let

$$\sigma^{(k+1)} = (\sigma^{(k)})^{\theta} \, \| [\nabla^2 \hat{q}]^{-1} \nabla F(x^k) \|^{1-\theta}$$

in order to obtain a value between the old radius and the corresponding unconstrained step, where θ is a parameter from [0,1].

Near the solution, σ converges to zero and the algorithm can be shown to be superlinearly convergent since the quadratic term is converging to the second derivative. The algorithm is also globally convergent in the sense that any limit point is a stationary point. Computational experiments indicate that this locally majorizing model algorithm is about as fast as Newton's method near the solution, and more likely to converge from poor starting points. An analogous algorithm for the Weber location problem has been devised, and it performs with similar efficiency.

2.3

Nondifferentiable Optimization

Claude Lemaréchal

INRIA, B.P. 105, 78153 Le Chesnay, France.

Nondifferentiable optimization deals with functions which are not continuously differentiable. Some good introductory papers are in Lemaréchal and Mifflin (1979). The given view of the subject is based on Lemaréchal (1978a, 1980a). For the underlying theory of convex analysis the book by Rockafellar (1970) is recommended.

The study is restricted to locally Lipschitz functions. These functions have a gradient $\nabla f(x)$ for almost every x, which is bounded on bounded sets. For each x one defines the *set of generalized gradients* to be

$$M(x) = \{g = \lim_{y\to x} \nabla f(y) \ : \ \nabla f(y) \text{ exists}, \ \ \nabla f(y) \text{ has a limit}\}. \tag{1}$$

This set describes the behaviour of f around x. If f has a continuous gradient at x, M(x) is the singleton $\nabla f(x)$. The *subdifferential*, or peridifferential, of f at x is the convex hull of M(x), i.e. the set

$$\partial f(x) = \{g = \Sigma\lambda_i g_i \ : \ \lambda_i \geq 0, \ \ \Sigma\lambda_i = 1, \ \ g_i \in M(x)\} . \tag{2}$$

A *necessary optimality condition* for x to minimize the locally Lipschitz function f is $0\in\partial f(x)$, which is a generalization of the condition $\nabla f(x) = 0$ when f is differentiable. Several optimality conditions are given in Clarke (1975).

Some important examples of nondifferentiable objective functions are as follows:

Example 1. ℓ_∞-functions. Let $f_i(x)$ be a smooth function for $i = 1,2,\ldots,m$, and let

$$f(x) = \max_{i=1,2,\ldots,m} f_i(x). \tag{3}$$

Then

$$M(x) \subset \{\nabla f_i(x) : f_i(x) = f(x)\}. \tag{4}$$

The optimality condition $0 \in \partial f(x)$ can be derived by reformulating the unconstrained problem of minimizing $f(x)$ as a standard NLP (nonlinear programming) problem with an extra variable v. The NLP calculation is to

$$\left.\begin{array}{l} \text{minimize} \quad v \\ \text{subject to } v \geq f_i(x), \quad i = 1,2,\ldots,m \end{array}\right\}. \tag{5}$$

Example 2. ℓ_1-functions. In this case f is the function

$$f(x) = \sum_{i=1}^{m} |f_i(x)| , \tag{6}$$

and M is the set

$$\left.\begin{array}{rl} M(x) \subset \{ \sum_{i=1}^{m} \varepsilon_i \nabla f_i(x) : & \varepsilon_i = 1 \text{ if } f_i(x) > 0, \\ & \varepsilon_i = -1 \text{ if } f_i(x) < 0, \\ & \varepsilon_i = \pm 1 \text{ if } f_i(x) = 0\} \end{array}\right\}, \tag{7}$$

where different choices of $\pm$ when $f_i(x) = 0$ generate different members of $M(x)$. Here again the minimization of $f(x)$ can be reformulated in NLP form:

$$\left.\begin{array}{ll} \text{minimize} & \sum_{i=1}^{n} v_i \\ \text{subject to } v_i \geq & f_i(x), \quad i = 1,2,\ldots,m \\ \text{and} \qquad v_i \geq & -f_i(x), \quad i = 1,2,\ldots,m \end{array}\right\}. \tag{8}$$

Example 3. Convex functions. If f is convex, then

$$\partial f(x) = \{g : f(y) \geq f(x) + g^T(y-x) \quad \forall y\}, \tag{9}$$

i.e. the subdifferential of a convex function can be defined equivalently by (2) or by (9).

Optimality conditions for these examples are considered by Clarke (1976), Hiriart-Urruty (1978, 1981), and Fletcher and Watson (1980).

The construction of minimization algorithms for locally Lipschitz functions is based on a study of the "differential quotient", which, for a given x and a direction d, has the value

$$q(t) = \frac{f(x+td) - f(x)}{t} \quad \text{for } t > 0. \tag{10}$$

Its behaviour when $t \to 0+$ is particularly important, because the following relations hold:

$$\left.\begin{aligned} \min \{g^T d : g \in \partial f(x)\} &\leq \liminf_{t \to 0+} q(t) \\ &\leq \limsup_{t \to 0+} q(t) \\ &\leq \max \{g^T d : g \in \partial f(x)\} \end{aligned}\right\}, \tag{11}$$

which shows that the definition of good directions of search implies a study of $\partial f(x)$, more specifically of its support function. In most applications (in particular in the three examples above) q(t) has a limit when $t \to 0+$, namely the *directional derivative*, which has the value

$$f'(x,d) = \lim_{t \to 0+} q(t) = \max\{g^T d : g \in \partial f(x)\} . \tag{12}$$

When directional derivatives exist, then d is defined to be a *direction of steepest descent* if it solves

$$\min_{|d| \leq 1} \max_{g \in \partial f(x)} g^T d, \tag{13}$$

or equivalently

$$\max_{g \in \partial f(x)} \min_{|d| \le 1} g^T d, \tag{14}$$

whose solution is collinear to

$$d = \text{-Projection of the zero vector into } \partial f(x). \tag{15}$$

If (12) does not hold, but only (11), then (15) gives at least a descent direction, if not the steepest.

Based on these developments, algorithms can be constructed, and there are essentially two different approaches.

Case 1. The complete set M(x) is available and is used, which is possible when f has the form (3) or (6). To every algorithm solving the equivalent form (5) or (8), there corresponds a Case 1 method solving the underlying unconstrained problem. Several types of algorithm are possible, e.g. penalty type methods (Bertsekas, 1975; Charalambous, 1979), feasible directions, active set strategies (Demjanov and Malozemov, 1974; Charalambous and Conn, 1978), and variable metric methods (Pschenichny and Danilin, 1975; Hald and Madsen, 1981; Han, 1981).

Case 2. Given x, only one element of M(x) is used, say g(x). There are two traditional approaches, both applicable to convex f:

a) Cutting plane methods, in which one minimizes at each iteration a piecewise linear underestimate of f.

b) Kiev methods. At each iteration, one tries to get closer to a solution, instead of trying to diminish f. The basic iteration is $x_{n+1} = x_n - t_n g(x_n)$, where the stepsize $t_n > 0$ is generally chosen beforehand (e.g. the n-th term of a divergent series whose elements tend to zero) (Poljak, 1967). These methods are closely related to the ellipsoid algorithm.

In a more recent class of methods, namely "bundle" methods, one tries to diminish f at each iteration. In addition to f being locally Lipschitz, one needs a hypothesis (Mifflin, 1977a) that is often met in practice (in particular in the three examples above), namely that

$$d^T g(x+td) \quad \text{has a limit when } t \to 0+. \tag{16}$$

The basic idea is to store the successive $g(x_i)$ to form a *bundle* approximating $\partial f(x_n)$. If, at some iteration n, the direction d_n is not downhill, the approximation of $\partial f(x_n)$ is enlarged to include a component of the true $\partial f(x_n)$ that caused d_n not to be downhill, and then a new direction is calculated from the same x_n. Algorithms of this type are proposed by Wolfe (1975), Lemaréchal (1975) and Mifflin (1977b); some modifications to these algorithms are given in Lemaréchal (1978b), Mifflin (1980) and Lemaréchal, Strodiot and Bihain (1981).

Convergence theory is based on the upper semi-continuity of the mapping $\partial f(x)$. If $x_n \to x$, $s_n \in \partial f(x_n)$ and $s_n \to 0$, then x satisfies the optimality condition.

It seems safe to say that nondifferentiable optimization is now well understood as far as first order terms are concerned, but a breakthrough is needed for higher order terms. Some results on the second order case are given by Lemaréchal and Nurminskii (1980) and Lemaréchal and Mifflin (1981). Nondifferentiable optimization may be more important for the new viewpoints that it brings to classical optimization, than for its applications.

2.4

Algorithms for Nonlinear ℓ_1 and ℓ_∞ Fitting

Michael L. Overton

New York University, 251 Mercer Street,
New York, NY 10012, USA.

1. Introduction

Let f(x) be a twice continuously differentiable function mapping $\mathbf{R}^n$ to $\mathbf{R}^m$. Consider the norm minimization problems:

$$\underset{x \in \mathbf{R}^n}{\text{minimize}} \ \| f(x) \|_1 \qquad (\ell_1 P)$$

and

$$\underset{x \in \mathbf{R}^n}{\text{minimize}} \ \| f(x) \|_\infty , \qquad (\ell_\infty P)$$

where

$$\| f(x) \|_1 = \sum_{i=1}^{m} | f_i(x) |$$

and

$$\| f(x) \|_\infty = \max_{1 \le i \le m} | f_i(x) | .$$

These problems usually arise in the form of choosing n parameters to fit m model functions to given data (normally with $n \ll m$). The choice of norm depends on the application. The ℓ_1 norm is often used to give a "robust" fit when some data are inaccurate, and the ℓ_∞ norm is used to minimize the maximum deviation of the model from the data.

The norm $\| f(x) \|_1$ is piecewise differentiable, with a probable derivative discontinuity at x if $f_i(x) = 0$ for some i. For $\ell_1 P$, the function f_i is called *active* if it equals zero at x. The norm $\|f(x)\|_\infty$ usually has a derivative discontinuity at x if $|f_i(x)| = |f_j(x)| = \|f(x)\|$ for some $i \neq j$. For $\ell_\infty P$, the function f_i is called *active* if it has modulus $\|f(x)\|_\infty$ at x. Let e denote the vector in $\mathbf{R}^m$ whose components are all equal to one. It is easily shown that $\ell_1 P$ and $\ell_\infty P$ are respectively equivalent to the following differentiable constrained optimization problems:

$$\left.\begin{array}{ll} \underset{x\in\mathbf{R}^n,\ u\in\mathbf{R}^m}{\text{minimize}} & e^T u \\ \text{subject to} & c_1(x,u) \geq 0 \end{array}\right\}, \qquad (EP_1)$$

where

$$c_1(x,u) = \begin{pmatrix} u - f(x) \\ u + f(x) \end{pmatrix},$$

and

$$\left.\begin{array}{ll} \underset{x\in\mathbf{R}^n,\ \tau\in\mathbf{R}^1}{\text{minimize}} & \tau \\ \text{subject to} & c_\infty(x,\tau) \geq 0 \end{array}\right\}, \qquad (EP_\infty)$$

where

$$c_\infty(x,\tau) = \begin{pmatrix} \tau e - f(x) \\ \tau e + f(x) \end{pmatrix} .$$

These problems could be solved by directly applying standard methods for nonlinearly constrained optimization. However, doing so discards the knowledge of the special structure of the objective function and constraints, and, in the case of EP_1, involves m extra variables. Instead it is possible to exploit the special structure of these problems.

2. Algorithms Based on Linear Programming

Let us first consider the case where all the $\{f_i(x)\}$ are linear, and hence EP_1 and EP_∞ are linear programs. By the fundamental theorem of linear programming it is clear that, for x to be a unique solution of $\ell_1 P$, at least n functions must be active (zero) at x, and for x to be a unique solution of $\ell_\infty P$, at least n+1 functions must be active (equal to τ in modulus) at x. Many algorithms have been designed to solve $\ell_1 P$ or $\ell_\infty P$ by linear programming techniques. The exchange algorithm of Stiefel (1960), which essentially solves the dual of EP_∞ by the simplex method, forms the basis of the most successful methods for the linear ℓ_∞ problem, although improvements have been suggested by many people. The best known linear ℓ_1 algorithms are those due to Barrodale and Roberts (1973) and Bartels, Conn and Sinclair (1978), which essentially solve variations of EP_1 by primal simplex methods taking account of the special structure for efficiency.

Let us now discuss methods for the nonlinear problems. The approach taken by many has been to solve a sequence of linear problems. For $\ell_\infty P$, Zuhovickii, Poljak and Primak (1963), Ishizaki and Watanabe (1968) and Osborne and Watson (1969) solve a linear program to obtain a direction of search at each iteration, following this by an exact line search to minimize $||f(x)||_\infty$ along the line. A similar method for $\ell_1 P$ was given by Osborne and Watson (1971). These algorithms are very effective if the number of active functions at the solution is n for $\ell_1 P$ and n+1 for $\ell_\infty P$. However, in the nonlinear case, these conditions may not hold even though the solution is unique, and the consequence is that the successive linear programming methods are extremely slow. Since these methods can be viewed as generalizations of the Gauss-Newton algorithm for nonlinear least squares to the ℓ_1 and ℓ_∞ norms, Madsen (1975) and Anderson and Osborne (1977) suggested regularized successive linear programming methods in the spirit of the Levenberg–Marquardt method for nonlinear least squares. The regularization can significantly improve the performance of these methods, but the rate of convergence, as for the Levenberg–Marquardt method, is often slow. Recently two-stage methods have been proposed by Hald and Madsen (1981) and G.A. Watson (1979) for $\ell_\infty P$ and by McLean and Watson (1980) for $\ell_1 P$. In the first stage a successive linear programming iteration is performed until it is thought that the set of functions which are active at the solution has been identified. In the second stage a Newton (or quasi-Newton) iteration is used to solve the nonlinear equations in the primal variables x and Lagrange multipliers λ which characterize the solution. The methods are normally superlinearly convergent and appear to be quite successful in practice. However, they have the drawback that only the residual of the nonlinear system and not $\|f(x)\|_1$ or $\|f(x)\|_\infty$ is necessarily reduced at each step of the second stage. Methods can be derived which reduce $\|f(x)\|_1$ or $\|f(x)\|_\infty$ at each step, by making fuller use of EP_1 and EP_∞.

3. Special Properties of EP_1 and EP_∞

One important property of the nonlinear programs EP_1 and EP_∞ is that feasibility can always be maintained. Given any x, the variables u and τ can be chosen by:

$$\left.\begin{array}{l} u_i = |f_i(x)|, \quad i = 1,2,\ldots,m, \text{ in } EP_1 \\ \tau = \|f(x)\|_\infty, \quad \text{in } EP_\infty \end{array}\right\} . \qquad (1)$$

Thus progress towards the solution can be measured by the objective function alone, which always has the value $\|f(x)\|_1$ or $\|f(x)\|_\infty$ respectively. This observation shows a significant simplification in the problems $\ell_1 P$ and $\ell_\infty P$ over the general nonlinear programming problem, since in the general case progress towards the solution must be measured by balancing reduction of the objective against violation of the constraints. Finding a suitable balance is sometimes difficult.

4. Further Notation

For both EP_1 and EP_∞ let t be the number of active functions at a point x, as defined earlier, and let $\hat{f}(x) \in \mathbf{R}^t$ be the vector of active functions at x. Let $\hat{V}(x) \in \mathbf{R}^{n \times t}$ be the matrix whose columns are $\{\nabla \hat{f}_i(x)\}$, and define $\hat{\Sigma}(x) \in \mathbf{R}^{t \times t}$ as the diagonal matrix whose entries are $\{\mathrm{sgn}(\hat{f}_i(x))\}$. Similarly let $\bar{f}(x) \in \mathbf{R}^{m-t}$ be the vector of inactive functions, let $\bar{V}(x) \in \mathbf{R}^{n \times (m-t)}$ be the matrix whose columns are $\{\nabla \bar{f}_i(x)\}$, and let $\bar{\Sigma}(x) \in \mathbf{R}^{(m-t) \times (m-t)}$ be the diagonal matrix whose entries are $\{\mathrm{sgn}(\bar{f}_i(x))\}$. Let $\hat{e} \in \mathbf{R}^t$ and $\bar{e} \in \mathbf{R}^{m-t}$ be vectors whose components are all equal to one. Now define the *active constraints* at a point x to be the constraints whose components in $c_1(x)$ or $c_\infty(x)$ are zero, where the choice between c_1 or c_∞ depends on which problem is being solved. Because of (1), the number of active constraints is m+t for EP_1 and t for EP_∞ (assuming $\|f(x)\|_\infty \neq 0$). Let $\hat{c}_1 \in \mathbf{R}^{m+t}$ and $\hat{c}_\infty \in \mathbf{R}^t$ be the vectors of active constraints for EP_1 and EP_∞ respectively, and let $A_1, \hat{A}_1, A_\infty$ and $\hat{A}_\infty$ be matrices whose columns are respectively the gradients of the components of c_1, $\hat{c}_1$, c_∞ and $\hat{c}_\infty$, with respect to x and u or x and τ.

We can order the variables $\{u_i\}$ and the active constraints so that:

$$\hat{A}_1 = \begin{pmatrix} -\bar{V}\bar{\Sigma} & -\hat{V} & \hat{V} \\ I_{m-t} & 0 & 0 \\ 0 & I_t & I_t \end{pmatrix}, \qquad \hat{A}_\infty = \begin{pmatrix} -\hat{V}\hat{\Sigma} \\ \hat{e}^T \end{pmatrix}.$$

Here I_t is the identity matrix of order t. Let Q_1 and Q_∞ be full

rank matrices whose columns span the null spaces of $\hat{A}_1^T$ and $\hat{A}_\infty^T$ respectively. Finally, let g_1 and g_∞ be respectively the gradients of the objectives of EP_1 and EP_∞, i.e.
$g_1^T = (0 \ldots 0\ 1 \ldots 1)^T \in \mathbf{R}^{n+m}$ and $g_\infty^T = (0 \ldots 0\ 1)^T \in \mathbf{R}^{n+1}$.

5. Optimality Conditions

It can be shown that the first order constraint qualifications always hold for EP_1 and EP_∞. Applying the standard theory of constrained optimization, we therefore get the following conditions for local minima of EP_1 and EP_∞, and hence for $\ell_1 P$ and $\ell_\infty P$. All matrices and vectors are to be evaluated at x.

First order necessary conditions. If (x,u) is a local minimum of EP_1, then $\exists\ \lambda \in \mathbf{R}^{m+t}$ s.t. $\lambda \geq 0$ and $\hat{A}_1 \lambda - g_1 = 0$. Some algebra shows that an equivalent consequence is that $\exists\ \pi \in \mathbf{R}^t$ s.t.

$$\left.\begin{array}{l} |\pi_i| \leq 1, \quad i = 1,\ldots,t, \quad \text{and} \\ \hat{V}\pi + \bar{V}\bar{\Sigma}\bar{e} = 0 \end{array}\right\} . \tag{2}$$

If (x,τ) is a local minimum of EP_∞, then $\exists\ \lambda \in \mathbf{R}^t$ s.t. $\lambda \geq 0$ and

$$\hat{A}_\infty \lambda - g_\infty = 0. \tag{3}$$

Second order necessary conditions. If (x,u) is a local minimum of EP_1 and the second order constraint qualifications hold, then

$$Q_1^T \begin{pmatrix} -\sum_{i=1}^{m+t} \lambda_i \nabla^2 (\hat{c}_1)_i & 0 \\ 0 & 0 \end{pmatrix} Q_1$$

is positive definite or semi-definite. Some algebra shows that this matrix can be written $Z_1^T W_1 Z_1$, where

$$W_1 = \sum_{i=1}^{t} \pi_i \nabla^2 \hat{f}_i + \sum_{i=1}^{m-t} \text{sgn}(\bar{f}_i) \nabla^2 \bar{f}_i ,$$

and where Z_1 is a matrix with full column rank spanning the null space of $\hat{V}^T$.

If (x,τ) is a local minimum of EP_∞ and the second order constraint qualifications hold, then $Z_\infty^T W_\infty Z_\infty$ is positive definite or semi-definite, where

$$W_\infty = \sum_{i=1}^{t} \lambda_i \, \text{sgn}(\hat{f}_i) \nabla^2 \hat{f}_i ,$$

and where Z_∞ is the matrix consisting of the first n rows of Q_∞.

Second order sufficient conditions. If (2) holds, if $|\pi_i| < 1$, $i = 1,2,\ldots,t$, and if $Z_1^T W_1 Z_1$ is positive definite, then (x,u) is a strong local minimum of EP_1. If (3) holds, if $\lambda > 0$, and if $Z_\infty^T W_\infty Z_\infty$ is positive definite, then (x,τ) is a strong local minimum of EP_∞.

Note that, in the linear case, if the solution is unique then the matrices Q_1, Q_∞, Z_1 and Z_∞ are all null matrices. Hence the second order conditions are also null.

6. Algorithms Based on EP_1 and EP_∞

By making full use of EP_1 and EP_∞, it is possible to design algorithms which (i) are locally superlinearly convergent, and (ii) reduce the norm $\|f\|_1$ or $\|f\|_\infty$ at every step of the iteration. Zangwill (1967b) and Charalambous and Conn (1978) proposed methods for $\ell_\infty P$ which have property (ii) but not (i) in general. These methods use projected steepest descent iterations. In order to have property (i) in general, it is clear from the optimality conditions that an algorithm must approximate the second order information in either the Lagrangian Hessian (W_1 or W_∞) or the projected Lagrangian Hessian ($Z_1^T W_1 Z_1$ or $Z_\infty^T W_\infty Z_\infty$). The natural way to do this is to solve successive quadratic programs, which incorporate a quadratic approximation to the Lagrangian of EP_1 or EP_∞ and linear approximations to the constraints. A direction of search can be obtained from the solution of the quadratic program, and then a line search can be used to reduce $\|f\|_1$ or $\|f\|_\infty$ along this direction. For a discussion of special line search algorithms for these problems see Murray and Overton (1979).

There are two basic possibilities for the choice of quadratic program. Either one can use an active set method and solve an equality constrained quadratic program (EQP) involving only the active constraints, or one can solve an inequality constrained quadratic program (IQP) involving all the constraints. The simplest form of the EQP to determine the search direction $p \in \mathbf{R}^n$ for $\ell_\infty P$ is:

$$\left.\begin{array}{ll} \underset{p\in\mathbf{R}^n,\ \rho\in\mathbf{R}^1}{\text{minimize}} & g_\infty^T\begin{pmatrix}p\\ \rho\end{pmatrix} + \tfrac{1}{2}p^T W_\infty^{(k)} p \\ \text{subject to} & \hat{A}_\infty(x^{(k)})^T\begin{pmatrix}p\\ \rho\end{pmatrix} = -\hat{c}_\infty(x^{(k)}) \end{array}\right\}, \qquad (EQP_\infty)$$

where $x^{(k)}$ is the current iterate, and where $W_\infty^{(k)}$ is an "estimate" of W_∞. This estimate usually depends on $x^{(k)}$ and on an approximation to the Lagrange multipliers, and it may be obtained from analytical Hessians, finite differences, or quasi-Newton updating. Because in practice one may treat f_i as active when $\|f\|_\infty - |f_i|$ is sufficiently small, the vector $\hat{c}_\infty$ may not be exactly zero. If certain conditions hold, it can be shown that p is a descent direction for $\|f\|_\infty$. EQP_∞ is essentially the basis of the methods of Conn (1979) and Murray and Overton (1980).

Alternatively, the IQP for $\ell_\infty P$ is:

$$\left.\begin{array}{l} \underset{p\in\mathbf{R}^n,\ \rho\in\mathbf{R}^1}{\text{minimize}} \quad g_\infty^T\begin{pmatrix}p\\ \rho\end{pmatrix} + \tfrac{1}{2}p^T W_\infty^{(k)} p \\ \\ \text{subject to } A_\infty(x^{(k)})^T\begin{pmatrix}p\\ \rho\end{pmatrix} \geq -c_\infty(x^{(k)}) \end{array}\right\} . \qquad (\mathrm{IQP}_\infty)$$

Again, if certain conditions hold, p is a descent direction for $\|f\|_\infty$. Methods of Han (1978, 1981), Wierzbicki (1978), and Charalambous and Moharram (1978) are based on this formulation. Solving the dual of IQP_∞ seems likely to be more efficient than solving the primal problem, given the experience with linear ℓ_∞ computations. We note that caution is required in solving either EQP or IQP, since $W_\infty^{(k)}$ may not be positive definite.

The situation for ℓ_1P is more complicated since there are n+m variables in EP_1. The EQP that determines the search direction is:

$$\left.\begin{array}{l} \underset{p\in\mathbf{R}^n,\ q\in\mathbf{R}^m}{\text{minimize}} \quad g_1^T\begin{pmatrix}p\\ q\end{pmatrix} + \tfrac{1}{2}p^T W_1^{(k)} p \\ \\ \text{subject to } \hat{A}_1(x^{(k)})^T\begin{pmatrix}p\\ q\end{pmatrix} = -\hat{c}_1(x^{(k)}) \end{array}\right\} , \qquad (\mathrm{EQP}_1)$$

where the Hessian approximation $W_1^{(k)}$ usually requires estimates of Lagrange multipliers. One can eliminate q analytically to obtain:

$$\left.\begin{array}{l} \underset{p\in\mathbf{R}^n}{\text{minimize}} \quad \bar{e}^T\bar{\Sigma}(x^{(k)})\bar{V}(x^{(k)})^T p + \tfrac{1}{2}p^T W_1^{(k)} p \\ \\ \text{subject to } \hat{V}(x^{(k)})^T p = -\hat{f}(x^{(k)}) \end{array}\right\} . \qquad (\mathrm{EQP}_1')$$

Thus the number of variables in the quadratic program has been reduced from n+m to m. Under certain conditions that are usually satisfied, p is a descent direction for $\|f\|_1$. The EQP formulation is used for nonlinear ℓ_1 calculations by Murray and Overton (1981) and by Bartels and Conn (1981). It should be clear how to express the IQP in n+m variables, but it is not possible in general to transform this to an IQP in n variables. Instead, just as in the Bartels, Conn and Sinclair (1978) algorithm for the linear ℓ_1 problem, it is preferable to solve the IQP by a method that takes account of the structure of the calculation, because using a standard method when there are n+m variables would be quite inefficient.

These QP-based algorithms are, of course, closely related to the well known successive quadratic programming methods for non-linearly constrained optimization discussed by many authors, including Wilson (1963), Murray (1969), Biggs (1972), Robinson (1974), Wright (1976), Han (1977a), Powell (1978a) and Coleman and Conn (1980b). Many aspects of the ℓ_1 and ℓ_∞ algorithms are different from the general case; for details see the original papers.

Other algorithms for ℓ_1P and ℓ_∞P have been suggested. For example, Charalambous and Bandler (1976) and El-Attar, Vidyasagar and Dutta (1979) suggest methods for ℓ_∞P and ℓ_1P respectively, which involve solving a sequence of unconstrained minimization problems. Unfortunately the subproblems become increasingly ill-conditioned as the solution is approached, and convergence is slow. It seems safe to say that eventually the best algorithms will prove to be those based on successive quadratic programming, by virtue of properties (i) and (ii) given at the beginning of this section. Although the two-stage linear programming/nonlinear equation methods show very good numerical results, there seems to be no reason why the IQP-based methods should not be equally successful, since both select the active set by solving inequality constrained subproblems. However, this has yet to be substantiated. There is still the question of whether an IQP or an EQP method is preferable. Han (1981) has proved global convergence for his IQP algorithm for ℓ_∞P, and no doubt a similar result could be proved for an IQP ℓ_1 method. Global convergence results also exist for EQP-based algorithms for constrained optimization (see Coleman and Conn, 1980b). The question of whether to solve IQP or EQP arises at all levels of inequality constrained optimization, and it is considered in the paper of Bartholomew-Biggs in this volume. As yet there does not appear to be enough numerical evidence to conclude which type of method is superior.

Discussion on "Nonlinear Fitting"

2.5 Tapia

It is well known that maximum likelihood problems, mentioned in Paper 2.1, can be ill-posed. Indeed there are several cases where the likelihood functional is unbounded. In density estimation statisticians usually deal with this difficulty by restricting the domain in a way that depends on the size of the random sample. For example, if a histogram is used as a maximum likelihood estimator, and if one allows arbitrarily small bin widths, then the likelihood functional is unbounded. However, when constructing a histogram, instead of allowing the bin width to be as small as one wants, the bin width should be chosen in accordance with the sample size. As the sample size goes to infinity the bin width may go to zero in a way that gives good convergence properties and stable calculations.

2.6 Dembo

In some cases maximum likelihood calculations can be formulated as geometric programming problems. This has the advantage that one is able to detect whether or not the problem is unbounded or otherwise ill-posed by examining the dual of the geometric programming problem. In general, I feel that too little use is made of duality, for it can be a powerful tool for analysing and sometimes solving mathematical programming models.

2.7 Himmelblau

The concept of uniqueness in obtaining *the* solution to a nonlinear programming problem rests on the concept of distinguishability. If one or more changes in the variables in a nonlinear programming problem have no influence on the value of the objective function, the particular variable cannot be uniquely estimated. More formally:

Definition 1 The pair of parameter values (θ,θ'), $\theta\epsilon S$, $\theta'\epsilon S$, where S is the set of parameter values, is said to be indistinguishable if the observations $y(t,\theta)$ and $y(t,\theta')$ are identical for all initial conditions $y(t_0)$ and all inputs $u(t)$, $0 \le t \le T$. Otherwise, the pair (θ,θ') is said to be distinguishable.

Definition 2 A parameter set S is globally identifiable in the range $0 \le t \le T$ if the pair (θ,θ') is distinguishable for all $\theta' \neq \theta$ where θ, $\theta'\epsilon S$. A parameter set S is said to be locally identifiable at θ' if there exists a δ such that the pair (θ,θ') is distinguishable for all $\theta\neq\theta'$ in the neighbourhood of θ', $\| \theta-\theta' \| < \delta$.

In nonlinear programming, the observations comprise the values of the objective function and constraints, the parameters are the values of the variables at the optimal solution, $\hat{\theta}$ say, and the inputs are the sequence of intermediate values of the variables.

Definition 3 A parameter set S is locally identifiable at $\hat{\theta}$ if a function $\phi(y(\theta))$ has an isolated local minimum at $\theta=\hat{\theta}$, and is globally identifiable if the minimum is a global one.

Definition 3 is equivalent to definition 2, but neither provide convenient guides for the analyst to establish identifiability for a given problem.

Whether or not one can estimate a unique set of parameters depends not only on the mathematical structure of the nonlinear programming problem, but also on the optimization method used. Previous work on the uniqueness of estimates can be classified as being based on (1) deterministic or (2) stochastic analysis; we are restricting our attention here to the former.

Bellman and Åström (1970) introduced the concept of structural identifiability that is independent of the identification method and the input used. Structural identifiability is the capability of ascertaining unique values of the parameters in the structural model from deterministic input and output data. Clearly, structural identifiability is a necessary condition to obtain a unique minimum.

How the concept of structural identifiability can be applied in a practical manner to problems to ascertain in advance of solving them whether or not they have unique solutions has not been examined in the nonlinear programming field. But it should be.

2.8 Lootsma

It is well known that statisticians are interested in second order information, in particular the Hessian matrix of the sum-of-squares function at the minimum solution in order to carry out some sensitivity analysis. Secant approximations, however, produce efficient search directions but not a good approximation to the Hessian or the inverse Hessian. What should be done in these cases?

2.9 Dennis

I assume that you are asking about the final Hessian approximation because of its utility in covariance estimation. If the Hessian is being updated by a variable metric approximation, then it should not be used for covariance estimation. The computer program NL2SOL gives the user a fresh finite difference or analytic estimate at the solution, and he has a choice of three forms since there does not seem to be a consensus which to use. This question is discussed in detail in Dennis, Gay and Welsch (1981). Perhaps Bob Schnabel would like to mention his experience of this problem.

2.10 Schnabel

Many statistical packages with which I am familiar use finite difference approximations for generating the components of the variance-covariance matrix. However it is important to note that each of the three common variance-covariance estimates for nonlinear least squares, given for instance in Dennis, Gay and Welsch (1981), can be misleading. For example, I know of an experiment, involving a highly nonlinear model, where all three estimators usually provide very similar estimates of variances, covariances and confidence-intervals, but they are often gross underestimates of the true quantities (which are known due to the design of the experiment). Basically, the difficulty is due to using only local information at the minimizer to estimate phenomena that are inherently global. Some statistical work is related to this problem, for example see Bates and Watts (1980).

2.11 Beale

In linear least squares the main use for the covariance matrix is to define confidence intervals for the individual regression coefficients (i.e. for the decision variables in usual optimization terminology). The theoretically best way to define approximate confidence intervals in nonlinear least squares is to use the likelihood ratio criterion. Let β denote the vector of regression coefficients, let $S(\beta)$ denote the residual sum of squares divided by the estimated residual variance, and let $\hat{\beta}$ denote a value of β that minimizes $S(\beta)$. To find a confidence interval for β_j, a component of β, consider the problem of minimizing $S(\beta)$ subject to the constraint that $\beta_j = \theta$, where θ is a parameter. A good approximate 95% confidence interval for β_j is the set of values of θ such that the objective function value in this constrained minimization problem is not greater than $S(\hat{\beta}) + 4$. This confidence interval cannot be computed using only local information near $\beta = \hat{\beta}$ for nonlinear models.

2.12 Robinson

A good example of an inverse problem that has received quite a lot of attention is that of factor analysis. The problem was analysed in the now classic paper of Rao (1955), and is described by Anderson and Rubin (1956) and in many other books and papers. Until the late 1960's, the preferred method of solution was to use simple iteration (successive substitution) on what is essentially a system of nonlinear equations in nonnegative variables (the "unique variances"). Not surprisingly this method was slow. What was not known at the time was that it was also often wrong, because people tended to stop computing when the iterates stabilized; they failed to realize that this does not guarantee convergence. When Jennrich and Robinson (1969) showed how to apply Newton's method to this problem, they pointed out that the illustrative example given in the paper of Rao (1955) had been incorrectly solved for just this reason, and that the true solution was quite different from that given by Rao. At about the same time K. Jöreskog (1967) showed in a series of papers how to apply the Davidon–Fletcher–Powell method to factor analysis. The introduction of superlinear methods into this area thus made a significant difference, not only in speed of solution but also in the ability to find the correct solution at all.

2.13 Lemaréchal

Concerning the fitting problem, is there a preference for the ℓ_1, the ℓ_2 or the ℓ_∞ norm?

2.14 Dennis

Different norms are suited to different problems, in particular it is certainly true that the ℓ_1 norm is better able to deal with gross errors in the observations.

2.15 Dixon

It is mentioned in Paper 2.1 that in many problems constraints of the form $x_i \geq 0$, $\Sigma_i x_i = 1$ occur naturally. My colleague S.E. Hersom recently had to solve a highly nonlinear problem in about 400 variables that were subject to such constraints. He transformed the problem into an unconstrained one by using generalized cylindrical polar coordinates θ_i, where

$$x_i = \cos^2\theta_i \prod_{j=1}^{i-1} \sin^2\theta_j, \quad i < n,$$

and

$$x_n = \prod_{j=1}^{n-1} \sin^2\theta_j.$$

Thus the constraints on x are satisfied naturally for all values of the variables θ_j, $j = 1,2,\ldots,n-1$. The unconstrained problem in the new variables was then easily solved by using a conjugate gradient algorithm. I can recommend this approach for problems with this type of constraint and highly nonlinear objective functions.

2.16 Meyer

For regression problems in which the objective function is convex, instead of using a majorizing quadratic function as in Paper 2.2, a majorizing approximating function may also be obtained via piecewise linear approximation. Such an approximation has the advantage that the computation of the associated trust region is trivial. Numerical experience in the separable case with large, constrained, data-fitting problems has shown that the piecewise linear approach is rapidly convergent.

2.17 Powell

Although the algorithms and theory that are given in Paper 2.2 have been developed for a general function ρ, is it usually more efficient in practical applications to take account of the particular ρ that occurs?

2.18 Byrd

The iteratively re-weighted least squares algorithm certainly takes advantage of the form of ρ in that it requires $\rho'(r)/r$ to be non-increasing as $|r|$ increases. Some of the other algorithms mentioned have similar requirements, but I know of none that exploits the form of ρ more explicitly.

2.19 Conn

We all know when to use least-squares, and there are some obvious examples where robust regression does exactly what we want. However, we should be careful in case the availability of "robust algorithms" causes practitioners to obtain "solutions" to insoluble or unsuitable problems. In other words we should not provide good answers to bad models.

2.20 Byrd

Statisticians who have proposed robust regression do not advocate automatic use of such estimators. Instead they examine a number of robust estimators in order to identify spurious data and to evaluate the validity of the chosen estimator.

2.21 Schnabel

It is not necessarily dangerous to provide software which gives solutions to data fitting problems in all cases, as long as the solution is followed by a statistical analysis to determine whether the solution is "dangerous", and, when this is the case, a warning is returned that the user cannot possibly miss.

For example, the nonlinear least squares software in STATLIB, from the National Bureau of Standards at Boulder, always examines the residuals and estimated confidence intervals at the end of the minimization, and, if something looks suspicious, it prints out a huge skull and crossbones before any of the results!

2.22 Tapia

If a particular statistics problem is ill-posed, it may be of little value to mention this to the statistician and ask him to reformulate the problem. The reformulation should be done jointly by the numerical analyst and the statistician. Let me illustrate my point with an example in an area of current concern, namely density estimation. In maximum likelihood density estimation it is possible to show that the optimization problem has a unique solution when we are optimizing over any finite dimensional subspace of integrable functions, but the optimization problem never has a solution when the domain is an infinite dimensional subspace. Therefore we must necessarily encounter numerical difficulties for large (but still finite) dimensions.

We, as numerical analysts, would say that the large finite dimensional case is ill-posed, but many statisticians would say that it is well-posed, because it has a unique solution. We must take an active role in helping model builders to reformulate their problems for they cannot do it adequately alone.

2.23 Gutterman

When working on a fitting problem posed by someone who does not have a strong statistical background, it is important to ensure that the work "makes good statistical sense". We should involve a statistician in the consultation in such cases to be sure that we are not, for example, using robust regression in situations that would be better handled by rejection of outliers.

2.24 Robinson

Users who are not trained in numerical analysis often have no idea of what they really want to do. One has to ask enough questions to understand what is their real problem, as opposed to the problem that they state. This is illustrated by an encounter I had once with a user who asked if I could help him to find good software for inverting a 100×100 matrix. I first asked him why he wanted to invert the matrix, and he replied that he wanted to solve a system of linear equations! This was bad enough, but not the worst: further questioning established that his 100×100 matrix was actually the direct sum of a 10×10 full matrix and a 90×90 tridiagonal matrix. In this case, therefore, asking enough questions beforehand saved a great deal of work, and that is typical of many other similar cases that I have seen.

2.25 Mangasarian

I presume that the statement in Paper 2.3, that there are no superlinearly or quadratically convergent algorithms for non-differentiable optimization yet, is not meant to include problems which can be converted to smooth optimization problems, such as

$$\min_{x \in \mathbf{R}^n} \max_{1 \le i \le m} f_i(x) \quad \text{or} \quad \min_{x \in \mathbf{R}^n} \sum_{i=1}^{m} |f_i(x)| ,$$

where each function f_i is from $\mathbf{R}^n$ to $\mathbf{R}$. In these cases superlinearly convergent algorithms exist.

2.26 Lemaréchal

This is true. Let me add that the Maratos effect exists, just as in the standard NLP case. If x^+ is a second order estimate coming from x, then, even when x is very close to the solution, it may happen that $f(x^+) \geq f(x)$. This was discovered long ago by Madsen, and motivated the paper by Hald and Madsen (1981).

2.27 Todd

In a survey paper Shor (1977) mentions that his generalized gradient method with space dilation in the direction of the difference between successive gradients performs better than bundle methods computationally for problems with dimension up to 200-300. Does Lemaréchal know of any relevant further testing?

2.28 Lemaréchal

I have myself conducted experiments on a series of academic test problems, and obtained the same conclusion (Lemaréchal, 1980b). Shor has defined two dilations: (1) along the gradient (convergence proved but bad behaviour in practice: it gives the Khachian algorithm), and (2) along the difference of two successive gradients (non-convergence proved but very nice numerical behaviour). The second dilation has a sensible numerical motivation and is close to the quasi-Newton method, but, like all Kiev methods, it lacks a stopping criterion and a good stepsize rule.

2.29 Meyer

For separable non-differentiable functions that are continuous and convex, piecewise linear approximation methods will converge to an optimal solution. Geometrically, piecewise linear secant approximations are just the opposite of subgradients in that they are "inner" approximations (overestimations) rather than "outer" approximations (underestimations) with respect to the epigraphs. A computational advantage of "inner" approximation is that improvement of the approximation function guarantees improvement of the original objective function, eliminating the need for line searches.

2.30 Griewank

Have there been any experiments with curvilinear searches for nondifferentiable optimization?

2.31 Mifflin

I had unfavourable numerical experience with a rather complex method for nonsmooth minimization which resulted in a curvilinear search. However, other types of curved searches, derived from those of smooth minimization, may be useful in the nonconvex case. Perhaps this type of research should wait until the convex case is better understood.

2.32 Robinson

G.P. McCormick has also reported unfavourable numerical results in experiments on curvilinear searches.

2.33 Beale

I should like to add to Paper 2.4 a historical note on linear ℓ_1 fitting. Barrodale and Roberts (1973) effectively treat the residuals as artificial variables, and find a true minimum of the sum of the absolute values of these variables by an extension of an efficient Phase 1 of the simplex method, but this algorithm had been proposed some years earlier by Davies (1967).

2.34 Mangasarian

Why is the second order constraint qualification needed in problem EP_1 of Paper 2.4? Most superlinearly convergent algorithms for this problem need the second order sufficiency conditions which do not require any constraint qualification.

2.35 Overton

I introduced the second order constraint qualification only to be able to state a theoretical second order necessary condition. I should perhaps add that, although the first order constraint qualification holds automatically for EP_1 and EP_∞, the second order one does not.

2.36 Conn

Nonlinear ℓ_1 problems are motivated by the usual criteria for robust regression — that is one is anticipating a *few* outliers. However, much of the data may be "good". Hence it is not unusual to have many residuals that are very small but that are not active at the solution. Thus degeneracy, near degeneracy, and identifying the active set are all difficult problems in this context. Therefore I suggest that nonlinear ℓ_1 problems are typically very difficult optimization problems.

2.37 Rosen

A program such as MINOS Augmented could be applied directly to solve EP_1 (or EP_∞) of Paper 2.4, often quite efficiently. Is there much difference between doing this and using the special ℓ_1 and ℓ_∞ algorithms?

2.38 Overton

There is a substantial difference between using a general nonlinearly constrained method to solve EP_1 or EP_∞, and using the special ℓ_1 and ℓ_∞ methods I have mentioned. In a general method, feasibility would not be maintained and $\|f\|_1$ or $\|f\|_\infty$ might not be reduced at each step. In the special methods $\|f\|_1$ or $\|f\|_\infty$ is reduced in the line search, and τ or $\{u_i\}$ are reduced so that equation (1) in Paper 2.4 holds at the end of each iteration.

2.39 Powell

Sometimes it is more efficient to regard ℓ_1 and ℓ_∞ calculations as general nonlinearly constrained problems in order to apply a general algorithm, instead of making use of the "natural objective function". In particular, if large weight is given to one term of the ℓ_1 or ℓ_∞ calculation, then the natural objective function forces this term to be very small. Thus one has the inefficiencies that occur if an algorithm has to follow closely a curved constraint boundary. However, the objective function that is formed automatically by some general algorithms depends on Lagrange multiplier estimates in such a way that scalings of constraint functions make no difference. Thus these algorithms can avoid the inefficiencies that tend to occur in special purpose algorithms.

2.40 Overton

It is a good point that one does not want to follow curved constraint boundaries too closely. However, I should point out that these algorithms do not follow curved boundaries too closely, in the sense that $\|f\|_1$ or $\|f\|_\infty$ is normally only reduced in the line search and not minimized exactly along the line. The latter strategy would generate points very near a curved boundary in your example. I would be surprised if making use of the natural merit function was not a useful thing to do, at least for general problems. However, the question cannot really be resolved without extensive numerical comparisons which have not yet been done.

2.41 Byrd

In terms of global convergence results, the nonlinear ℓ_∞ problem seems easier than the general NLP. For the general NLP Han has proved convergence under assumptions which can only be guaranteed *a priori* if the problem is convex. For the nonlinear ℓ_∞ problem he shows global convergence without such requirements on the functions.

2.42 Han

I think the special structure of the minimax problem is very suitable for the sequential quadratic programming approach. In this case the quadratic programming subproblem is always feasible and its Lagrange multipliers remain bounded.

2.43 Conn

The successive quadratic programming methods that have been mentioned are asymptotically similar. Therefore they have comparable scaling invariance of the constraints in the neighbourhood of the stationary point (since they all approximate the "correct" Lagrangian). However, because of the following three remarks, I am not convinced of the power of this invariance globally, as is suggested in Contribution 2.39. (1) The Lagrangian may be the "wrong" model and the invariance may reinforce this model. (2) Globally, some methods do not approximate the Lagrangian, which makes them free of the above invariance, and they can be protected by the globally consistent merit function. (3) Sometimes it is desired that one stays close to a constraint, even when one is far from the solution, and then the natural merit function may be much more convenient.

2.44 Overton

I should mention that another special feature of problems ℓ_1P and $\ell_\infty P$ in Paper 2.4 is that the $\{f_i\}$ are "naturally scaled", in the sense that the solution is changed if the $\{f_i\}$ are scaled. In general constrained optimization problems, scaling the constraints does not change the solution.

2.45 Coleman

In my opinion it is unnecessary, in the approach of Paper 2.4 to the minimax problem, to introduce an artificial variable τ. In that formulation τ is used in the computation of the search direction, but not in the line minimization procedure. However, the search direction problem can be restated by simply replacing τ with any one of the 'active' functions (it does not matter which one is chosen). This modification has an educational advantage: the magical τ (here it is, here it isn't) suggests *too strongly* that the minimax problem is just another nonlinear programming problem. There is no intrinsic computational advantage to either formulation. In particular, any computational procedure used in one formulation (to estimate dual variables, for example) can be easily transformed to the other formulation.

2.46 Overton

I include τ in EP_∞ simply as a device to obtain the search direction in $\mathbf{R}^n$ by solving a quadratic programming problem with equality constraints in $\mathbf{R}^{n+1}$. One can rearrange the calculation so that one factorizes a matrix with n rows instead of n+1 rows, but this leads to a bias in first order multiplier estimates (Murray and Overton, 1980).

2.47 Dixon

Many methods attempt to ensure feasibility by means of penalty terms. If an equality constraint is nonlinear, and if the points $x^{(k)}$ and $x^{(k+1)}$ are far apart and are approximately feasible, then the penalty term introduces a "hump" in the penalty function on the straight line joining $x^{(k)}$ to $x^{(k+1)}$. If the algorithm is to find such a point $x^{(k+1)}$ on the other side of the "hump", then the line search must be able to cope with several minima. Few algorithms include a suitable line search, and it could be a promising area for research. Note that, to pass through the hump, the search direction should not be tangential to the non-linear constraint at $x^{(k)}$.

2.48 Overton

Although true for general constrained problems, this difficulty is not relevant to the special ℓ_1 and ℓ_∞ algorithms. The reason is that $\|f\|_1$ or $\|f\|_\infty$ is reduced directly in the line search in n dimensions, without any introduction of penalty terms. The "hump" would occur only if the line search were done in n+m or n+1 dimensions respectively, as suggested in Contributions 2.37 and 2.39.

2.49 Steihaug

The term "global convergence" has so far been used in three different ways. *Convergence* is either that the gradients are converging to zero or that the iterates are converging. *Global* is either "convergence" from any starting point or "convergence" to a global minimizer. I suggest that we are more precise and use, for example, "potential convergence" when the gradients are converging to zero, and let "global" be used when the convergence result is obtained without restriction on the starting point.

PART 3

Linear Constraints

3.1

Linearly Constrained Optimization*

Philip E. Gill, Walter Murray,
Michael A. Saunders and Margaret H. Wright

Systems Optimization Laboratory,
Department of Operations Research,
Stanford University, Stanford, California 94305, USA.

1. Introduction

The problem to be considered in this paper is the minimization of a smooth nonlinear function subject to the requirement that certain *linear* functions of the variables must be exactly zero, nonnegative, or nonpositive. The general form for each linear function is $\ell(x) = a^T x - \beta$, for some row vector a^T and scalar β. By convention, the linear constraints will be written in the form $a^T x \geq \beta$. (Clearly, constraints of the form $a^T x - \beta \leq 0$ can equivalently be stated as $-a^T x + \beta \geq 0$.)

We shall begin by considering the following form of the linearly constrained problem:

$$\left.\begin{array}{ll} \underset{x \in \mathbf{R}^n}{\text{minimize}} & F(x) \\ \text{subject to} & Ax \geq b \end{array}\right\} , \qquad \text{(LIP)}$$

where F is twice continuously differentiable, the matrix A is

*This research was supported by the U.S. Department of Energy Contract DE-AC03-76SF00326, PA No. DE-AT03-76ER72018; National Science Foundation Grants MCS-7926009 and ECS-8012974; the Office of Naval Research Contract N00014-75-C-0267; and the U.S. Army Research Office Contract DAAG29-79-C-0110.

$m \times n$, and the i-th row of A contains the coefficients corresponding to the i-th constraint. For simplicity we shall not treat the case where the constraints are a mixture of equalities and inequalities; algorithms for such a problem can be developed in a straightforward manner from those to be described. We shall assume that LIP has a bounded solution and that F is bounded below in the feasible region. We shall denote the gradient vector and Hessian matrix of $F(x)$ by $g(x)$ and $G(x)$ respectively.

The problem format LIP is generally used to define problems that do not have a large number of variables, i.e. problems that are not *large scale*. We shall consider a problem format for large scale problems in Section 3.5.

In Section 2 we shall give a brief overview of methods for nonlinear programming subject to linear constraints. In Section 3 we consider some specific problem categories that we believe will be the subject of future research.

2. A Brief Overview of Methods

A very general class of methods for such problems is that of *feasible point active set methods* (see, e.g., Gill and Murray, 1977a; Gill, Murray and Wright, 1981). This name reflects two important properties of the methods. Firstly, all iterates are feasible after an initial feasible point has been found by a "phase 1" procedure. Secondly, a subset of the constraints — which we shall call the *working set* — is used to define the search direction at each iteration; typically, the working set includes all the constraints that are exactly satisfied at the current iterate.

We shall be concerned initially with "step-length-based" methods, i.e., methods that generate iterates that satisfy the relationship $x_{k+1} = x_k + \alpha_k p_k$, where p_k denotes the direction of search and α_k the scalar step length ("trust-region-based" methods will be discussed briefly in Section 3.2).

Assume that the working set contains t_k constraints, and let $\hat{A}_k$ denote the matrix of coefficients of the constraints in the current working set. The search direction p_k is constrained to lie within the subspace of vectors that are orthogonal to the rows of $\hat{A}_k$; in effect, the constraints in the working set are treated as *equalities* for the purpose of defining the search direction p_k.

Depending on the nature of the algorithm, there are various strategies for adding and deleting constraints from the working set. Typically, a constraint is *added* to the working set when its presence restricts the step length to be taken along p_k. An inequality constraint is *deleted* from the working set when a sufficiently accurate Lagrange multiplier estimate predicts that the constraint is unlikely to be active at the solution.

Let Z_k denote a matrix whose columns form a basis for the set of vectors orthogonal to the rows of $\hat{A}_k$, so that $\hat{A}_k Z_k = 0$; the search direction p_k is then of the form

$$p_k = Z_k p_Z,$$

where p_Z is an $(n-t_k)$-vector. This definition ensures that $\hat{A}_k p_k = 0$, and hence moves along p_k do not alter the values of any of the constraints in the working set.

In some of the most successful methods, the search direction is the solution of the quadratic programming subproblem

$$\left.\begin{array}{ll} \underset{p \in \mathbf{R}^n}{\text{minimize}} & g_k^T p + \tfrac{1}{2} p^T H_k p \\ \text{subject to} & \hat{A}_k p = 0 \end{array}\right\} . \tag{1}$$

The quadratic objective function is usually viewed as an approximation to the objective function $F(x)$, and H_k represents the Hessian matrix of second derivatives or an approximation to it.

The solution of (1) is found by solving the equations

$$Z_k^T H_k Z_k p_Z = -Z_k^T g_k, \tag{2}$$

and setting $p_k = Z_k p_Z$.

In the unconstrained case, Newton's method may be regarded as an "ideal" or "natural" strategy that is usually to be preferred if there are no restrictions on storage or the availability of second derivatives. If H_k is G_k, the exact Hessian at x_k, a method can be devised based on solving (2) that is a direct generalization of Newton's method to the linear constraint case. An important feature of this "ideal" algorithm is that an *orthogonal* matrix Z_k is recurred using an orthogonal factorization of the matrix of constraints in the working set (see, e.g., Gill and Murray, 1977a). Note that the projected Hessian matrix $Z_k^T G_k Z_k$ must be at least positive semi-definite at the solution, whereas the full Hessian matrix may be indefinite at every iteration.

If it is not possible to provide second derivatives, other methods, such as quasi-Newton methods or conjugate gradient methods, must be used that emulate Newton's method without the associated overhead. To illustrate how the equations (2) can be used to define a particular method, consider the construction of a quasi-Newton method. In this case an $(n-t_k)$-dimensional quasi-Newton approximation to the projected Hessian is updated and used to compute the vector p_z from (2). The standard quasi-Newton update formulae can be adapted to reflect curvature within the subspace defined by Z_k by using only projected vectors and matrices in the computation of the updated matrix. This approach to applying quasi-Newton methods to LIP has the attractive feature that the property of hereditary positive definiteness can be carried over to the linear constraint case. The projected Hessian matrix must be at least positive semi-definite at the solution, whereas the full Hessian matrix may be indefinite at every iteration. Furthermore, the search directions always satisfy the constraints of (1) with the same accuracy (depending on the form of Z_k), and the quasi-Newton updates thus do not affect feasibility.

In many problems, all the constraints of LIP are simple bounds on the variables:

$$\ell_i \le x_i \le u_i, \quad i = 1,2,\ldots,n,$$

where either bound may be omitted.

The special form of bound constraints leads to considerable simplifications in the application of equations (2). Since a bound constraint in the working set indicates that a variable lies on one of its bounds, the matrix $\hat{A}_k$ will contain only signed columns of the identity matrix; the corresponding components of x are termed the *fixed* variables. Furthermore, the matrix Z_k can be taken as a set of the columns of the identity matrix corresponding to the variables that are not on their bounds (the *free* variables, denoted by x_{FR}). Thus, the projected gradient $Z_k^T g_k$ is simply the subvector of g_k corresponding to free variables, which we denote by g_{FR}; similarly, the rows and columns of the Hessian matrix G_k corresponding to the free variables comprise the projected Hessian, which we shall denote by G_{FR}.

All the algorithms for linear constraints have a particularly simple form when applied to a bound constrained problem. For example, the Newton equations become

$$G_{FR}p_{FR} = -g_{FR}.$$

At each iteration, the variables of interest are simply a subset of the original variables; hence, methods for bound constrained minimization are more closely related to algorithms for unconstrained problems than to problems with general linear constraints. An interesting consequence of this relationship is that, though it may be necessary to pursue *research* into methods for unconstrained minimization, it is not absolutely necessary to have *software* for unconstrained minimization. Even when the problem of interest is *unconstrained*, it is often helpful to introduce unnecessary but reasonable bounds on the variables, and then to solve the problem by means of a bound constraint algorithm.

3. Some Topics for Future Research

3.1. Quadratic programming If $F(x)$ is a positive definite quadratic function, there are only two possibilities for the step length. A step of unity along p_k is the exact step to the minimum of the function restricted to the null space of $\hat{A}_k$. If a step of unity can be taken, the next iterate will be a constrained stationary point with respect to the equality constraints defined by $\hat{A}_k$, and exact Lagrange multipliers can be computed to determine whether a constraint should be deleted. Otherwise, the step along p_k to the nearest constraint is less than unity, and a new constraint will be included in the working set at the next iterate.

As with linear programming, advantage can be taken of the special features of the quadratic objective function, so that the needed quantities are computed with a minimum of effort (see, e.g., Gill and Murray, 1978a).

Recently, there has been considerable interest in methods for solving quadratic programs — primarily because of the renewed interest in methods for nonlinearly constrained minimization that solve a quadratic programming subproblem. If a quadratic program is used in this application, care should be taken to avoid the possibility that the point computed during the first phase of the QP algorithm is far from the optimum of the new subproblem. Several quadratic programming algorithms have been devised that do not fall precisely within the class of two-phase active set methods discussed in Section 2 (see, e.g., Conn and Sinclair, 1975, and Goldfarb's research seminar at this meeting). For example, in the single phase algorithm of Conn and Sinclair, each iterate satisfies a set of constraints exactly, but is not necessarily feasible. In this case, the method is one of descent with respect to a nondifferentiable penalty function involving the linear constraints.

In the context of nonlinear programming, it remains to be seen whether a single phase QP algorithm will prove to be more efficient than a two phase QP algorithm in which the initial working set can be specified for the first phase (see Section 3.4 for more discussion concerning the choice of initial working set).

3.2. The form of the QP subproblem The search direction for a linearly constrained method does not have to be defined using the precise QP subproblem (1). Consider the more general problem

$$\left.\begin{array}{lll} \underset{p\in \mathbf{R}^n}{\text{minimize}} & g_k^T p + \tfrac{1}{2} p^T H_k p \\ \text{subject to} & \hat{A}_k p = d_k \\ \text{or} & \hat{A}_k p \geq d_k \end{array}\right\} .$$

In this case, the matrix $\hat{A}_k$ includes some or all of the constraints of the original problem.

There are two extremes of subproblem that might be posed, with other variants in between. At one extreme, an equality constrained (EQP) subproblem is solved whose constraints are those in a working set. At the other extreme, the subproblem is an inequality constrained QP (IPQ), in which *all* the original inequality constraints are represented as inequalities. The approaches differ in several ways: the effort required to solve the subproblem; the original information required about F; the theoretical properties of the resulting search direction; and the information returned after the problem has been solved. All these differences affect the performance of the outer algorithm.

Any algorithm for linearly constrained minimization must include some procedure — the active set strategy — for determining which constraints hold with equality at the solution. The procedure for the determination of the active set is one of the most important differences between the pure EQP and IQP approaches. Algorithms based on solving an EQP include a *pre-assigned* active set strategy that specifies which constraints are to be treated as equalities in the QP; the term "pre-assigned" signifies that the decision about the working set is made before posing the QP subproblem. With the pure IQP approach, on the other hand, a *QP-assigned* active set strategy is used, in that the set of constraints active at the solution of the QP will be taken as a prediction of the active set. The pre-assigned strategy is tacitly assumed to involve solving a problem that is less complex than a general IQP. However, the pre-assigned strategy associated with an EQP approach may itself require the solution of another EQP.

When a pure EQP approach is used, the search direction is the solution of the subproblem (1). When a pure IQP approach is used, one possible subproblem is

$$\left.\begin{array}{ll} \underset{p\in\mathbf{R}^n}{\text{minimize}} & F_k + g_k^Tp + \tfrac{1}{2}\, p^TH_kp \\ \text{subject to } Ap \geq b - Ax_k & \end{array}\right\}. \tag{3}$$

If the solution of (3) is used as the search direction, complications arise because it may be necessary to store or represent the full matrix H_k. Although only the matrix $Z_i^TH_kZ_i$ is required at iteration i of the qudratic subproblem, it is not known *a priori* which set of constraints will define Z_i as the iterations proceed. Hence, most inequality QP methods assume that the full matrix H_k is available. In contrast to the positive definiteness of the projected Hessian in the equality constraint case, the Hessian of F need not be positive definite, even at the solution. If H_k is indefinite, (3) may not have a bounded solution; furthermore, even if a solution exists, p_k is not necessarily a descent direction for F. Even if $Z_i^TH_kZ_i$ is positive definite during *every* iteration of the subproblem, descent is not assured.

A number of approaches have been proposed to overcome the difficulties of solving an inequality constrained QP. Fletcher (1972) has suggested computing the search direction from a QP subproblem that includes all the original inequality constraints as well as additional bounds on each component of the search direction; this method is similar to that of Griffith and Stewart

(1961), in which a linear programming subproblem is solved. The purpose of the extra constraints in Fletcher's method is to restrict the solution of the QP to lie in a region where the current quadratic approximation of F is likely to be reasonably accurate. The bounds on p are adjusted at each iteration if necessary to reflect the adequacy of the quadratic model. Fletcher's algorithm effectively includes the "trust region" idea that is used in other areas of optimization.

In the situation where the exact second derivatives of F are known, it is likely that methods for linear constraints based on an EQP subproblem and an IQP subproblem are each best for some problem categories. For example, if the objective function is very expensive to evaluate compared to the cost of updating the factorizations associated with the constraints in the working set, the IQP approach may prove to be superior. However, it must be emphasized that it is not yet clear that the extra work required within the IQP will lead to an overall reduction in the number of function evaluations needed to solve the problem.

The prospects for IQP methods are more uncertain for problems for which second derivatives are not available. For example, if the BFGS method is used to approximate the full $n \times n$ Hessian, extra precautions are necessary to ensure positive definiteness since the usual conditions that guarantee hereditary positive definiteness are less likely to apply. If positive definiteness is enforced, it is not clear how this will affect the algorithm at arbitrary points where the ability of the IQP approach to predict the correct working set is critically affected by the curvature of the quadratic approximation.

3.3. Problems with few constraints in the working set Methods based upon computing p_k from (2) tend to improve in efficiency as the number of constraints in the working set increases. However, standard methods of this type may not be the best choice when the number of linear constraints active at the solution is *small* compared to the number of variables, since the dimension of the null space will be much larger than that of the range space. Methods for the solution of this type of problem still require further research. Of particular interest are methods that are applicable to large scale problems.

One approach is to store Z_k and the factors of $Z_k^T H_k Z_k$ in a compact form. Such an approach is possible only if the matrix of constraint coefficients and the Hessian matrix have a special structure. For example, if the constraints are all generalized upper bound (GUB) constraints, and B_k is tridiagonal, a matrix Z_k can be found such that $Z_k^T H_k Z_k$ has band-width five.

An alternative approach is to design a *range space method* for which the work per iteration is reduced as the number of constraints in the working set decreases. Let t_k denote the number of active constraints. Range space methods are based upon solving the $n + t_k$ linear equations

$$\begin{pmatrix} H_k & -\hat{A}_k^T \\ \hat{A}_k & 0 \end{pmatrix} \begin{pmatrix} p_k \\ \lambda_k \end{pmatrix} = \begin{pmatrix} -g_k \\ 0 \end{pmatrix} . \tag{4}$$

Range space methods for various categories of linearly constrained problem have been suggested by several authors, including Goldfarb (1969), Murtagh and Sargent (1969) and Fletcher (1971). All these methods are based upon forming some part of the partitioned inverse of the matrix in (4):

$$\begin{pmatrix} H_k^{-1} - H_k^{-1}\hat{A}_k^T \Phi_k \hat{A}_k H_k^{-1} & H_k^{-1}\hat{A}_k^T \Phi_k \\ -\Phi_k \hat{A}_k H_k^{-1} & \Phi_k \end{pmatrix}, \text{ where } \Phi_k = (\hat{A}_k H_k^{-1} \hat{A}_k^T)^{-1} .$$

Range space methods tend to be less numerically stable than null space methods (especially when the Hessian of the objective function is not positive definite), and several theoretical and practical difficulties need to be resolved. The most important problem is that, unless H_k is positive definite, the factorizations of H_k and Φ_k do not provide information concerning whether or not $Z_k^T H_k Z_k$ is positive definite. As a result, it is more difficult to distinguish between a constrained stationary point and a constrained minimum. To some extent, this problem can be resolved in the case of indefinite quadratic programming. If x_k is a constrained minimum, and if a constraint is deleted from the working set, the point $x_k + p_k$ will be a constrained minimum unless p_k is a direction of negative curvature within the subspace defined by the reduced working set. Thus, if x_0 is a vertex of the feasible region, it is possible to detect whether or not subsequent iterates are constrained minima by examining the curvature of F along each search direction. If negative curvature is detected, a constraint must be added to the working set before any further constraints are deleted. (This exchange scheme is the basis of Fletcher's, 1971, range space QP algorithm.) A feature of this approach is that n constraints may need to be in the working set at x_0. Thus the problem may be temporarily converted into one for which range space methods are inefficient.

3.4. The active set strategy There is no universal agreement today about the best possible active set strategy (see, e.g., Lenard, 1979). Different methods require different strategies. With null space methods, the row dimension of $\hat{A}_k$ should be as large as possible; for range space methods, it should be as small as possible. Since constraints are usually added and deleted one at a time, the size of the working set at the initial point of the feasible point algorithm greatly influences the efficiency of the subsequent computation.

In order to initiate the first phase of the optimization, an initial working set (the *crash set*) must be determined by a procedure that is called a *crash start*. The only mandatory feature of the initial working set is that all the linearly independent equality constraints should be included. The selection of any other constraints is determined primarily by the nature of the second phase of the optimization, i.e. whether a null or range space method is employed. If the second phase uses a null space method, the initial working set should include as many constraints as possible. If a range space method is employed, the initial working set should include the linearly independent equality constraints only.

<u>3.5. Large scale problems</u> If the number of variables is large and the matrix of constraint coefficients is sparse, it is advantageous to consider the constraints in the same form as that used in large scale linear programming:

$$\left.\begin{aligned} &\underset{x\in\mathbf{R}^n}{\text{minimize}} \quad F(x) \\ &\text{subject to } Ax = b, \\ &\text{and} \qquad\quad \ell \le x \le u \end{aligned}\right\} . \tag{5}$$

As a rule, there are *fewer* algorithmic options for large problems, since many computational procedures that are standard for small problems become unreasonably expensive in terms of arithmetic and/or storage. However, in another sense the options for large problems are less straightforward because of the critical effect on efficiency of special problem structure and the details of implementation.

At a typical iteration of an active set method applied to (5), the matrix $\hat{A}_k$ will contain all the rows of A and an additional set of rows of the identity that correspond to variables on one of their bounds. Let r denote the number of fixed variables at the current iteration. Then the matrix A is (conceptually) partitioned as follows:

$$A = (\; B \quad S \quad N \;).$$

The m×m "basis" matrix B is square and nonsingular, and its columns correspond to the *basic* variables. The r columns of N correspond to the *nonbasic* variables (those fixed on their bounds). The $n - m - r$ columns of the matrix S correspond to the remaining variables, which will be termed *superbasic* (Murtagh and Saunders, 1978). The number of superbasic variables indicates the number of degrees of freedom remaining in the minimization (since there are $m + r$ constraints in the working set).

At a given iteration, the constraints in the working set (rearranged for expository purposes) are given by

$$\hat{A}_k x = \begin{pmatrix} B & S & N \\ 0 & 0 & I \end{pmatrix} \begin{pmatrix} x_B \\ x_S \\ x_N \end{pmatrix} = \begin{pmatrix} b \\ b_N \end{pmatrix},$$

where the components of b_N are taken from either ℓ or u, depending on whether the lower or upper bound is binding.

A matrix Z_k that is orthogonal to the rows of $\hat{A}_k$ is given by

$$Z_k = \begin{pmatrix} -B^{-1}S \\ I \\ 0 \end{pmatrix}. \tag{6}$$

The matrices B^{-1} and Z_k need not be computed explicitly. An advantage of (6) is that Z_k or Z_k^T may be applied to vectors using only a factorization of the square matrix B.

The form (6) of Z_k means that the partitioning of the variables into basic, nonbasic, and superbasic sets carries directly over to the calculation of the search direction p_k. If p_k is partitioned as $(p_B\ p_S\ p_N)$, it follows that $p_N = 0$ and

$$Bp_B = -Sp_S. \tag{7}$$

Equation (7) shows that p_B can be computed in terms of p_S. Thus, the only variables to be adjusted are the superbasic variables, which act as the "driving force" in the minimization. The determination of the vector p_S is exactly analogous to that of p_Z in the methods discussed earlier, in that it completely specifies the search direction.

The motivation that underlies the definition of a "good" choice of the vector p_S is similar to that discussed in Section 2 for defining p_Z. We wish to specify p_S so that the search direction "solves" an equality constraint quadratic problem of the form

$$\left.\begin{array}{ll}\underset{p\in\mathbf{R}^n}{\text{minimize}} & g_k^Tp + \tfrac{1}{2}\, p^TH_kp \\ \text{subject to } \hat{A}_kp = 0 & \end{array}\right\}, \tag{8}$$

where H_k is an approximation to the Hessian matrix of $F(x)$. The vector p_S that corresponds to the solution of (8) can be computed using only the *projected matrix* $Z_k^TH_kZ_k$, by solving the linear system

$$Z_k^TH_kZ_kp_S = -Z_k^Tg_k, \tag{9}$$

where Z_k is given by (6). Note that $Z_k^Tg_k = g_S - S^T\pi$, where π is the solution of $B^T\pi = g_B$. When Z_k has the form (6), $Z_k^Tg_k$ is called the *reduced gradient*.

Given that we can obtain a representation of B^{-1} (and hence of Z_k), an important unresolved issue in implementing a method for large scale problems is the best way to solve the equations (9) for p_S. In general, the projected matrix $Z_k^TH_kZ_k$ will be dense even if A and H_k are sparse (although there may be exceptions when the constraints have a very special structure). Moreover, there may be inadequate storage to retain a full version of H_k.

In many cases, the dimension of the projected Hessian matrix $Z_k^TH_kZ_k$ will be relatively small at every iteration, even when the problem dimension is large. If $Z_k^TH_kZ_k$ is small enough to be stored, standard approaches from the dense case may be used. For example, a quasi-Newton approximation of the projected Hessian may be maintained using update procedures similar to those in the unconstrained case. Any questions concerning such procedures apply generally to linearly constrained optimization, and are not particular to large scale problems. However, the technique of computing finite differences along all the columns of Z_k, which is very successful for small problems, is too expensive in the large scale case because of the effort required to form the columns of Z_k explicitly. Furthermore, even if the exact Hessian G_k itself is available, it is probably too costly to form $Z_k^T G_k Z_k$.

When limitations of storage and/or computation preclude an explicit representation of $Z_k^T H_k Z_k$, a linear conjugate gradient method can be used to solve (9), since the product of $Z_k^T H_k Z_k$ and a vector v can in some circumstances be computed efficiently even when $Z_k^T H_k Z_k$ is not available. For example, if it is possible to store a sparse Hessian G_k and a sparse representation of Z_k, the necessary matrix vector products of the form $Z_k^T G_k Z_k v$ can be computed relatively cheaply by forming, in turn, $v_1 = Z_k v$, $v_2 = G_k v_1$ and $v_3 = Z_k^T v_2$ (a similar procedure can be used if a sparse Hessian approximation is available). If a sparse Hessian is not available, the vector v_2 may be approximated using a single finite difference of the gradient along v_1.

There are several possibilities for the application of preconditioning, depending on the available information and on the speed with which the iterates of the linear conjugate gradient method converge.

An *inexact Newton method* (see Dembo, Eisenstat and Steihaug, 1980) may prove to be an important tool for the solution of (9). For example, a *truncated* Newton method "solves" (9) by performing a *limited number* of iterations of the linear conjugate gradient method. The final iterate of the truncated sequence is then taken as an approximate solution of (9). The hope with a truncated Newton method is for the required number of linear conjugate gradient steps to be small, and the use of preconditioning would therefore seem to be essential. A preconditioned truncated Newton method will be particularly effective when the *projected* Hessian matrix is small enough to be stored explicitly. In this case, a quasi-Newton approximation to $Z_k^T H_k Z_k$ can be used as a preconditioning matrix.

4. Conclusions

The basic strategy for the solution of linearly constrained problems is to use the constraints in order to effect a reduction in the dimensionality of the minimization. The resulting "reduced" problem has no constraints and can be solved using any one of the myriad of methods for unconstrained minimization. An immediate consequence is that many improvements in methods for the unconstrained case often carry over to the linear constraint case.

In the unconstrained case, Newton's method may be regarded as an "ideal" or "natural" strategy that is usually to be preferred if there are no restrictions on storage or the availability of second derivatives. If such restrictions apply, other methods, such as quasi-Newton methods or conjugate gradient methods, must be used that emulate Newton's method without the associated overhead. The "ideal" algorithm for the linear constraint case has two components: (i) a procedure to reduce the dimensionality of the minimization using an orthogonal factorization of the matrix of active constraints; and (ii) a second order procedure, analogous to Newton's method, for minimizing the reduced problem (see, e.g., Gill, Murray and Wright, 1981).

This method has several important features: the condition of the reduced problem is no worse than that of the original problem; the method allows the computation of both first and second order multiplier estimates; and finally, the method will (under mild conditions) exhibit quadratic convergence to a local minimum.

However, as in the unconstrained case, it may be inconvenient or difficult to compute second order information, and therefore quasi-Newton approximations of the projected Hessian matrix are appropriate. In the linear constraint case there is the additional complication that storage limitations may make it impossible to compute the orthogonal factorization of the matrix of active constraints. In this case, a less satisfactory unconstrained subproblem of the "reduced gradient" type must be formulated.

Recent research in linearly constrained optimization has primarily been directed towards solving problem categories that the "natural" methods cannot handle efficiently.

3.2

Numerically Stable Approaches to Linearly Constrained Optimization*

D. Goldfarb

Department of Computer Sciences,
The City College of New York, New York, N.Y. 10031, USA.

Most algorithms for solving linearly constrained optimization problems that have appeared in the literature employ an active set strategy. Whether such a strategy is combined with an EQP (equality constrained quadratic programming) or an IQP (inequality constrained quadratic programming) approach for determining the direction of search, it is important to implement these approaches in a numerically stable manner. (This has already been mentioned by Gill in his Paper 3.1.) When one uses an IQP approach, the quadratic programming algorithm usually solves its subproblems by an active set EQP algorithm. Consequently, both the EQP and IQP approaches are quite similar when considered from the point of view of their numerically stable and efficient implementation.

Such implementations are invariably based upon one or more of the following factorizations: (i) a symmetric LDL^T factorization of the Hessian matrix G of the objective function or a variable metric approximation to it (assumed to be positive definite), (ii) an orthogonal QR factorization of the matrix N of active constraint normals, (iii) an LU factorization of the square and nonsingular basis matrix when the problem constraints are considered in the standard LP form, (iv) variants of these factorizations. For example, the Cholesky factorization $\hat{L}\hat{L}^T$ and the

*Partially supported by the National Science Foundation under Grant MCS-8006065.

QR factorization of $\hat{L}^{-1}N$ are often used in place of (i) and (ii) respectively. The symmetric factorization of Z^TGZ, where the columns of Z provide a basis for the null space of N, or $N^TG^{-1}N$ or something quite similar to it, is also frequently used in place of, or in addition to, (i).

It is not possible to single out which factorization or factorizations are best, as such a judgement depends upon the average cardinality of the active set during a problem run, whether or not some of the constraints are simple bounds, whether or not the matrices G and N are large and sparse, how the optimization algorithm performs the calculations which involve the factorizations, and which data structures are used to store the factorizations, to name just a few influencing factors.

Efficient and numerically stable methods for updating these factorizations when a constraint is either added to or deleted from the active set or is replaced by another constraint, or when G is modified by a symmetric rank-one or rank-two correction, are well known, see for example (Gill, Golub, Murray and Saunders, 1974; Gill, Murray and Saunders, 1975; Goldfarb, 1976).

Future research in this area will probably be directed towards customizing such methods to matrices with special structure, and in dealing with the case of a nonpositive definite Hessian or approximate Hessian matrix G. Here the factorizations of Bunch and Parlett (1971), Aasen (1971), Dax and Kaniel (1977), Bunch and Kaufman (1977), and Dax (1980) can be used. Thus far, ways to update only the first of these factorizations have been studied when G is modified by a symmetric rank-one term (Sorensen, 1977).

How one represents a problem is very important in that it significantly influences the factorizations used and other aspects of the implementation of an algorithm. If a problem is stated in the standard LP form, it is quite natural to use an LU factorization of the basis matrix, whereas a QR factorization of N is more natural for problems given in inequality form, although LU factorizations can also be used in this case (e.g. Goldfarb,

1976). Which of these representations is preferable is very problem dependent, and it would be foolish to make any all-inclusive recommendations. One advantage of the standard LP form, which becomes very important for large sparse problems, is that one can readily use all of the basis matrix machinery developed for the simplex method. The best example of this is the MINOS code of Murtagh and Saunders (1978).

Another factor which is partially within the control of an algorithm developer is the cardinality of the active set. As already mentioned, the advisability of keeping the active set small or large depends heavily on the factorizations used and consequently on the problem representation. Although it is again not possible to make any general statement, my own experience with primal quadratic programming is that a small active set is better, mainly because it results, on the average, in a reduction in the number of iterations required to obtain an optimal solution. Let me now turn to a discussion of some promising directions for future research, which are concerned less directly with numerically stable methods for modifying factorizations.

One important topic that has not been dealt with very much in the past is the problem of degeneracy and linear dependence. From a numerical standpoint we are, of course, also interested in "near" degeneracy and "near" linear dependence. Typically, in primal active set algorithms, the need (for reasons of numerical stability) to keep a "sufficiently" linearly independent active set is balanced against the need to keep the iterates "sufficiently" feasible. The concept of ε-active constraints has been used, but mainly to ensure convergence by preventing zigzagging. One approach that is possible is to make the requirement of feasibility very loose initially, and then gradually to tighten it as the solution of the problem is approached. This would allow one to ensure a "sufficiently" linearly independent active set until very close to the solution. Here we are assuming that some infeasibility in the constraints can be tolerated. Without theoretical criteria to relate near linear dependence to near feasibility, the above approach is *ad hoc*. Results on the perturbations of systems of linear inequalities may be useful in developing such criteria (Robinson, 1974).

An alternate approach to this problem is provided by duality. The key observation is that in a dual algorithm there is a choice of which constraint to add to the active set (but usually there is no choice of which constraint to drop), while in a primal algorithm the constraints that are added to active sets are selected by feasibility conditions (usually, however, there is a choice of which constraint to drop). Consequently, one has more control of the linear independence of the active set in dual methods than one does in primal methods.

To illustrate further the advantages of a dual method, consider the positive definite quadratic program:

$$\text{minimize} \quad f(x) = a^T x + \tfrac{1}{2} x^T G x,$$

$$\text{subject to } C^T x \geq b.$$

It is easy to derive the following dual problem (Lemke, 1962):

$$\text{minimize} \quad w(u) = -d^T u + \tfrac{1}{2} u^T \hat{G} u,$$

$$\text{subject to } u \geq 0,$$

where $\hat{G} = C^T G^{-1} C$ and $d = b + C^T G^{-1} a$. Clearly degeneracy and linear dependence problems do not occur in the dual constraints $u \geq 0$. Notice that, because x and u are related by $x = G^{-1}(Cu - a)$, it is possible for x to remain unchanged when u changes. In the dual problem the matrix $\hat{G}$ is only positive semi-definite, and a potentially serious disadvantage is that a principal submatrix of $\hat{G}$ may be ill-conditioned, due to near linear dependence of a subset of columns of C.

Let us consider a dual quadratic programming method from the point of view of the original primal problem. When primal feasibility is tested, one expects to find substantial constraint violations until the solution is reached, but the current x is in theory optimal with respect to an active subset **A** of the primal constraints. In practice, however, because of rounding errors, x will not lie on the linear manifold **M**, which is the intersection of the boundaries of the constraints in **A**. Now suppose that there are some primal constraints, that are not in **A**, whose bounding hyperplanes also contain **M**. We call this set of constraints **D**. Due to computer rounding errors, some of the constraints of **D** may be violated by the current x. If **A** is the *optimal* active set, then these violations should fall within a prescribed tolerance for termination, and thus cause no difficulty for the algorithm. A similar statement holds for the case where the manifold **M** is only "nearly degenerate"; i.e. the constraints in $\mathbf{A} \cup \mathbf{D}$ are only "nearly" linearly dependent, and are very close to being satisfied as equalities at the point x. Alternatively, if **A** is not the optimal active set, and if x is so infeasible that the algorithm does not terminate, then we expect a constraint that is not in $\mathbf{A} \cup \mathbf{D}$ to be violated by x. This constraint can be chosen as the constraint to be added to the active set, which avoids an ill-conditioned intermediate active set constraint matrix when the given constraints are consistent. In fact, choosing the most violated constraint is a good strategy to help the prevention of numerical instabilities. However, as mentioned above, in primal methods one cannot choose which constraint to add to the active set; therefore ill-conditioned intermediate calculations are far more likely to occur.

Another area of research that appears to be worth investigating is the use of the ellipsoid method (Shor, 1970) and related relaxation methods (Agmon, 1954; Motzkin and Schoenberg, 1954) for solving convex programming problems. In these methods iterative procedures reduce the region in which the optimal solution lies, by using the constraint and objective functions, where linear approximations are made to any nonlinear functions in the calculation. It appears that near linear dependence of a set of constraints is less damaging than in active set algorithms, but a thorough analysis remains to be done.

A research topic, that is relevant to linearly constrained optimization and that could lead to significant algorithmic improvements, is the problem of converting general linear constraints $Ax = b$ into network constraints $\bar{A}\bar{x} = \bar{b}$, where $\bar{A}$ is a node-arc incidence matrix for some network. Working with $\bar{A}$ can be orders of magnitude more efficient than working with A, and, because each element of $\bar{A}$ is in the set $\{0, 1, -1\}$, numerical instability problems do not occur. Some results in this area have been obtained by Bixby and Cunningham (1980), which are based upon an algorithm (due to Tutte) for determining whether or not a given binary matroid is graphic. Further algorithmic development, however, is necessary for such an approach to become practicable; also the numerical stability of the transformation algorithm needs to be investigated.

In this paper we have discussed new approaches to the problem of developing numerically stable algorithms for linearly constrained optimization, as well as outlining extensions of more traditional approaches. It is hoped that the paper will stimulate fresh ideas that will be even more fruitful than any of those presented here, and that will lead to new algorithms which are more efficient and more robust than those that are currently in use.

3.3

Fixed-point Methods for Linear Constraints*

Michael J. Todd

School of Operations Research, Cornell University,
Ithaca, New York 14853, USA.

1. Introduction

We are concerned with the application to nonlinear programming problems of the class of so-called fixed-point algorithms (also known as simplicial or piecewise-linear (PL) homotopy methods). These algorithms are designed primarily for zero-finding problems. They have been called fixed-point algorithms because their first application was to problems where existence could be guaranteed only by appealing to a topological fixed-point theorem, and convergence could be assured for such problems. For general references on these methods, see Allgower and Georg (1980), Eaves (1976), Scarf and Hansen (1973) and Todd (1976). Since the algorithms may not be familiar to many readers, we will outline briefly their approach in Section 2.

Besides their attractive global properties (for certain problems), fixed-point algorithms can be implemented on smooth problems so as to converge quadratically to a solution. However, it appears that the computational work in applying these methods increases rather rapidly with the dimension. (A rough rule is that $O(n^2)$ function evaluations and $O(n^4)$ algebraic operations

*Research supported in part by a Guggenheim Fellowship and by National Science Foundation Grant ECS-7921279.

are required in dimension n.) There has therefore been some interest in taking advantage of the special structure of certain problems; here we will discuss this work in the context of nonlinear programming problems — see Section 3. While PL homotopy methods can be applied to general inequality constrained problems, to discuss all the approaches that have been suggested we will confine ourselves to the case of linear constraints.

Consider the problem:

$$\min\ \theta(x) \quad \text{subject to } A^T x \le b, \tag{NLP}$$

where $\theta : \mathbf{R}^n \to \mathbf{R}$ is continuously differentiable, $A = [a_1, a_2, \ldots, a_m] \in \mathbf{R}^{n\times m}$ and $b = (\beta_1, \beta_2, \ldots, \beta_m)^T \in \mathbf{R}^m$. Since the algorithms with which we are concerned apply to zero finding problems, we proceed to the first order necessary conditions of F. John:

$$\left.\begin{array}{l} w_0\ \nabla\theta(x) + Aw = 0, \\ w_0 \ge 0, \quad (w_0, w) \ne 0, \\ w \ge 0, \quad A^T x \le b, \quad w^T(A^T x - b) = 0. \end{array}\right\} \tag{1.1}$$

The Karush–Kuhn–Tucker conditions further require that $w_0 = 1$. Both our zero finding formulations arise from (1.1) and thus lead to direct methods for solving the optimality conditions. The algorithms therefore make no use of function values of θ, and while this property is a consequence of their general nature, it must be viewed as a drawback in their application to optimization problems.

Our first formulation seeks a zero of a point-to-set mapping defined on $\mathbf{R}^n$. Define

$$X_0 = \{x : A^T x \le b\}, \qquad X_i = \{x : a_i^T x - \beta_i = \max_j [a_j^T x - \beta_j] \ge 0\}, \tag{1.2}$$

$$f_0 = \nabla\theta \ , \qquad f_i \equiv a_i \tag{1.3}$$

and

$$\Gamma(x) = \text{conv}\{f_i(x) : i \quad \text{such that } 0 \le i \le m \text{ and } x \in X_i\}\ . \tag{1.4}$$

Then Γ takes as values nonempty convex compact subsets of $\mathbf{R}^n$ and is upper semi-continuous (see e.g., Todd, 1976). Our problem is to find $x^* \in \mathbf{R}^n$ satisfying

$$0 \in \Gamma(x^*). \tag{1.5}$$

Note that, if x^* with some w_0 and w solves (1.1), then x^* satisfies (1.5); the converse is true if the feasible region X_0 is nonempty. This approach has been taken by Scarf and Hansen (1973), Merrill (1972), Todd (1976) and Saigal (1979).

The second formulation gives rise to a continuous function but in a space of increased dimension n+m. For $y \in \mathbf{R}^m$ define the positive and negative parts of y by $y = y_+ - y_-$, so $y_+ \geq 0$, $y_- \geq 0$, and $y_+^T y_- = 0$. Then the Karush–Kuhn–Tucker conditions are

$$g(x,y) \equiv \begin{pmatrix} \nabla\theta(x) + Ay_+ \\ \\ b - A^T x - y_- \end{pmatrix} = 0. \tag{1.6}$$

This approach has been followed by Fisher and Gould (1974) and Kojima (1974).

2. An Outline of PL Homotopy Algorithms

Suppose we seek a zero of a continuous function $f : \mathbf{R}^k \to \mathbf{R}^k$. Let z_0 be a guess for a zero and let $r : \mathbf{R}^k \to \mathbf{R}^k$ be one-to-one and onto with z_0 its unique zero (e.g. $r(z) = z - z_0$). Define the homotopy $h : \mathbf{R}^k \times [0,1] \to \mathbf{R}^k$ by

$$h(z,t) = tf(z) + (1-t)r(z). \tag{2.1}$$

Early continuation (or embedding) methods solved $h(z,t_i) = 0$ for $0 = t_0 < t_1 < \ldots < t_p = 1$ (see e.g. Ortega and Rheinboldt, 1970); more recent methods allow t to regress. PL homotopy methods make a piecewise-linear approximation ℓ to h using a triangulation T of $\mathbf{R}^k \times [0,1]$. That is, ℓ agrees with h on the vertices of T and is linear on each simplex of T.

Assuming r is linear, $(z_0,0)$ is a zero of ℓ, and the algorithm traces a path of zeroes of ℓ starting from this point. Tracing this path requires, for each simplex through which it passes, one evaluation of h and one linear programming pivot. Of course, the homotopy h (i.e., the functions f and/or r) is only evaluated where necessary to continue the path. Under suitable conditions (see below), the path must lead to a point $(z_1,1)$ and z_1 is then an approximate zero of f — this is one major cycle of the algorithm. If z_1 is insufficiently accurate, a new major cycle is initiated with a triangulation of finer mesh and a new function r_1 (e.g. $r_1(z) = z - z_1$) replacing r.

If a zero of a point-to-set mapping Γ is required, we proceed as above with $f : \mathbf{R}^k \to \mathbf{R}^k$ an arbitrary selection from Γ; i.e., we just require $f(x) \in \Gamma(x)$ for all x — we write $f \in \Gamma$. Of course f is usually discontinuous, but the properties listed below (1.4) ensure that any limit point generated by the algorithm satisfies (1.5).

The suitable conditions guaranteeing convergence stipulate that f and r bear some relationship far from the origin. For instance, if $r(z) = z - z_0$, then it is sufficient if there exists $\varepsilon > 0$, $c \in \mathbf{R}^k$, and a bounded set $C \subset \mathbf{R}^k$ such that, for all z not in C and all z' with $\|z' - z\| \le \varepsilon$, the function f satisfies

$$f(z)^T(z' - c) > 0.$$

This condition holds for $f \in \Gamma$, where Γ is defined in (1.4), if (see Merrill, 1972; Todd, 1976)

$$X_0 \textit{ is bounded, or } \theta \textit{ is convex and (NLP) has a bounded non-empty set of optimal solutions.} \tag{2.2}$$

If f is sufficiently smooth (Lipschitz continuous derivative) and regular (all zeroes have nonzero Jacobians), then the algorithm can be implemented to converge quadratically (Saigal, 1977). Basically this requires that the meshes of the triangulations shrink sufficiently rapidly, and that asymptotically we choose $r_i(z) = D_i(z - \hat{z}_i)$ with $D_i \approx \nabla f(z_i)$ (such approximate derivative matrices are generated automatically), where $\hat{z}_i = z_i$ or $z_i - D_i^{-1} f(z_i)$. For large i, each major cycle traverses exactly n+1 simplices.

The amount of work required for each major cycle depends directly on the number of simplices traversed. This number can increase rather rapidly with the dimension and (for a fixed starting point z_0) with decreases in mesh size. Thus generally a fairly large mesh (of the order of an estimate of $\| z_0 - z^* \|$ where $f(z^*) = 0$) is selected initially, and usually the formulation (1.5) has been preferred to (1.6).

3. Exploiting the Structure of (NLP)

First we discuss the formulation (1.6). Note that $g(x,y) = g_1(x) + g_2(y)$ with g_2 piecewise-linear — indeed the pieces of linearity are the orthants of $\mathbf{R}^m$. It makes sense, therefore, to choose an artificial function r of the same form, i.e. $r(x,y) = r_1(x) + g_2(y)$, where

$$r_1(x) = \begin{pmatrix} D_0(x - x_0) - Aw_0 \\ \\ b_0 - A^T x \end{pmatrix},$$

with D_0 positive definite, $w_0 \geq 0$, $b_0 - A^T x \geq 0$ and $w_0^T(b_0 - A^T x) = 0$. We can then use the homotopy

$$h(x,y,t) = tg_1(x) + (1-t)r_1(x) + g_2(y). \tag{3.1}$$

Since h is already PL in y, it is only necessary to triangulate $\mathbf{R}^n \times [0,1]$ (rather than $\mathbf{R}^{n+m} \times [0,1]$) to make a PL approximation to h. Thus we deal with a sequence of pieces each of which is a product of an (n+1)-simplex in $\mathbf{R}^n \times [0,1]$ and an orthant in $\mathbf{R}^m$.

This idea was first proposed for the nonlinear complementarity problem in which some variables appear linearly by Kojima (1974). Later, the author (Todd, 1980) described the method in the present context, proved global convergence under condition (2.2), and showed how quadratic convergence could be achieved. Asymptotically, given smoothness of θ and the strong second order sufficiency conditions, the orthant in $\mathbf{R}^m$ does not change (the correct active set is identified), and each major cycle requires only n+1 evaluations of $\nabla\theta$.

Recently, Awoniyi (1980) and Awoniyi and Todd (1981) have considered acceleration using the formulation (1.5). Note that there are several applications other than optimization that lead to problem (1.5) with Γ as in (1.4); some examples are given by Awoniyi (1980) and by Awoniyi and Todd (1981). Here application of the "crude" algorithm for several major cycles leads to a guess of the active constraints — suppose there are q. Let $\bar{A}$ and $\bar{b}$ be the corresponding submatrix of A and subvector of b. We then triangulate $\mathbf{R}^n$ in such a way that the (n-q)-dimensional affine set $S = \{x : \bar{A}^T x = \bar{b}\}$ is triangulated with (n-q)-simplices, for example by using the orthogonal transformation in a QR factorization of $\bar{A}$. For points x in $X \equiv X_0 \cap S$, we make a particular choice of f(x) from $\Gamma(x)$; if possible, pick $f(x) \in \{z : \bar{A}^T z = 0\}$. In fact, if the Lagrange multiplier vector $w = -(\bar{A}^T\bar{A})^{-1}\bar{A}^T\nabla\theta(x)$ is nonnegative, the scaled projected gradient $(1 + e^T w)^{-1}(\nabla\theta(x) + \bar{A}w)$ is such a vector (e is the vector of ones). Thus f is a continuous function on a subset of X. One can then perform a major cycle of a PL homotopy algorithm, applied in $\mathbf{R}^n \times [0,1]$, in such a way that the algorithm remains within the (n-q+1)-dimensional subset $X \times [0,1]$ if possible. The result is an algorithm which enjoys global convergence under condition (2.2) and converges quadratically when appropriate smoothness properties hold. Moreover, the algorithm has an automatic active set strategy and uses the active constraints to reduce the dimensionality of the space which is searched.

4. Remarks on Computation

Unfortunately, there has been only limited computational experience with PL homotopy algorithms applied to nonlinear programming problems, particularly with the accelerated versions of Section 3. Merrill (1972) applied his algorithm with the formulation (1.5) and no acceleration to the Colville (1968) nonlinear programming test problems. He successfully solved all but the second of these, and his times ranged from equal to up to twenty times those achieved with GRG for those problems solved by both codes. More computational experience with similar algorithms is reported in Gochet, Loute and Solow (1974), Saigal (1979) and Solow (1980). Fisher and Gould (1974) give some results with formulation (1.6). All the results indicate that fixed-point methods become quite slow as the dimension increases.

Only one numerical example is presented in Kojima (1974), and to my knowledge no other computation has been performed with the homotopy (3.1). Recently, a few experiments have been conducted using the approach of Awoniyi and Todd. While the acceleration clearly improves the algorithm, a moderately large number of iterations still seem necessary. For details, see Awoniyi and Todd (1981). To give some idea of the computational experience, however, we give results for three problems. The so-called Post Office Problem is to minimize $-x_1x_2x_3$ subject to $0 \le x_j \le 42$, $j = 1,2,3$, and $0 \le x_1+2x_2+2x_3 \le 72$. With the standard starting point of $(10,10,10)^T$, the accelerated algorithm required 23 evaluations of $\nabla\theta$ and 18 linear programming pivots; without acceleration, 60 evaluations of $\nabla\theta$ and 91 linear programming pivots were required to get a solution of far lower accuracy. The Shell primal problem in the Colville (1968) study has five variables and fifteen constraints. The accelerated algorithm required 19 evaluations of $\nabla\theta$ and 67 linear programming pivots, while without acceleration the figures would be 47 and 180 respectively. Finally, the Gauthier problem in the Colville study was modified by solving for the last eight variables in terms of the first eight. The resulting inequality constrained problem has eight variables and thirty two constraints. We used the starting point $(10,10,\ldots,10)^T \in \mathbf{R}^8$, which does not correspond exactly with the usual point $(10,10,\ldots,10)^T \in \mathbf{R}^{16}$. The resulting figures were 137 evaluations of $\nabla\theta$ and 968 linear programming pivots for the accelerated algorithm, and 193 and 1197 respectively without acceleration. These numbers indicate that PL homotopy algorithms are not hopelessly slow, especially if any available structure is fully exploited, but that the associated computational burden seems to increase rather rapidly with the dimension. Further research is needed to improve the behaviour of the algorithm in switching active sets. At present, identifying the correct active set may necessitate a rather small mesh, thus slowing the algorithm, particularly in high dimensions; once the correct set is found, convergence appears to be very rapid.

3.4

Computational Testing of a Large Scale Linearly Constrained Nonlinear Optimization Code

David F. Shanno

Department of Management Information Systems,
University of Arizona, Tucson, Arizona 85721, USA.

The purpose of this paper is to discuss problems which have been encountered during computational testing of the COSMOS system for large scale linearly constrained nonlinear optimization problems, and to discuss the ramifications of these problems for future testing of such codes.

COSMOS is a new program developed at the University of Arizona by Roy Marsten and the author. It is a reduced gradient code based upon Marsten's (1978) XMP large scale linear programming code. It uses the basic, nonbasic, superbasic strategy for handling variables introduced by Murtagh and Saunders (1978). Minimization of the nonlinear object function in the reduced set of variables which remain when the basic variables are set to satisfy the constraints is done by either a conjugate gradient or a variable metric algorithm, the choice being made by the user.

Mathematically, the problem can be stated as

$$\text{minimize } f(x), \quad x = (x_1, x_2, \ldots, x_n), \tag{1}$$

$$\text{subject to } Ax = b, \tag{2}$$

and

$$\ell \leq x \leq u, \tag{3}$$

where A is an $m \times n$ matrix, $m < n$.

The reduced gradient method of Murtagh and Saunders partitions the matrix A so that

$$Ax = \left(\overset{m}{B} \;\middle|\; \overset{s}{S} \;\middle|\; \overset{n-m-s}{N} \right) \begin{pmatrix} x_B \\ x_S \\ x_N \end{pmatrix} = b \;, \tag{4}$$

where the basic variables x_B are used to satisfy the constraint set, the superbasic variables x_S are allowed to vary to minimize $f(x)$, and the nonbasic variables x_N are fixed at their bounds. Here B is a square, nonsingular matrix. At each step, the problem then becomes determining a step vector $\Delta x = (\Delta x_B, \Delta x_S, \Delta x_N)$ which reduces the value of $f(x)$ while forcing $x + \Delta x$ to continue to satisfy the constraints (2). As the values x_N are fixed, we get immediately

$$\Delta x_N = 0, \tag{5}$$

and allowing Δx_S to be chosen freely means that Δx_B is determined by

$$\Delta x_B = -B^{-1} S \; \Delta x_s. \tag{6}$$

Considering the unconstrained problem of minimizing f as a function of the current set of superbasic variables x_S, we note that the reduced gradient h of $f(x)$ in the variables x_S is given by

$$h = g_S - S^T \pi, \tag{7}$$

$$\pi = (B^T)^{-1} g_B, \tag{8}$$

where g_B, g_S, and g_N are the components of the gradient vector $\nabla f(x)$ corresponding to x_B, x_S, and x_N, respectively. Theoretically, the algorithm continues until $\|h\| = 0$, in which case $\Delta x_S = 0$

and x_S is a stationary point of f on the manifold determined by the choice of basic, superbasic and nonbasic variables. The Lagrange multipliers λ, defined by

$$\lambda = g_N - N^T \pi , \tag{9}$$

are computed, and, if $\lambda \geq 0$ for each x in x_N at its lower bound and if $\lambda \leq 0$ for each x in x_N at its upper bound, then the point x is a stationary point of f. Otherwise one or more nonbasic variables are allowed to become superbasic, and a further reduction in f is made by solving a new unconstrained problem. The algorithm continues until a stationary point is found.

In Shanno and Marsten (1979) a new conjugate gradient algorithm for driving the reduced gradient h to zero for a fixed set of superbasic variables is proposed. Although we do not go into the mathematical details, which are fully developed by Shanno and Marsten, we note that the innovative feature of the new method is an attempt to preserve information concerning past search directions when either a superbasic or a basic variable reaches a bound. This is accomplished by extending Shanno's (1978) definition of the conjugate gradient direction

$$d_{k+1} = -Q_{k+1} g_{k+1}, \tag{10}$$

where Q_{k+1} is the positive definite matrix defined by

$$Q_{k+1} = I - \frac{p_k y_k^T + y_k p_k^T}{p_k^T y_k} + \left(1 + \frac{y_k^T y_k}{p_k^T y_k}\right) \frac{p_k p_k^T}{p_k^T y_k} , \tag{11}$$

where $p_k = \alpha_k d_k = x_{k+1} - x_k$, and where y_k is the change in the gradient due to the step from x_k to x_{k+1}. The extension preserves directional information by modifying the matrix Q_{k+1} to Q^*_{k+1}, following the variable metric transformations of Murtagh and Saunders, and setting

$$d_{k+1} = -Q^*_{k+1} h_{k+1} . \tag{12}$$

A salient point here is that, although the transformation (12) may well prove effective in practice, it has none of the nice theoretical properties of maintaining conjugacy which variable metric methods possess. Hence, the potential value of using (12) rather than restarting with

$$d_{k+1} = -h_{k+1} \tag{13}$$

whenever a bound is encountered must be verified experimentally.

To this end, the literature was searched for test problems, and disappointingly few of any interest were found. The original tests were the three most interesting suitable test problems of this form from Himmelblau's (1972) book. While indications are that the transformation (12) reduces work by about twenty per cent over nontransformed conjugate gradient algorithms, the test set is too small to be definitive.

Further, the test set has the disadvantage that bounds are encountered quickly, thus eliminating superbasic or basic variables before good search directions have been developed. Typically, if k of the variables that are superbasic initially have to become nonbasic during the minimization, then all these bounds are set in fewer than 2k major iterations of the conjugate gradient algorithm. The unconstrained minimization then continues unhindered.

Therefore the questions arise whether this behaviour is typical of linearly constrained problems, thus mitigating the need to preserve direction information, or whether the test problems are uninteresting. Some computational evidence suggests the latter, and clearly more work is needed to provide large problems with interesting constraints.

It would therefore be highly valuable to adopt a standard format for data input to test problems, and for data that define interesting classes of test problems to be generated and distributed; alternatively there is a need for test problems whose data can be generated easily. We cannot continue to draw conclusions about the efficiency of algorithms devised for very large general problems, by testing with very small problems or very specially structured large problems simply because this is all that is available in the literature.

As another example of the inadequacy of the currently published test problems, we consider a variable metric algorithm that is also implemented in COSMOS. In particular we consider the behaviour of the algorithm when $\|h\| = 0$, but the λ's defined by (9) do not indicate convergence. In this case, new superbasic variables are added to those already in the active set. Following Murtagh and Saunders, COSMOS attempts to update the current approximation B to the Hessian matrix by appropriate differencing in the new variables. When the differencing maintains positive definiteness, all is well.

Sometimes, however, other measures must be taken because negative curvature in the objective function makes the new approximation indefinite. Murtagh and Saunders suggest that, if B is the existing approximation, then a suitable initial approximation $\hat{B}$ to the new Hessian is given by

$$\hat{B} = \begin{pmatrix} B & 0 \\ 0 & 1 \end{pmatrix} . \tag{14}$$

It can be shown, however, that this leads in many cases to such a poor scaling of the problem, that computationally it is better to set $\hat{B} = I$, i.e. a complete resetting of $\hat{B}$ to the identity matrix.

The reasons for this poor scaling can be easily demonstrated. Measures can be adopted that preserve the information so painstakingly gathered in B while preserving scaling. As in the conjugate gradient algorithm, these measures are heuristic, and thus can only be evaluated by numerical testing. Unfortunately, again the available set of test problems has proved inadequate to demonstrate any real differences between heuristically altering $\hat{B}$ or resetting $\hat{B}$ to I, but it is clear that both of these techniques are preferable to (14).

I personally cannot believe that in general it is useful to discard information about the true Hessian, so must submit again that our limited set of test problems is badly at fault.

In conclusion, I feel that, for linearly constrained algorithms, we are little beyond the stage in testing where we were in unconstrained optimization when Rosenbrock's function and Wood's function characterized virtually our entire test set. We must arrive at some vehicle for making a varied and interesting set of problems of varying size available in order to effectively evaluate new and existing algorithms.

Discussion on "Linear Constraints"

3.5 Dembo

The "Newton" feasible search direction, p, for linearly constrained problems is given by the calculation:

$$\underset{p \in \mathbf{R}^n}{\text{Minimize}} \quad \tfrac{1}{2} p^T H p + g^T p \tag{Q1}$$

$$\text{subject to } Ap = 0, \tag{Q2}$$

where H and g are the Hessian and gradient of the objective function and A is the coefficient matrix of the current active constraint set.

If the Hessian is *singular* (which can even occur at a strong local minimum) a question arises as to what is a suitable dual-Newton (range-space) algorithm, since dual methods involve the Hessian inverse and therefore it appears that they are not defined for such cases.

An appropriate dual method can be derived by transforming the above problem as follows. Make the change of variables $\bar{p} = Ep$ where E is some nonsingular matrix such that

$$\bar{H} \equiv E^T H E = \begin{pmatrix} \bar{H}_N & 0 \\ 0 & 0 \end{pmatrix}$$

with $\bar{H}_N$ nonsingular. Note that E may be obtained from some suitable factorization of H. The key point is that $\bar{H}_N$ should be nonsingular and, if possible, easy to invert.

If $\bar{p}$ is partitioned into $(\bar{p}_N, \bar{p}_L)^T$, corresponding to the above partition of $\bar{H}$, then problem (Q) becomes

$$\underset{\bar{p}\in\mathbf{R}^n}{\text{Minimize}} \quad \tfrac{1}{2}\bar{p}_N^T\bar{H}_N\bar{p}_N + \bar{g}_N^T\bar{p}_N + \bar{g}_L^T\bar{p}_L \tag{P1}$$

$$\text{subject to } \bar{A}_N\bar{p}_N + \bar{A}_L\bar{p}_L = 0 , \tag{P2}$$

where $\bar{g} \equiv (\bar{g}_N, \bar{g}_L)^T = E^{-T}g$ and $\bar{A} \equiv [\bar{A}_N \ \bar{A}_L] = AE^{-1}$ and the subscripts N and L denote the "nonlinear" and "linear" components of the transformed search direction vector.

It is often the case that E is some permutation of the identity matrix. This occurs, for example, in problems in which only some of the variables appear nonlinearly.

A *primal-Newton* algorithm (null-space method) would solve problem (P) using a reduced gradient method to get the search direction.

A *dual-Newton* algorithm solves the dual of problem (P) above which may be written as the following *constrained* quadratic programming problem.

$$\text{Minimize} \quad \tfrac{1}{2}\lambda^T(\bar{A}_N\bar{H}_N^{-1}\bar{A}_N^T)\lambda + (\bar{A}_N\bar{H}_N^{-1}\bar{g}_N)^T\lambda \tag{D1}$$

$$\text{subject to } \bar{A}_L^T\lambda = -\bar{g}_L. \tag{D2}$$

The relationship between the primal and dual based methods is as follows. The multipliers associated with the primal constraints (P2) are the dual variables λ. The multipliers associated with the dual constraints (D2) are the "linear" primal variables $\bar{p}_L$. Finally, the "nonlinear" primal variables can be computed from:

$$\bar{p}_N = -\bar{H}_N^{-1}(\bar{A}_N^T\lambda + \bar{g}_N).$$

Note that, if there are no linear variables (i.e. when H is nonsingular), then the dual is unconstrained.

3.6 Gill

When E is a permutation matrix, the scheme you suggest is equivalent to defining an alternative partitioned inverse of the n+t matrix

$$\begin{pmatrix} H & -\hat{A}^T \\ \hat{A} & 0 \end{pmatrix}$$

(see Section 3.3 of Paper 3.1).

It is interesting to note that the matrix Z used in the reduced gradient and orthogonal factorization methods can be interpreted as being part of a transformation E. Suppose that the matrix E is defined as

$$E = \begin{pmatrix} V \\ \hat{A} \end{pmatrix}, \qquad (**)$$

where V is any $(n-t)\times n$ matrix such that E is nonsingular. It can be shown that the first n-t rows of E^{-1} span the null space of the rows of $\hat{A}$ and therefore constitute a suitable definition for the matrix Z. If the rows of V are selected from rows of the $n\times n$ identity matrix, the corresponding matrix Z is equivalent to that used in the "reduced gradient" formulation. If the rows of V are selected from rows of the orthogonal factor of $\hat{A}$, the corresponding matrix Z is orthogonal (see Wolfe, 1963; Gill and Murray, 1974a; Buckley, 1975). If the projected Hessian matrix is nonsingular, the matrix E defined by (**) is such that $\bar{H}_N$ is equal to the projected Hessian Z^THZ. In this case, problem (P) gives the usual primal method with $Z^THZ\bar{p}_N = -Z^T\bar{g}_N$.

On the question whether it is more efficient to use a range-space or a null-space method when G_k is positive definite, range-space methods tend to be advantageous only when the number of constraints in the working set is less than about ten per cent of the number of variables (assuming all the matrices are dense). However, range-space methods become more attractive if G_k is a sparse matrix (see Contribution 3.9).

3.7 Sargent

I wish to make two points.

First, in connection with IQP versus EQP, it seems to me that the answer is fairly clear if the objective function and its gradient are expensive to evaluate. One then wishes to make maximum use of each evaluation, and hence it is desirable to find the IQP optimum at each iteration. This conclusion is reinforced if there are also nonlinear constraints which are expensive to evaluate. On the other hand, if function and gradient evaluations are cheap, it could be advantageous to re-evaluate these before the optimum is reached, and whether an EQP or IQP strategy is more efficient seems a more open question, which is probably problem dependent.

Secondly, I want to mention an extension to the "range-space" methods which answers most of the objections to these methods. Instead of updating the matrices as the constraint set changes, one can solve the system as a linear complementarity problem using Lemke's algorithm (Sargent, 1974). This can be implemented using stable factorizations (Sargent, 1978), does not require the Hessian (or even the projected Hessian) to be positive definite, is not upset by degeneracy, and will find a solution if one exists, or otherwise indicate that there is no solution. Further, it does not require an initial feasible point, which is an advantage in the initial phase for a linearly constrained problem, but is a more important advantage at every step with nonlinear constraints. Nevertheless, it is probably not the ideal solution, since Lemke's algorithm makes no use of the value of the objective function, and it is likely that solving the dual QP by a descent method, as will be described in the research seminar by Goldfarb, can do this and also retains the other advantages of the linear complementarity approach.

3.8 Gill

Firstly, I would like to re-emphasize the point that it is yet to be proven that solving an IQP subproblem as opposed to an EQP subproblem reduces the *overall* number of iterations.

Secondly, it is my understanding that the published version of Lemke's method is not guaranteed to converge to a solution when G is indefinite (see Mylander, 1974, for an example).

3.9 Tanabe

In sparse problems it is often best to solve the complete system of linear equations that occurs in the range-space method without any partitioning.

3.10 Gill

I agree. This avoids the requirement that G_k be nonsingular.

3.11 Goldfarb

I also agree with Tanabe that in the sparse case it may be best to solve the equations

$$\begin{pmatrix} B_k & -\hat{A}_k^T \\ \hat{A}_k & 0 \end{pmatrix} \begin{pmatrix} p_k \\ \lambda_k \end{pmatrix} = \begin{pmatrix} -g_k \\ 0 \end{pmatrix}$$

directly. These equations are reminiscent of the system of

equations

$$\begin{pmatrix} I & -A \\ A^T & 0 \end{pmatrix} \begin{pmatrix} r \\ x \end{pmatrix} = \begin{pmatrix} -b \\ 0 \end{pmatrix},$$

which are solved by Hachtel's method (see Duff and Reid, 1976) for the sparse linear least squares problem: minimize $\|r\|_2$, where $r = Ax-b$. Therefore I believe that the experience gained in recent years in solving large sparse linear least squares problems can be quite relevant to large sparse nonlinear programming problems.

3.12 Griewank

What is the optimal conditioning of the reduced gradient basis that can be achieved by suitable pivoting?

3.13 Gill

The condition number of Z depends upon the column permutation P that gives the partition $AP = (B \quad S)$. If the whole of A is factorized using Gaussian elimination with partial or complete pivoting, the resulting P will provide a Z that is acceptably close to optimal. However, it is seldom practical to take advantage of this. In the sparse case (particularly), it is more convenient to accept the partition from a previous iteration and factorize only the corresponding B. This also allows an existing projected Hessian approximation to be retained.

3.14 Powell

The suggestion that it is redundant to have an algorithm for unconstrained optimization as well as an analogous method for linearly constrained optimization does not apply to conjugate gradients, because many difficulties occur in conjugate gradient methods for linear constraints if one wishes to preserve quadratic termination, due to the need for restarts when there are changes to the active set.

3.15 Gill

I agree. My comment was concerned with software for small dense problems. Much more research needs to be done to find an efficient conjugate gradient method for the bound constraint case.

3.16 Bertsekas

In connection with large scale problems, I find that it is a severe limitation if the technique for handling constraints allows the addition of at most one constraint to the working set per iteration. So, for example, if there are a thousand active constraints at the solution that are not active at the starting point, then the methods of Paper 3.1 will require at least one thousand iterations to identify these constraints alone, and probably quite a few more iterations to obtain an adequate approximation to the solution. Since there is no systematic way to guess which constraints are active at the solution, these methods will typically require an unreasonably high number of iterations. Therefore I think that one will not be able to solve efficiently truly large problems unless one completely abandons the given philosophy of handling constraints.

3.17 Wolfe

Why are several thousand iterations "a lot"? I would think that for a problem with several thousand constraints it would be a reasonable number.

3.18 Bertsekas

I thoroughly disagree with the notion that a thousand iterations is an acceptable number to solve a linearly constrained problem for which Hessian approximations can be meaningfully employed. This is particularly so if the problem happens to be large in which case iterations are more time consuming. I have been working for example on nonlinear multicommodity flow problems arising in data communication networks and involving many thousands and even millions of variables and linear constraints for which an iteration can take of the order of minutes of computer time. In practice one may have to provide reasonable solutions to these problems in about five to ten iterations, and methods for accomplishing this are presently available (Bertsekas, 1979; Bertsekas, Gafni and Vastola, 1979; Bertsekas and Gafni, 1981).

3.19 Gill

We must always assume that the cost of factorizing the optimal active set is not excessive. Even if the starting point is optimal, the first order necessary optimality conditions cannot be verified unless the $t \times n$ active set is factorized. The cost of this factorization is comparable to t iterations in which a constraint is added to the basis one at a time. This defines what I mean by an "iteration".

The only situation in which a large number of constraint

changes may lead to an unacceptable cost is if a function or gradient evaluation must be performed each time the working set is changed. In this case, an IQP approach may be appropriate since the function and gradient are only evaluated during the outer iteration. I think that the work involved in one of the "iterations" of Contribution 3.18 must be comparable to the work required to solve a single IQP.

3.20 Mangasarian

I would like to point out that there exist SOR (successive overrelaxation) methods which can solve huge linearly constrained problems of special types. In particular quadratic (Mangasarian, 1977; Han and Mangasarian, 1981; Mangasarian, 1981b) and linear (Mangasarian, 1980; Mangasarian, 1981a) programming problems can be solved using SOR methods. The numerical experience of Mangasarian (1977) and Kreuser (1981) with these methods has been very encouraging. These SOR methods can take real advantage of sparsity (Mangasarian, 1981b; Kreuser, 1981) and, unlike pivotal methods, sparsity can be completely preserved throughout these algorithms. SOR methods are also natural candidates for the solution of large nonlinearly constrained problems, because they can be used to solve the quadratic programming subproblems of iterative quadratic programming methods. Recently, with a code written by my student David P. Anderson for the VAX 11/780, we solved a sparse linear program with 10,000 nonnegative variables, 5,000 inequality constraints (with 4 nonzero coefficients in each constraint) in less than 2.5 minutes to an accuracy of 10 figures in the objective function and 10^{-7} in primal feasibility.

3.21 Overton

Why would you use SOR to solve these systems rather than the conjugate gradient method, particularly if they have no special properties such as Property A?

3.22 Mangasarian

Both SOR methods (see references in Contribution 3.20) and conjugate gradients (Poljak, 1969; O'Leary, 1980), adapted for problems with nonnegatively constrained variables, can be used for solving linear and quadratic programs. However it seems that conjugate gradient methods are not as effective as SOR methods.

3.23 Lemaréchal

This gives me an opportunity to raise a point which I feel strongly about. There is no clear division between "linear" and "nonlinear". For example, linear constraints define a domain whose boundary is piecewise linear. However, if the number of constraints is enormous, this boundary may seem very smooth. In other words, it may sometimes pay to treat a linear problem as though it were nonlinear.

3.24 Fletcher

I would like to comment on the difficulties caused by linear dependence in active constraint normals and the 'spurious vertex' problem mentioned in Paper 3.2. An advantage of the 'single phase' approach to quadratic programming (see Conn and Sinclair, 1975, for example) using an ℓ_1 penalty term, is that the spurious vertex difficulty is avoided. The search along the almost common line passes through the spurious vertex since there is only an ε-change in the slope, and the search usually continues to a true vertex further along the line. At no stage in the process therefore does near-dependence in the active constraint normals arise.

3.25 Conn

Whereas I agree that penalty functions may increase the opportunity to avoid degenerate vertices (for example one may go through them without regard to feasibility), the really difficult degenerate situations are degenerate optimal points, and in this case the penalty function is no help at all.

3.26 Beale

Degeneracy is a less serious problem when the objective function is nonlinear than in linear programming, because degeneracy means having basic variables at their bounds. In nonlinear programming such variables can usually be exchanged for independent (or superbasic) variables.

3.27 Jittorntrum

I wish to draw attention to the distinction between degeneracy caused by near linear independence and that caused by zero Lagrange multipliers. A simple one-dimensional example where the constraint gradient is linearly independent but has zero Lagrange multiplier is that of minimizing x^2 subject to $x \geq 0$. We should distinguish between these two kinds of situations. There is now more understanding about the nature of degeneracy caused by zero Lagrange multipliers, and it is possible to handle this kind of degeneracy without loss in efficiency.

3.28 Wright

In a feasible point method it may not be necessary to add a new (possibly nearly dependent) constraint to the working set simply because it restricts the step length. One of the active set strategies considered by Lenard (1979) (which she termed the "add when necessary" strategy) is based on restricting the step length if a new constraint is encountered, but adding the constraint to the working set only if the next search direction violates the relevant constraint. With such a strategy, it might be possible under some circumstances to avoid adding nearly dependent constraints to the working set.

3.29 Gill

Methods for linearly constrained optimization that are based upon minimizing the ℓ_1 penalty function, mentioned by Fletcher in Contribution 3.24, are less attractive when $F(x)$ is a general nonlinear function because of the need to evaluate F outside the feasible region.

3.30 Goldfarb

I agree, but there is a place for dual feasible algorithms, especially because of their possibly superior ability in avoiding near linear dependence.

3.31 Gutterman

Maintenance of feasibility is important in practical problems for two reasons. First, the objective function (and, in the case of nonlinear constraints, other constraints) may be undefined when a constraint is violated. Second, if the computer fails or the client runs out of budget before the optimum point is reached, a feasible point can be of considerable practical use.

3.32 Wolfe

Please extend the discussion of refinement methods, given in Paper 3.3, and say how they relate to the rate of convergence.

3.33 Todd

For fast convergence on smooth problems the grid size must be shrunk at a quádratic rate. This is performed using a finite difference approximation to the Jacobian matrix that is a natural byproduct of the methods. The next grid size is then chosen to be of the order of the length of the corresponding discrete Newton step $\| D_i^{-1} f(z_i) \|$. For nonsmooth problems, the grid size is usually shrunk at a linear rate, and at best R-linear convergence is attained.

3.34 Steihaug

I think that the numerical results may not be fair to the piecewise linear homotopy methods, because they can treat discontinuities of first derivatives and all the problems of Paper 3.3 are smooth.

3.35 Todd

For general point-to-set mappings, the piecewise linear paths generated can be extremely chaotic and the algorithms are sometimes very slow. Most mappings that arise in practice exhibit some sort of structure, and this should be exploited to the full. My computational results were mainly to illustrate that such structure could be exploited with considerable savings. I compared them with good special purpose optimization codes to indicate that these methods are not hopelessly slow on simple optimization problems. Note also that the homotopy methods would probably not be significantly slower on similar stationary point problems over a polyhedron, where the function whose stationary point is sought is not necessarily the gradient of a scalar function.

3.36 Boggs

Can one apply adaptive procedures to the simplices?

3.37 Todd

In order to have simple rules for moving between simplices, the triangulations used are almost always nonsingular linear transformations of various standard subdivisions. To obtain consistent approximations to derivative matrices it is advisable to use orthogonal transformations, and these are employed to line up the triangulations with the active set manifold in the algorithm outlined in Section 3 of my paper.

3.38 Wright

What contributions are homotopy methods likely to make to future research on efficient algorithms? Do you consider that they are likely to be effective compared to other methods?

3.39 Todd

In their application to optimization problems, piecewise linear homotopy methods ignore function values and symmetry of Hessians, and for these and other reasons are unlikely ever to provide a practical alternative to more specialized methods. They do seem to have value in other nonlinear problems. For example, for a 13-dimensional economic equilibrium problem my code was faster than a fairly sophisticated successive linear programming approach developed by Manne, Chao and Wilson (1980). Layne T. Watson (1979) has also reported success on some hard engineering problems.

3.40 Dembo

Are the global convergence properties of homotopy methods particularly valuable?

3.41 Todd

I should add a note of caution to the global convergence results. They do require that one finds a suitable function r that is compatible with f far from the origin. For some real problems and test problems constructing such an r can be very hard. Global convergence of the methods is guaranteed only for the class of problems whose behaviour far from the origin is understood, for example functions satisfying certain coercivity conditions.

3.42 Moré

My impression is that homotopy methods are quite appropriate for fixed point problems, but that their application to general smooth nonlinear optimization problems is not justified. Do you agree?

3.43 Todd

I believe that the main value of these methods will be in solving nonsmooth problems. On the other hand, the applications of which I am aware that give rise to such nonsmooth problems generally lead to mappings that share the property of being built up from smooth functions in much the same way as the mapping Γ of equation (1.4) in my paper. Thus it is important to exploit this structure in solving these problems, and I have tried to indicate how this might be done in the simple context of linearly constrained optimization.

3.44 Toint

I would like to support the statement in Paper 3.4 on the need for interesting large dimensional tests. As was said, we not only need such problems, but we also need them in a form that allows easy communication and testing in different settings. This is also true for unconstrained large dimensional problems.

3.45 Schnabel

One rich source of linearly constrained optimization problems is the path constraint problem in software validation. A path through a computer program can be described as a system of linear and nonlinear inequality constraints, and the problem is to determine whether the system of constraints is feasible. Using an exponential penalty function, this problem is easily recast as a sequence of linearly constrained minimization problems (Schnabel, 1981b).

3.46 Davidon

Though we have been discussing the need for good overall test problems for evaluating algorithms, there is also a need for developing special tests to evaluate selectively various parts of algorithms under development. These would give more detailed information about each aspect of an algorithm rather than just an overall evaluation, and since they need not be from typical applications, they could eliminate the need for providing large amounts of data by computing the data instead.

3.47 Shanno

I agree that such test problems would be very useful, and in particular I think there is a need for test problems in linearly constrained optimization that would require much pivoting to achieve numerical stability.

3.48 Rosen

Good test problems for a linearly constrained, nonlinear objective algorithm can easily be constructed from any good LP test problem. One way to do this is to use the objective function

$$f(x) = c^T x + \theta\phi(x) \quad ,$$

where $\theta \geq 0$ is a parameter and where $\phi(x)$ is any suitable nonlinear function. For sufficiently small θ, the behaviour will be similar to the LP obtained when $\theta = 0$, while as θ is increased the nonlinear aspects will dominate.

3.49 Goldfarb

I doubt that simply perturbing linear test problems would give useful results, because it would be very difficult to control how many steps a nonlinear programming algorithm would need to take for a given active set before adding or deleting a constraint.

3.50 Dembo

There appears to be some confusion between the number of pivots and the number of active set changes. In nonlinear optimization the pivots determine the basis, which is used for projection onto the current active set. There may be many active set changes for any given basis and typically there will be.

In response to the suggestion that one may construct NLP problems with many basis changes by simply adding a small nonlinear term to the objective function of an LP, it is most likely that the resulting NLP will go through few basis changes, depending on the algorithm used. This is because the role of the basis in a nonlinear programming code is completely different from its role in an LP code. Basis changes are made in reduced gradient codes only to improve conditioning or when a basic variable hits a bound.

3.51 Wolfe

In Paper 3.4 it is mentioned that people who have *real* large scale problems tend to describe them as "proprietary" and will not release them to the public as test problems. We encountered that difficulty in some degree in the early 1960's when assembling the SHARE set of linear programming problems. I felt that in most cases the "proprietary" complaint was an easy way to avoid the considerable work of (1) producing a portable version of the problem and (2) obtaining management approval of an activity not of obvious benefit to the firm involved.

We overcame the difficulty, and ultimately collected a useful problem set. On point (1), the SHARE Linear Programming Committee adopted a standard problem format (fortunately similar to that of the most used existing software) for all future software. On point (2), it emphasized the importance of such problems in developing techniques that would benefit users. It further took care to lavish praise on any contributor, while not seeking information about the nature of the particular application.

Perhaps the fact that the request for problems came from a (prestigious?) group of researchers, in which the firms were already represented, was helpful. In your case, perhaps the Committee on Algorithms of the Mathematical Programming Society could be of help.

3.52 Shanno

My experience is limited, but both my collaborators on this work and I have found "proprietary rights" to be a real difficulty.

3.53 Gutterman

I have found no suitable test problems in our company for this type of algorithm. When our analysts go through the effort of linearizing constraints, they have also been linearizing the objective function. If software for nonlinear objective functions were more readily available and more robust and reliable, this might not be so. We are in a circular situation — the formulators will not do anything until the software is better and the software cannot become better until the formulators provide test problems.

3.54 Jackson

First, let me say that I endorse the call for more and better test problems and test problem collections. The Committee on Algorithms is very concerned about it and is working to try to improve the situation. I will say more about this in my Paper 7.2.

Secondly, for almost 10 years, we too have been trying to obtain proprietary problems from the big industrial firms, but have been repeatedly turned down. The situation may change in the future, but I doubt it. In the meantime, let me encourage all participants here to make test problems available. This is very important.

Lastly, I would like to ask Shanno why he feels that varying starting points in his code testing would be an unrealistic approach. It seems to me that this would satisfy better the tenets of designing statistical experiments well, which is a key point of the discussion.

3.55 Shanno

Standard starting points can be very useful in testing whether a change to a component of an algorithm gives an improvement over previous work. Eventually, however, we wish to compare the efficiencies of entire codes, which includes the relative ways of finding initial feasible points when one is not provided by the user. Hence standard points should be provided for certain aspects of testing, but should not always be used.

3.56 Gay

In Shanno's comparisons of conjugate gradient algorithms in the context of linearly constrained optimization, was the same basic computer code used, the changes being only to conjugate gradient details? The concern for standardizing starting points (and bases) prompts me to ask this.

3.57 Shanno

In all my published results, the experiments were carefully controlled to ensure that all features stayed the same except the specific algorithmic changes being tested. However, the high cost of running large problems is eventually going to force us to compare results of new algorithms to already published results, which makes it imperative to be able to duplicate many elements of the published tests, such as starting points.

3.58 Steihaug

Several speakers have pointed out an instability of the nonlinear conjugate gradient method. I would like to make the remark that Hestenes and Stiefel (1952) discuss the propagation of roundoff errors in the (linear) conjugate gradient method, and we apparently have no complete error analysis for the *standard* conjugate gradient method.

3.59 Shanno

There is a Ph.D. thesis (Bollen, 1980) that analyses the stability of the conjugate gradient method.

3.60 Lootsma

It has been said many times that we are not sure which is the best variable metric method and which is the best conjugate gradient method, although we have excellent experience with the BFGS updating formula, for instance. Perhaps the reason for the uncertainty is that quadratic models do not tell us very much (sometimes these models only distinguish methods with the property of quadratic termination from those that do not terminate). Would theoretical research on the performance of these methods on nonquadratic models, such as conic functions or homogeneous functions (when the nonquadratic behaviour is described by a few parameters), lead to a better understanding?

3.61 Shanno

I do not know the answer to this, but Nocedel's (1980) computational results on conjugate gradient methods based on conic models appears to be encouraging. Consequently, analysis of these models may provide additional insight.

PART 4

Nonlinear Constraints

4.1

Methods for Nonlinear Constraints

R. Fletcher

Department of Mathematics, The University,
Dundee DD1 4HN, Scotland.

1. Introduction

The general nonlinear programming problem can conveniently be expressed as finding a local solution x^* of

$$\left.\begin{array}{lll} \text{minimize} & f(x), & x \in \mathbf{R}^n \\ \text{subject to} & c_i(x) = 0, & i \in \mathbf{E} \\ & c_i(x) \geq 0, & i \in \mathbf{I} \end{array}\right\} , \qquad (1.1)$$

where **E** and **I** are finite index sets of equality and inequality constraints respectively, and where the functions $f(x)$ and $c_i(x)$ are smooth but in general nonlinear. This subject is of great practical importance and so has attracted considerable research interest. In view of this it is difficult to be exhaustive in surveying the area; however this paper attempts to provide a brief introduction to the subject and to set out some of the main criteria which must be kept in mind when developing or assessing methods. Then the existing methods are divided into six broad categories, the merits and demerits of these general approaches are discussed, and current areas of research are identified; thus an assessment is given of the "state of the art" in the subject. Finally there is a brief description and summary of my recent work into ℓ_1 exact penalty functions, which I regard as an attractive approach to the solution of (1.1) in many circumstances. In fact this is a personal assessment of the whole subject and is to some extent biased towards those areas in which I have worked. Fortunately the other papers and the discussion on "Nonlinear Constraints" present independent points of view, so I hope that a balanced picture emerges overall. I have tried to cross-reference this material with that in this paper where possible. More detailed information can be obtained from source papers or books. There are relatively few of the latter, but two which adopt a similar presentation to that given here are Gill and Murray (1974b) and Fletecher (1981b). I shall make frequent references to the latter to clarify and extend some of the points raised here.

An important preliminary observation is that, for any given problem, there may be various ways of presenting it in a form corresponding to (1.1). For example there exist possible pre-transformations of the problem which might be considered in order to simplify or to improve the conditioning of the problem. These ideas include elimination of variables, adding extra variables (e.g. slack variables), and making transformations of variables or constraints, especially diagonal scaling. More substantial transformations can involve the use of duality theory or geometric programming for example. A review of many of the possibilities is given by Fletcher (1981b). Notice that some simplifying transformations may have undesirable side effects, for example the use of quadratic slacks for replacing inequality constraints by equations.

In fact (1.1) is often an oversimplified representation of a nonlinear programming problem. For example there often exist some linear constraints in the problem, and for maximum efficiency it can be desirable to give these constraints special treatment (see discussion contributions 4.21 and 4.22). In particular simple bounds $\ell_i \le x_i \le u_i$ frequently arise, and it is usually important to treat these directly to eliminate a variable if the bound is active, rather than indirectly in a penalty term say. Another possibility is that f(x) (and also some of the functions $c_i(x)$) may not be defined for all x but only for points which are feasible (with respect to the remaining constraints). This places some restrictions on how a method can operate (see discussion contributions 4.33—4.39). It is my impression, however, that this problem can largely be dealt with by adding extra variables and equations if necessary. For example, if $f(x) \equiv (x_1^3 + x_2^3)^{\frac{1}{2}}$, then by introducing a new variable x_3 and a new equation $x_3 = x_1^3 + x_2^3$, f(x) can be replaced by $f(x) \equiv x_3^{\frac{1}{2}}$. Thus it is only necessary in the method for f to be defined on $x_3 \geq 0$, and it is usually straightforward to modify a method so that all iterates are feasible with respect to the simple bounds in the problem. All this discussion illustrates that the provision of efficient and useful packaged computer software is a difficult and demanding task, and in fact much yet remains to be done in this area. Nonetheless the underlying methods can be relatively straightforward and are better understood if a fairly simple representation like (1.1) is considered. In fact it is often convenient for purposes of exposition to consider even more simple cases than (1.1), such as the "equality problem" in which the set **I** is empty, or the "inequality problem" in which the set **E** is empty.

It is assumed that the problem functions are smooth. The notation $g = \nabla f$ and $G = \nabla^2 f$ denotes the gradient vector and Hessian matrix of f, likewise $a_i = \nabla c_i$ and $G_i = \nabla^2 c_i$. It is convenient to write $g(x^*)$ as g^* and so on. Active constraints at a fixed point x' are defined by the set

$$\mathbf{A}' = \mathbf{A}(x') = \{i: c_i(x') = 0\}. \qquad (1.2)$$

First order necessary conditions for a feasible point x^* to solve an equality problem are that g^* is a linear combination of the vectors $\{a_i^*;\ i \in \mathbf{E}\}$, that is there exist Lagrange multipliers λ_i^* (collectively λ^*) such that

$$g^* = \sum_{i \in \mathbf{E}} a_i^* \lambda_i = A^* \lambda^*, \tag{1.3}$$

where A^* is the Jacobian matrix of $c(x)$ at x^* with columns $\{a_i^*; i \in \mathbf{E}\}$. For an inequality problem the corresponding conditions are that (1.3) must hold with respect to the active constraints at x^*

$$g^* = \sum_{i \in \mathbf{A}^*} a_i^* \lambda_i^*, \tag{1.4}$$

and in addition the corresponding multipliers are nonnegative

$$\lambda_i^* \geq 0, \quad i \in \mathbf{I} \cap \mathbf{A}^*. \tag{1.5}$$

These conditions express the fact that at a local solution of (1.1) there are no feasible descent directions (that is no feasible arcs with negative slope). Strictly speaking it is required that a regularity assumption holds in order to eliminate certain unlikely situations, but these situations are fairly pathological and can be ignored for all practical purposes. Each constraint has an associated Lagrange multiplier (at the solution) which is conventionally zero if the constraint is inactive. The Lagrange multiplier λ_i^* can be interpreted (to first order) as the rate of change in $f(x^*)$ that would result from a perturbation in the constraint function c_i. It is often convenient to restate these results by introducing a "Lagrangian function"

$$L(x, \lambda) = f(x) - \lambda^T c(x). \tag{1.6}$$

For an equality problem, condition (1.3) and the feasibility condition $c(x^*) = 0$ can be expressed as

$$\nabla_x L(x^*, \lambda^*) = 0, \quad \nabla_\lambda L(x^*, \lambda^*) = 0, \tag{1.7}$$

or equivalently that (x^*, λ^*) is a stationary point of the Lagrangian function. For an inequality problem the first order conditions can be restated as

$$\left.\begin{aligned} \nabla_x L(x^*,\lambda^*) &= 0 \\ c(x^*) &\geq 0 \\ \lambda^* &\geq 0 \\ \lambda^{*T} c^* &= 0 \end{aligned}\right\} , \tag{1.8}$$

and are often referred to as Kuhn–Tucker (KT) conditions. The final condition is often referred to as the "complementarity condition", and it expresses the fact that λ_i^* and c_i^* cannot both be nonzero, or equivalently that inactive constraints have zero multipliers. If there exists no $i \in \mathbf{I}$ such that $\lambda_i^* = c_i^* = 0$ then "strict complementarity" is said to hold.

It is also possible to look at second order effects caused by feasible changes in the solution. Second order necessary conditions for the solution of an equality problem are that the Lagrangian function has nonnegative curvature for all feasible directions, that is

$$s^T W^* s \geq 0 \quad \text{for all } s: A^{*T} s = 0, \tag{1.9}$$

where

$$W^* = G^* - \sum_i \lambda_i^* G_i^* \tag{1.10}$$

is the Hessian $\nabla_{xx}^2 L$ of the Lagrangian function at (x^*,λ^*). Second order sufficient conditions for an isolated local solution of an equality problem are that x^* is feasible, (1.3) holds, and

$$s^T W^* s > 0 \quad \text{for all } s: \|s\| = 1,\ A^{*T} s = 0. \tag{1.11}$$

Similar conditions hold for an inequality problem in which (1.9) has to hold for all feasible directions of zero slope: usually this is the set $\{s: a_i^{*T} s = 0,\ i \in \mathbf{A}^*\}$. More details of the theory of nonlinear programming can be found in Fiacco and McCormick (1968) and in Fletcher (1981b).

2. General Considerations

Methods for the solution of (1.1) are usually iterative so that a sequence $x^{(1)}, x^{(2)}, x^{(3)}, \ldots$ is generated from a given point $x^{(1)}$, hopefully converging to x^*. Often estimates $\{\lambda^{(1)}, \lambda^{(2)}, \lambda^{(3)}, \ldots\}$ of λ^* are also required in the method. In devising a method to solve nonlinear programming problems, three main areas must be kept in mind, namely how to deal with the equations, the inequalities, and the minimization aspects of the problem. Later in this section other general considerations are considered relating to the comparative assessment of methods.

There is no well-accepted choice of a best general purpose method for solving systems of nonlinear equations, and the same is true for handling any nonlinear equations which arise in (1.1). One direct approach is to use elimination, either to eliminate specified variables, or in a more general way. Another is the *method of Lagrange multipliers* in which a solution (x^*, λ^*) to the first order conditions $g = A$ and $c = 0$ is sought. Both these approaches require the solution of a system of non-linear equations. Successive linearization (i.e. some form of the Newton-Raphson method) is usually the most efficient way of dealing with this, although there are difficulties in guaranteeing global convergence. The alternative approach is to use some form of *penalty function* (see Section 3). Here the aim is to reduce the problem to one of unconstrained optimization by adding to the objective function a term which weights constraint violations. There are two main types of penalty function; *sequential*, in which the solution x^* is located as the limit point of a sequence of unconstrained penalty function minimizations, and *exact* in which x^* is (usually) the minimizer of a single penalty function. Efficiency depends on the details of implementation and on how the penalty function is used, although an exact penalty function is generally more efficient. It seems inevitable, however, that some sort of penalty function must be present to get good global convergence properties. An overview of the state of the art for penalty functions is given in Conn's paper (contribution 4.4). The use of both linearization and penalty functions is likely to occur in a successful practical method.

Early methods for handling inequality constraints involve the inclusion of a barrier term which restricts the domain of the problem to the interior of the feasible region. These methods are now largely out of favour, so most modern methods try to identify the active constraints at the solution and treat them as equations. Thus the problem is seen as a combinatorial one of correctly selecting a discrete index set. In some methods the selection of active constraints is obtained by taking the active constraints of an inequality subproblem which approximates the main problem (the IQP approach). In other methods a working set of constraint indices is kept and corresponding equality subproblems are solved on each iteration (the EQP or "active set" approach). The working set is changed whenever it becomes clear that certain constraints should be added or deleted; usually at most one addition or deletion takes place on each iteration. There was considerable discussion at the conference on the merits of these two alternatives, see for example contributions 4.2, 4.16–4.20, 4.31 and 4.32. My opinion is that IQP methods are more effective in locating the active set quickly but require somewhat more computation per step. In the usual case, where function evaluation costs are dominant and it is not easy to guess the active constraints at the solution, then I would expect an IQP method to perform better.

Many good methods exist for solving unconstrained minimization problems (see the papers and discussion of Part 1 of these proceedings) and most of their techniques carry over to nonlinear programming, with Newton or quasi-Newton methods foremost. However, because of the interaction of constraint curvature effects, it is not yet clear how best to update curvature information, although some good possibilities exist. Conjugate gradient methods become important in large sparse problems in conjunction with nonlinear elimination and reduced gradients (see Section 3.3). Levenberg–Marquardt or trust region strategies can also be useful (see Section 4), but one should be sure that they have the desired effect of forcing global convergence.

In choosing a good general purpose method for nonlinear programming it is necessary to assess how well the method will solve typical problems of a general nature. One aspect of this is examining the convergence properties of the method. Concerning the ultimate rate of convergence, methods exist with second order convergence (Sections 3.3 and 4), assuming that second derivative information is available. For first derivative methods superlinear convergence can be obtained, and there is no good reason to settle for less. However, analyses of this sort present a considerable challenge and there are a number of cases where the theory is incomplete. Turning to the global convergence properties (i.e. convergence to a local solution from any $x^{(1)}$), it would be nice to guarantee to solve all problems every time. So-called global convergence proofs exist for a number of algorithms, and, although these proofs are often of intrinsic interest, they unfortunately do not always give a guarantee of acceptable behaviour. The catch is that the proofs are consequent upon certain assumptions about the problem or the iterative sequence. If it is not possible to verify (or accept as reasonable) that these assumptions hold, then the value of a proof is doubtful. I prefer to see a more positive attitude to global convergence (and rate of convergence) proofs, in which the algorithm designer seeks out those situations that may cause difficulties, and then produces an algorithm and an associated convergence proof that avoids the difficulties. Of course a distinction has to be made between difficulties that are infrequent but possible, and those which are entirely pathological and therefore can be ignored. This is a matter for individual judgement, and the ultimate long term test is the success or otherwise of the resulting software package. Apropos of global convergence, typical situations which cause difficulty include:

a) the least singular value of the Jacobian matrix of the active constraints tends to zero,
b) Lagrange multiplier estimates tend to infinity,
c) the problem has no feasible points, and
d) first order or second order sufficient conditions may not hold at a solution.

Situations (a) and (b) are related and are often "assumed away" in convergence proofs, whereas they do occasionally cause trouble in practice. Thus it is preferable to seek methods in which (a) and (b) can be proved not to cause trouble — see Section 4 for example. Situation (c) can easily occur if the user has set up his problem wrongly; the algorithm should identify this situation, rather than just failing to converge. However, it is not possible to make a 100% certain identification without resorting to expensive global optimization techniques (see discussion contribution 4.13), and this illustrates that some situations exist which are unlikely to be resolved by any practical algorithm. Situation (d) is a difficulty which will trouble many (but not all) methods, but I do not know of any *practical* example in which it has caused difficulty. Therefore I think this is a case where a prior assumption to exclude the situation is reasonable. Returning to rate of convergence proofs, another example of a difficult situation is

e) the Maratos effect: $x^{(k)}$ can be arbitrarily close to x^* but a unit step of the method may not reduce the exact penalty function.

My experience is that inefficiencies from this effect are rare, especially if parameters are carefully chosen. Clearly, however, it is preferable to avoid relying on the choice of parameters, and recent research has now produced a number of suggestions which avoid the effect.

In my opinion the paramount way of assessing methods should be by observing their performance in solving practical problems. Clearly *efficiency* can only be assessed in this way, since this is a rather vague concept related to consistently good performance over a wide range of problems. Likewise for *reliability*, whilst realistic global and rate of convergence proofs are helpful, only extensive numerical experience can indicate that steady progress is made remote from the solution and that no unforeseen pitfalls are likely to arise. Thus, when developing a method, it is important to assess the results of numerical experiments. This gives rise to many difficult decisions, for instance what test problems to use and how to compare different methods (see Parts 6 and 7 of these proceedings). Ease of programming must be considered, and a method which lends itself to modular construction, e.g. sequential QP, gives more confidence than a long and complex routine in which many *ad hoc* features interact. Unfortunately codes of the latter type are not infrequent in nonlinear programming, and developments which enable the structure of these codes to be simplified are welcome.

The importance of special treatment for linear constraints and bounds, and of pre-scaling constraints and/or variables (see discussion contributions 6.43–6.50 on whether this should be done by the user or automatically) has already been mentioned. Another similar question is whether the penalty parameter should be fixed by the user or automatically adjusted by the software. Some discussion of this question appears in contributions 4.14, 4.15, 4.53 and 4.54. Another important consideration is that most methods currently assume that at least the first derivatives of f and c_i can be calculated: the no-derivative situation would probably be handled by finite difference approximations which might well be less reliable, except on long word-length computers. I suspect that suitable software is non-existent, so there is plenty of scope for developments here. Large scale problems in which sparsity is taken into account also require special treatment (see Part 5). Special methods are also merited in particular cases of (1.1) when the objective function is identically zero, firstly when **I** is empty (solution of systems of nonlinear equations) and secondly when **I** is non-empty (feasible points of equations and inequalities). Developments in solving these problems have implications for the solution of (1.1) and vice versa.

3. Methods for Nonlinear Programming

It is impossible to do justice to this subject in a short paper, so I intend to classify the methods into a few main types and give a prototype method. Various ways in which each prototype method can be developed are then briefly described, with some indication of the relative merits and demerits, guided by the general discussion of the previous section. More details are given in Fletcher (1981b).

3.1. Sequential multiplier-free penalty functions

For the equality problem a simple penalty function (Courant, 1943) is

$$\phi(x,\sigma) = f(x) + \tfrac{1}{2}\sigma \sum_{i\in \mathbf{E}} [c_i(x)]^2 . \tag{3.1}$$

The penalty is formed from a sum of squares of constraint violations, and the parameter σ determines the amount of the penalty. If $x(\sigma)$ denotes the minimizer of (3.1) for fixed σ, then it can be proved under mild conditions that the solution x^* of (1.1) is the limiting value of $x(\sigma)$ as $\sigma\to\infty$. Thus the technique of solving a sequence of unconstrained problems numerically is suggested. A prototype implementation for this is as follows:

i) Choose a fixed sequence $\{\sigma^{(k)}\}\to\infty$, typically $\{1,10,100,\ldots\}$.
ii) For each $\sigma^{(k)}$ find a local minimizer $x(\sigma^{(k)})$ of $\phi(x,\sigma^{(k)})$.
iii) Terminate when $c(x(\sigma^{(k)}))$ is sufficiently small.

In early work $x(\sigma^{(k)})$ is used as the initial approximation when minimizing $\phi(x,\sigma^{(k+1)})$, but Fiacco and McCormick (1968) (in a slightly different context) obtain significant improvements using an extrapolation scheme. The method is easy to program and theoretically robust (only difficulty (c) in Section 2 can cause problems). In practice, however, severe numerical difficulties arise because the solution of step (ii) becomes increasingly badly determined as $\sigma\to\infty$. This is expressed mathematically by the condition number of the Hessian matrix $\nabla^2\phi$ approaching infinity. Although the choice of the sequence in step (i) represents an attempt to ameliorate these effects, in practice it is only possible to obtain quite low accuracy with these algorithms. It can be seen that a good estimate of x^* is not helpful, since $x(\sigma^{(1)})$ is usually remote from x^*, and this also is disadvantageous. Because the rate of convergence is only linear (about one decimal place per minimization with the above sequence), the method is quite slow. The method can be modified for inequalities in an obvious way, but further problems then arise. An alternative approach is to use barrier functions to handle inequalities (e.g. Fletcher, 1981b), but the numerical difficulties are no less severe and there are two additional problems: the barrier function is not defined outside the feasible region, and an initial strictly feasible point is required. In view of all these disadvantages, methods in category 3.1 currently attract little interest (but see discussion contributions 4.35 and 4.38).

<u>3.2. Sequential multiplier (augmented Lagrangian) penalty functions</u> The way in which (3.1) is used to solve (1.1) can be envisaged as an attempt to create a local minimizer at x^* in the

limit $\sigma^{(k)} \to \infty$. However, x^* can be made to minimize ϕ for finite σ by including in (3.1) a term which is linear in $c(x)$, giving

$$\phi(x,\lambda,\sigma) = f(x) - \lambda^T c(x) + \tfrac{1}{2}\sigma\, c(x)^T c(x) \tag{3.2}$$

(Hestenes, 1969; Powell, 1969). This is the augmented Lagrangian penalty function, and it can be used sequentially to give a way of solving (1.1) which is as easy as that in Section 3.1. There exists a value of λ for which x^* minimizes $\phi(x,\lambda,\sigma)$, and this is in fact λ^*, the Lagrange multiplier vector associated with the solution x^*. This result is usually true independent of σ (provided σ is sufficiently large), so it is convenient to denote the minimizer of (3.2) by $x(\lambda)$ and to regard σ as being fixed during the algorithm. The basic approach is therefore:

i) Determine a sequence $\{\lambda^{(k)}\} \to \lambda^*$.
ii) For each $\lambda^{(k)}$ find a local minimizer $x(\lambda^{(k)})$ to minimize $\phi(x,\lambda^{(k)},\sigma)$.
iii) Terminate when $c(x(\lambda^{(k)}))$ is sufficiently small.

The main difference between this algorithm and that in Section 3.1 is that, because λ^* is not known in advance, the sequence in step (i) cannot be predetermined. However, by examining the function $\psi(\lambda) \equiv \phi(x(\lambda),\lambda,\sigma)$, it follows by the optimality of $x(\lambda)$ and by $c^* = 0$ that

$$\psi(\lambda) \le \phi(x^*,\lambda,\sigma) = \psi(\lambda^*). \tag{3.3}$$

Thus λ^* is a maximizer of $\psi(\lambda)$, and a method for generating a sequence $\lambda^{(k)} \to \lambda^*$ can be determined by applying unconstrained minimization methods to $-\psi(\lambda)$. After evaluating derivatives it follows that Newton's method for this problem can be written

$$\lambda^{(k+1)} = \lambda^{(k)} - (A^T W_\sigma^{-1} A)^{-1} c \Big|_{x(\lambda^{(k)})}, \tag{3.4}$$

where W_σ is the Hessian of (3.2). For first derivative applications, W_σ can be replaced by the approximate Hessian calculated by a quasi-Newton method applied in step (ii) of the basic method. For no-derivative applications the approximate form

$$\lambda^{(k+1)} = \lambda^{(k)} - \sigma c(x(\lambda^{(k)})) \tag{3.5}$$

of (3.4) is often suitable (Powell 1969; Hestenes 1969). Again more details of this development are given in Fletcher (1981b).

Equation (3.4) induces a second order rate of convergence in the basic method (albeit in terms of the number of unconstrained minimizations required) and (3.5) can be used to obtain an arbitrarily fast rate of linear convergence. However, to ensure global convergence the basic algorithm must be modified, and Powell (1969) suggests a technique based on increasing σ if necessary. This has proved useful and theoretically avoids difficulties (a) and (b) in Section 2. In practice, however, increasing σ can cause ill-conditioning as in Section 3.1, and difficulty (c) is also troublesome. There is therefore scope for research into other methods of inducing global convergence.

The basic method is readily adapted to handle inequality constraints. Another good feature is that a good estimate of λ^* is valuable for starting off the method. In practice local convergence is often rapid, and high accuracy can be achieved with (3.4) in about four to six minimizations. By carrying forward the Hessian matrix of (3.2) from one step to the next, the computational effort required in each minimization goes down rapidly. Nonetheless the total number of function and derivative calls is often much higher than with methods of the type described in Section 3.3. Low accuracy minimization is possible, and can be used to include some features from these methods in order to improve efficiency, but currently these research possibilities are not attracting much interest. More discussion on augmented Lagrangian functions is given in Bertsekas's paper in this session and in discussion contributions 4.43 and 4.44.

<u>3.3. Lagrange-Newton (Sequential QP) methods</u> A direct and effi-

cient approach to nonlinear programming is to iterate on the basis of certain approximations to the functions $f(x)$ and $c(x)$. For an equality problem the method can be explained as Newton's method applied to solve the nonlinear equations which arise in the method of Lagrange multipliers. Thus estimates $x^{(k)}$ and $\lambda^{(k)}$ of both x^* and λ^* are iterated, and a linearization of the equations $g = A\lambda$ and $c = 0$ about $(x^{(k)}, \lambda^{(k)})$ gives

$$\begin{pmatrix} W^{(k)} & -A^{(k)} \\ -A^{(k)T} & 0 \end{pmatrix} \begin{pmatrix} \delta^{(k)} \\ \lambda^{(k+1)} \end{pmatrix} = \begin{pmatrix} -g^{(k)} \\ c^{(k)} \end{pmatrix} \tag{3.6}$$

after some rearrangement, where $W^{(k)} = \nabla^2 f^{(k)} - \Sigma_i \lambda_i^{(k)} \nabla^2 c_i^{(k)}$. This system of linear equations is solved for $\delta^{(k)}$ and $\lambda^{(k+1)}$, and then $x^{(k+1)} = x^{(k)} + \delta^{(k)}$ is set. An equivalent derivation is to say that, in (1.1), $c(x)$ is approximated by the linear function $\ell^{(k)}(\delta)$ defined by

$$c(x^{(k)} + \delta) \approx \ell^{(k)}(\delta) \equiv c(x^{(k)}) + A^{(k)T}\delta, \tag{3.7}$$

and $f(x)$ is approximated by the quadratic function $q^{(k)}(\delta)$ defined by

$$f(x^{(k)} + \delta) \approx q^{(k)}(\delta) \equiv f(x^{(k)}) + g^{(k)T}\delta + \tfrac{1}{2}\delta^T W^{(k)}\delta. \tag{3.8}$$

Expression (3.7) is a first order (linear) Taylor series approximation for $c(x)$, and (3.8) is a quadratic Taylor series approximation for $f(x)$ together with terms in $W^{(k)}$ which account for constraint curvature. By analogy, (1.1) is replaced by a sequence of quadratic programming (QP) subproblems of the type

$$\left.\begin{array}{lll} \text{minimize} & q^{(k)}(\delta), & \delta \in \mathbf{R}^n \\ \text{subject to} & \ell_i^{(k)}(\delta) = 0, & i \in \mathbf{E} \\ & \ell_i^{(k)}(\delta) \geq 0, & i \in \mathbf{I} \end{array}\right\} . \tag{3.9}$$

Given initial approximations the basic method is:

For $k = 1, 2, 3, \ldots$

i) Solve (3.9) to get $\delta^{(k)}$ and let $\lambda^{(k+1)}$ denote the corresponding vector of Lagrange multipliers of the functions $\{\ell_i^{(k)}(\delta);\ i \in \mathbf{E} \cup \mathbf{I}\}$.

ii) Set $x^{(k+1)} = x^{(k)} + \delta^{(k)}$.

This is known as the sequential QP method, and was first suggested by Wilson (1963) in his SOLVER algorithm.

The resulting method is easy to program if good QP software is available, it is efficient, and it can advantageously use a good estimate of x^* (and λ^* if available). The rate of convergence of the method is second order (Fletcher, 1981b), and this is in terms of the number of function calls required (in contrast to Section 3.2). However, in some applications the expense of solving a QP problem at each iteration might be significant. As described above the method requires both first and second derivatives of f and c to be available, but an important aspect of research into this method is how best to update the matrix $W^{(k)}$ by using only first derivative information. Powell (1978a) updates the full matrix by a BFGS-like update, with a modification to ensure positive definiteness. However, this is not suitable for large problems in which curvature information over a relatively small subspace is required. Methods which update a reduced Hessian matrix are currently being researched, for example Coleman and Conn (1980a) and Womersley (1981), but keeping positive definiteness is not as easy as in unconstrained optimization.

The main difficulty of sequential QP is in how to modify the method to obtain good global properties. The basic method may not converge, and indeed becomes undefined if the QP (3.9) is either infeasible or unbounded. Levenberg–Marquardt or trust region modifications circumvent the latter difficulty, but they do not have the same effect in forcing convergence as in unconstrained optimization. Indeed a trust region modification makes the infeasibility problem worse. One idea (Han, 1977a) is to attempt to induce global convergence by using a line search. The solution to (3.9) is used to define a search direction along which an exact ℓ_1 penalty function (see Section 4) is minimized. This method often works well in practice and gives significantly better performance as against methods in Sections 3.1 and 3.2. Han gives a global convergence proof, but the assumptions exclude the difficult cases given in Section 2, and in practice difficulties (a), (b), (c) and (d) can all cause trouble (Fletcher, 1981a). Exactly how serious this is is a matter of some contention, but I believe that the difficulties can be avoided at no significant cost by preferring the method described in Section 4. This retains all the advantageous features of the sequential QP approach, but avoids the difficulties, and is globally convergent under very mild assumptions. Another difficulty of sequential QP when used in conjunction with an exact ℓ_1 penalty function is that the Maratos effect ((e) in Section 2) can occur. There is a similar difficulty in following steep sided curved grooves caused by the nondifferentiability of the penalty function, especially if the problem is badly scaled, or the penalty parameter has been chosen badly. This can be avoided by using an additional projection step (Coleman and Conn, 1980a) analogous to that in Section 3.4, or by relaxing the requirement that every iteration of the sequential QP method must reduce the penalty function (Chamberlain, Lemaréchal, Pedersen and Powell, 1980).

<u>3.4. Feasible direction methods</u> An apparently attractive approach to nonlinear programming is to try to produce direct methods which generalize the ideas for linear constraints considered in Part 3. These methods attempt to maintain feasibility by moving from one feasible point to another along feasible arcs. They can also be regarded as elimination methods in which the nonlinear constraints are used to eliminate variables (Fletcher, 1981b). The prototype algorithm (Rosen, 1961) has the form:

For $k = 1,2,3,\ldots$

i) Given a feasible point $x^{(k)}$ generate a feasible descent direction $s^{(k)}$.

ii) Define a feasible arc $x(\alpha)$ by projecting the point $x^{(k)} + \alpha s^{(k)}$ into the feasible region of the active constraints at $x^{(k)}$.

iii) Carry out a line search along $x(\alpha)$ to determine the best feasible point $x^{(k+1)}$.

Note that, for every value of α that is considered in the line search, it is necessary to solve a nonlinear system of equations to determine the projection into the feasible region. The actual details of how $s^{(k)}$ is calculated or what sort of projection is used differ from one algorithm to another; more details are given by Fletcher (1981b), for example, and more discussion and references to many recent methods are given by Lasdon (contribution 4.5).

Advantages of the method are that no penalty function is required and hence there are no penalty parameters to choose. Also the objective function need only be defined on the feasible region. It is possible to define $s^{(k)}$ by a Newton method giving a second order rate of convergence, or to use quasi-Newton or conjugate gradient schemes if only first derivatives are available (see also discussion contribution 4.64). Unfortunately there are also a number of difficulties, which become apparent when a completely reliable scheme is sought. The problem of finding an initial feasible point is that of solving a system of nonlinear equations (and inequalities). The same is true for finding the projected point described above in step (ii). In both cases the basic Newton—Raphson method is not guaranteed to converge, and more elaborate alternatives would unduly complicate the basic method. Convergence can be proved if difficulty (a) in Section 2 is excluded, but it is not attractive to make this assumption. A similar difficulty is that the feasible arc $x(\alpha)$ may not be defined for sufficiently large α.

Additional complications arise when attention is switched to consider inequality constraints. The most obvious approach is an active set strategy similar to that used for linear constraints. Unfortunately the calculation of the value of α at which a new constraint becomes active in the search along $x(\alpha)$ (step (iii)) is now nonlinear, and therefore requires a further level of iteration which makes the whole scheme very inefficient. It may be possible to judge which constraint will become active and to zero this constraint along with the other active constraints in the projection iteration, but there are difficulties in writing down a strategy which covers all possibilities in a satisfactory way. It is also necessary to include additional heuristics to avoid zigzagging. In view of all these complications the methods become cumbersome and unappealing, and require a long and complex computer code. Some combinations of these ideas with penalty functions or with a projection iteration involving the solution of QP-like problems have been suggested, and indeed have potential, but then the methods become more closely related to those in other sections. Further discussion of feasible direction methods appears in discussion contributions 4.57–4.69.

3.5. Other methods Many other ingenious ideas have been suggested for the solution of nonlinear programming problems, and four of these are mentioned here. One possibility is to attempt to define an exact penalty function which is minimized locally by the solution x^* to (1.1). One way to do this is the ℓ_1 exact penalty function; this function and a suitable minimization method are described in Section 4. Nonetheless the ℓ_1 function is non-differentiable, and its minimization requires special techniques. Thus other research has concentrated on trying to find *smooth* exact penalty functions which can therefore be minimized by conventional unconstrained minimization techniques. Methods of this type are described in more detail by Bertsekas (Paper 4.3) and in discussion contributions 4.41, 4.42, and 4.49—4.52. The main difficulty with this approach is that calculating the objective function, $\phi(x)$ say, requires first derivatives of f and c_i to be available. Thus $\nabla\phi$ requires second derivatives which is a serious disadvantage. There may also be other difficulties when the methods are generalized to account for inequality constraints.

Many ideas have involved the use of linearizations of the constraint functions such as those in Section 3.3. Another interesting method of this type (recursive QP) can be derived from the penalty function (3.1), but turns out to have much more in common with the methods of Section 3.3. The method has been widely and successfully used in practice, and is described in more detail by Bartholomew-Biggs (Paper 4.2) and is mentioned in discussion contributions 4.28 and 4.30. My only reservation concerns the extent to which global convergence can be proved, especially the extent to which difficulties (a) and (b) in Section 2 can be avoided.

Another class of methods, which is also closely related to those in Section 3.3, makes the linear approximation described by (3.7), but does *not* make a quadratic approximation to f(x). Instead f(x) is replaced by a shifted Lagrangian function, giving the optimization problem

$$\left.\begin{array}{ll} \text{minimize} & f(x^{(k)}+\delta) - \lambda^{(k)T}\{c(x^{(k)}+\delta) - \ell^{(k)}(\delta)\} \\ \text{subject to} & \ell_i^{(k)}(\delta) = 0, \quad i\in\mathbf{E} \\ & \ell_i^{(k)}(\delta) \geq 0, \quad i\in\mathbf{I} \end{array}\right\} \qquad (3.10)$$

Robinson (1972). Then $x^{(k+1)} = x^{(k)} + \delta$ and $\lambda^{(k+1)}$ are taken as the solution and the multipliers of this subproblem. A list of references to this idea is given in discussion contribution 4.17 (see also 4.21). The main feature is that (3.10), although linearly constrained, has a general nonquadratic objective function in contrast to (3.9). Thus in general more computational effort must be put into solving the subproblem, especially if function calls are expensive, and this is disadvantageous. On the other hand the method does avoid the difficulties associated with the sequential QP method over the representation and updating of the matrix $W^{(k)}$, and it handles linear constraints more efficiently. It might also be argued that the more general form of the objective in (3.10) might somehow lead to a more meaningful solution to the subproblem. However, difficulties over convergence and infeasible or unbounded subproblems still remain, and it is not clear to me that (3.10) can be readily modified to avoid these difficulties in a satisfactory way.

Finally it is pointed out that special methods have been suggested for the case when the nonlinear functions in the problem are separable ($f(x) = \Sigma_i f_i(x_i)$). These methods replace each function of one variable $f_i(x_i)$ by a piecewise linear approximation, and so reduce the problem to a piecewise linear programming problem. In general these methods are only capable of low accuracy solutions. However it is possible to improve the methods by making local piecewise linear approximations over suitably chosen neighbourhoods, and in this form it is possible to solve successfully some very large network problems: see the paper by Meyer in Part 5.

4. An Exact ℓ_1 Penalty Function Method

The exact ℓ_1 penalty function for solving (1.1) is the function

$$\phi(x) = \nu f(x) + \sum_{i \in \mathbf{E}} |c_i(x)| + \sum_{i \in \mathbf{I}} \max[-c_i(x), 0], \qquad (4.1)$$

in which the penalty term is simply a sum of constraint violations, and a penalty parameter ν weights the relative contributions of the objective and constraint functions (Pietrzykowski, 1969). Necessary conditions for x^* to minimize $\phi(x)$ are that $\nu g^* = A^*\lambda^*$ and that

$$\left.\begin{array}{l}\left.\begin{array}{ll}|\lambda_i^*| \le 1 & \\ \lambda_i^* = -\text{sign } c_i^*, & \text{if } c_i^* \neq 0\end{array}\right\} i\in\mathbf{E} \\ \left.\begin{array}{ll}0 \le \lambda_i^* \le 1 & \\ \lambda_i^* = 1, & \text{if } c_i^* < 0 \\ \lambda_i^* = 0, & \text{if } c_i^* > 0\end{array}\right\} i\in\mathbf{I}\end{array}\right\} \quad . \tag{4.2}$$

A simple explanation of this result is the following. Because derivative discontinuities in (4.1) occur only when $c_i = 0$, define the set $\mathbf{Z}^* = \{i\colon c_i^* = 0\}$ and define λ_i^* as in (4.2) when $i\notin\mathbf{Z}^*$. It follows that x^* also solves

$$\left.\begin{array}{ll}\text{minimize} & \nu f(x) - \sum_{j\notin\mathbf{Z}^*} \lambda_j^* c_j(x) \\ \text{subject to } c_i(x) = 0, & i\in\mathbf{Z}^*\end{array}\right\} \quad . \tag{4.3}$$

Now let $\{\lambda_i^*;\ i\in\mathbf{Z}^*\}$ be defined as the Lagrange multipliers associated with the constraints in (4.3). It follows for arbitrary perturbations $\{\varepsilon_i;\ i\in\mathbf{Z}^*\}$ in the constraints of (4.3) that, if x solves the perturbed form of (4.3), then

$$\frac{d}{d\varepsilon_i}\left(\nu f(x) - \sum_{j\notin\mathbf{Z}^*} \lambda_j^* c_j(x)\right) = \lambda_i^*, \quad i\in\mathbf{Z}^* . \tag{4.4}$$

Consider $i\in\mathbf{Z}^*\cap\mathbf{I}$. If $\varepsilon_i > 0$ then $d(\max[-c_i,0])/d\varepsilon_i = 0$ and hence $d\phi^*/d\varepsilon_i = \lambda_i^*$. If $\varepsilon_i < 0$ then $d(\max[-c_i,0])/d\varepsilon_i = -1$ and hence $d\phi^*/d\varepsilon_i = \lambda_i^* - 1$. Minimality of ϕ^* requires that $d\phi^*/d\varepsilon_i \ge 0$ when $\varepsilon_i > 0$ and $d\phi^*/d\varepsilon_i \le 0$ when $\varepsilon_i < 0$, and so it follows that $0 \le \lambda_i^* \le 1$, $i\in\mathbf{Z}^*\cap\mathbf{I}$. A similar development gives $|\lambda_i^*| \le 1$, $i\in\mathbf{Z}^*\cap\mathbf{E}$, which justifies (4.2). From (1.3) and (1.8) it follows that the conditions (4.2) will hold at the solution x^* to (1.1) if

$$\nu < 1/\|\lambda^*\|_\infty, \tag{4.5}$$

where λ^* refers to the multiplier vector associated with (1.1). In fact, if in addition second order sufficient conditions hold at the solution to (1.1), then x^* both solves (1.1) and minimizes (4.1); thus the "exact" property of the penalty function (4.1) is demonstrated, provided that the penalty parameter is selected below a fixed threshold value. There are some cases when local solutions of (1.1) and (4.1) do not correspond (see Fletcher, 1981b, and also discussion contributions 4.12—4.15), but these are infrequent and, when these cases occur, they cause difficulties in most penalty function methods. The simple form of (4.1) requires that the constraint functions are mutually well-scaled: whether this should be done automatically or by the user is considered in the discussion contributions 4.14, 4.15, 4.53 and 4.54. Also the choice of penalty parameter may require some care (see contribution 4.15); in this respect, because I have noticed that an ℓ_1 penalty may not be as effective as a sum of squares penalty for large constraint violations, a possible remedy may be to use a hybrid loss function which combines the best features of both, although I do not think this has been investigated (perhaps some suitable ideas are given in Part 2 of these proceedings).

The main difficulty in using (4.1), however, is that its non-differentiability does not allow effective minimization by conventional smooth unconstrained techniques. Recently (Fletcher 1981a, 1981c, 1981d) I have been researching a direct method for minimizing (4.1), which also includes ideas from sequential QP and from trust regions in unconstrained minimization. Indeed the method is very easily explained along the lines of Section 3.3, by replacing f and c_i in (4.1), using (3.8) and (3.7) respectively, to give the function

$$\psi^{(k)}(\delta) \equiv \nu q^{(k)}(\delta) + \sum_{i\in \mathbf{E}} |\ell_i^{(k)}(\delta)| + \sum_{i\in \mathbf{E}} \max[-\ell_i^{(k)}(\delta),0], \quad (4.6)$$

which can therefore be regarded as an approximation to $\phi(x^{(k)}+\delta)$ local to $x^{(k)}$. This differs from (3.9) in that there are no side conditions, and so there are no difficulties due to an infeasible subproblem. Also it readily allows the introduction of a trust region constraint as a means of inducing global convergence. Thus the subproblem which is solved on iteration k is

$$\left.\begin{aligned} &\text{minimize} \quad \psi^{(k)}(\delta) \\ &\text{subject to } \|\delta\|_\infty \leq \rho^{(k)} \end{aligned}\right\} . \quad (4.7)$$

If $\delta^{(k)}$ denotes the solution of this subproblem, then $x^{(k+1)} = x^{(k)} + \delta^{(k)}$ is set and $\lambda^{(k+1)}$ becomes the multiplier vector associated with the functions $\ell_i^{(k)}(\delta)$ in (4.6). The radius $\rho^{(k)}$ of the trust region is adjusted adaptively by comparing actual to predicted reductions in $\phi(x)$ in an analogous way to unconstrained optimization (see Sorensen, contribution 1.4). A discussion of which norms to use in (4.1) and (4.7) is given in contribution 4.7. The convergence properties of the method are very good. Global convergence to a stationary point can be proved under very mild conditions (Fletcher, 1981c and see also 4.8—4.11). Asymptotically, if the trust region is inactive, then the method is equivalent to the sequential QP method, and hence converges at a second order rate. Likewise any modifications to sequential QP, such as updating $W^{(k)}$ or automatic scaling, can equally be applied to this method. Far from the solution, however, the methods differ, and the trust region is seen to play an important part. Because the constraint functions c(x) are represented in (4.7) only through a penalty term, not all the linearized constraints may be zeroed by the subproblem — even if they are equality constraints. Those that are zeroed can be interpreted as being "locally active", and this gives some indication why difficulty (a) (loss of rank in the

Jacobian of the active constraints) causes no trouble. In this method only the locally active constraint normals need be independent, which is achieved automatically by solving (4.7). Another important property of the method is that the Lagrange multipliers are always bounded (see (4.2)) as a consequence of solving (4.7), so this feature need not be assumed *a priori*. The method handles difficulty (c) by converging to a best ℓ_1 solution (more accurately a minimizer of (4.1)) if no feasible point exists.

It is important to realize that the subproblem (4.7) is no more difficult to solve than a conventional QP problem, and might be referred to as an ℓ_1QP problem. An active set method is used in which the currently zeroed constraints are active. The line search is replaced by a line search on $\psi^{(k)}(\delta)$ (see (4.6)), and a constraint becomes active if it is zeroed in the line search. At the solution of an equality problem for the active set, a constraint is made inactive if it violates the multiplier conditions (see (4.2)). The techniques for handling matrices are the same as in QP.

Numerical experience with (4.7) on small dense problems is good (Fletcher, 1981a), and comparable to sequential QP, although a number of problems are solved which sequential QP does not solve. It will be interesting to study the method on larger problems, and to find to what extent structure (e.g. sparsity, linear constraints) can be taken into account, as compared with other methods. No evidence of the Maratos effect is noticed, although this is probably because the problem constraints are reasonably well scaled, and because the penalty parameter is chosen not too far below its threshold value. However, because it is preferable not to have to rely on this being done, a further paper (Fletcher, 1981d) describes modifications which avoid the Maratos effect, and which enable the method to follow steep sided curved grooves much more rapidly. In fact both the basic method (4.7) and these modifications are applicable to a wide range of nondifferentiable optimization problems (finite min-max, ℓ_1, ℓ_∞ best approximation, etc.), and are presented as such in Fletcher (1981d).

The main idea of the modifications in Fletcher (1981d) is based on the projection iteration in feasible direction methods (Section 3.4), but only the first step of this iteration is required; it is referred to as the second order correction. If the basic step (4.7) gives a point $\hat{x} = x^{(k)} + \delta^{(k)}$, then the second order correction, $\hat{\delta}$ say, is obtained by re-linearizing the constraints about $\hat{x}$, but using the Jacobian at $x^{(k)}$. Thus $\hat{\delta}$ is most readily calculated by letting it be the value of δ that solves:

$$\left.\begin{array}{ll} \text{minimize} & \nu q^{(k)}(\delta + \delta^{(k)}) + h(\hat{c} + A^{(k)T}\delta) \\ \text{subject to} & \|\delta + \delta^{(k)}\| \le \rho^{(k)} \end{array}\right\}, \tag{4.8}$$

where h(c) denotes the penalty terms in (4.1). It is convenient that this is also an ℓ_1QP problem, because it can share the same software as (4.7). In fact, if this software permits parametric programming, then the solution of (4.8) is calculated rapidly whenever the locally active constraints remain unchanged, or are only changed a little. Therefore because no extra derivative evaluations occur (but $c(\hat{x})$ and $f(\hat{x})$ are required), the second order correction is usually easy and cheap to compute.

It is fairly straightforward to maintain the convergence properties of the method when using the second order correction. In the algorithm either $x^{(k)} + \delta^{(k)}$ or $x^{(k)} + \delta^{(k)} + \hat{\delta}$ may be accepted to define $x^{(k+1)}$. The motivation is that, if the basic step fails to improve ϕ, then the second order correction may be tried if it is predicted to do well. This enables the method to follow a curved groove more rapidly. If, however, the basic step improves ϕ significantly, then the second order correction is not made, because it is disadvantageous to force the iterates into a groove unnecessarily. Theoretical results show that asymptotically the second order correction provides a higher order approximation to the reduction $\phi^{(k)}-\phi^*$, and hence the Maratos effect can be avoided. Some numerical results (Fletcher, 1981d), in which the steepness of a curved groove is proportional to a parameter σ, show that the effort required by the basic method (4.7) to follow the groove is of order σ, whereas, when the second order correction (4.8) is also used, then the effort is only of order $\sigma^{\frac{1}{2}}$; thus a considerable improvement is obtained. For well scaled problems no significant differences between the two algorithms are noticed. Thus the use of (4.8) extends the basic method to solve badly behaved problems, and to avoid the Maratos effect at little extra cost and complexity.

The algorithms described so far in this section represent perhaps the most straightforward way of minimizing (4.1), and are of IQP type (see Section 2). In fact an EQP type of algorithm is also described in Fletcher (1981a), in which the locally active set is adjusted in the outer iteration by rules analogous to those in ℓ_1QP, but including anti-zigzagging precautions. The 2-norm is used to define the trust region, and the change in the variables is calculated by a Hebden–Moré type of iteration. The method can be implemented in $n^2 + O(n)$ storage using straightforward linear algebra techniques (no quadratic programming). Numerical experience suggests that the method takes too many iterations to locate the correct active set on large inequality problems, but might be useful for solving equality problems, and particularly for systems of nonlinear equations. Modification of the method to include the second order correction step would be advantageous.

4.2

Recursive Quadratic Programming Methods for Nonlinear Constraints

M.C. Bartholomew-Biggs

School of Information Sciences, The Hatfield Polytechnic, Hatfield, Herts., England.

The recursive quadratic programming approach to the general nonlinear programming problem is regarded as one of the most promising at present. This paper gives a brief review of the state of the art in this area, together with an indication of some lines of research aimed at producing better algorithms.

We consider the general nonlinear programming problem with both inequality and equality constraints:

$$\text{minimize } F(x), \quad x\in\mathbf{R}^n ,$$

$$\text{subject to} \begin{Bmatrix} \text{either } c_i(x) \geq 0 \\ \text{or } \quad c_i(x) = 0 \end{Bmatrix}, \quad i = 1,2,\ldots,m, \tag{1}$$

where some, at least, of the functions F and $\{c_i;\ i = 1,2,\ldots,m\}$ are nonlinear. For the purposes of discussion we shall make the (usual) assumptions that the functions are twice differentiable, and that a solution x^* exists for problem (1), at which the normals to the binding constraints are linearly independent.

We shall discuss methods which approach the solution x^* iteratively, each iteration involving the solution of a quadratic programming subproblem. Methods of this type will be called "recursive quadratic programming" techniques. At the start of an iteration an estimate, x, of x^* is known; a direction of search, p, for a better solution estimate, is then found by solving a subproblem of the general form:

$$\text{minimize } QF(p), \quad p \in \mathbf{R}^n,$$

$$\text{subject to} \left\{ \begin{array}{ll} \text{either } Lc_j(p) \geq 0 \\ \text{or } \quad Lc_j(p) = 0 \end{array} \right\}, \quad j = 1,2,\ldots,k \ (\leq m). \tag{2}$$

In (2) QF is a quadratic function which can be viewed, for the moment, as being an approximation of F about the point x. Similarly, each Lc_j is a linearization of a constraint in (1).

Subproblems of this form have been studied by many authors; recent work suggests that the recursive QP approach is very promising for minimization under nonlinear constraints. However, a number of different forms have been proposed for the subproblem (2), and it is not entirely clear which ideas will lead to the most successful algorithm. A common form for the function in (2) is

$$QF(p) = p^T \nabla F(x) + \tfrac{1}{2} p^T Bp, \tag{3}$$

where B is a (positive definite) approximation to $\nabla^2 L$, the Hessian matrix of the Lagrangian function associated with (1). Very briefly, B is defined in this way because the second order conditions for optimality of x* are expressed in terms of $\nabla^2 L$. This in turn means that superlinear convergence can be proved if $B \approx \nabla^2 L$ (rather than $B \approx \nabla^2 F$, which was used in some earlier algorithms). Of course $\nabla^2 L$ is not usually positive definite, but the computational convenience of using a positive definite approximation can be justified by the observation that the superlinear convergence result requires B to agree with $\nabla^2 L$ only in the subspace where L must have positive curvature. Some authors, however, are considering how to devise stable algorithms where B is not forced to be positive definite.

In the treatment of the constraints in (2) there are two distinct schools of thought. Some authors pose (2) as an *inequality* constrained problem (IQP) with k = m, so that linearizations of *all* the constraints in (1) are included. Others propose that a preliminary "active set" strategy be used, so that (2) involves only those constraints which are judged to be binding at x*. This use of an "active" set suggests the further simplification that all the constraints in (2) should be regarded as *equalities*. Algorithms employing an equality constrained subproblem (EQP) have been discussed by Biggs (1972), Biggs (1975), Murray and Wright (1978) and Van der Hoek (1980). IQP methods (which seem to have received more attention) are described by Wilson (1963), Fletcher (1973, 1975), Han (1977a), Powell (1978a) and Tapia (1977).

It is important at this stage to show the connections that exist between QP subproblems and *penalty functions*. IQP subproblems are conceptually quite simple, amounting to quadratic/linear approximations to the original problem. Most EQP subproblems, however, are based on ideas that were developed originally (Murray, 1969) to overcome some difficulties of minimizing the classical penalty function

$$P(x,r) = F(x) + r^{-1} \sum_{i=1}^{m} \phi(c_i(x)) \tag{4}$$

for small values of the penalty parameter r. Here $\phi(c_i) = c_i^2$ if the i-th constraint is an equality, and $\phi(c_i) = (\min[0,c_i])^2$ if the i-th constraint is an inequality. Murray showed that it is possible to approximate the minimum of P(x,r) by the solution to a certain quadratic programming problem. He further suggested that (1) might be solved very efficiently via a sequence of such problems so as to approach x* along a *trajectory* resembling the trajectory of penalty function minima. This idea proved very successful in practice, and it was further developed into an EQP method by Biggs (1972, 1975) and Murray and Wright (1978) independently. A QP subproblem based on barrier functions is also described by Murray and Wright (1978). More recently, Coleman and Conn (1980a, 1980b) have considered the relationship between QP methods and the exact ℓ_1 penalty function, defined by

$$Q(x,r) = F(x) + r^{-1} \sum_{i=1}^{m} \psi(c_i(x)),$$

where $\psi(c_i) = |c_i|$ if the i-th constraint is an equality, and $\psi(c_i) = \max[0,-c_i]$ if the i-th constraint is an inequality. Although considerable interest is being shown in the function $Q(x,r)$, and although it is the basis of several QP methods that will appear soon, we shall be more concerned with existing work related to the penalty function (4).

From now on we consider the IQP subproblem in the form

$$\begin{aligned} &\text{minimize } p^T\nabla F + \tfrac{1}{2}p^T Bp \\ &\text{subject to} \begin{Bmatrix} \text{either } c_i + p^T\nabla c_i \geq 0 \\ \text{or } \quad c_i + p^T\nabla c_i = 0 \end{Bmatrix}, \quad i = 1,2,\ldots,m. \end{aligned} \tag{5}$$

The EQP subproblem, based on penalty functions, will be written

$$\begin{aligned} &\text{minimize } p^T\nabla F + \tfrac{1}{2}p^T Bp \\ &\text{subject to } g_j + p^T\nabla g_j = -\tfrac{1}{2}ru_j, \quad j = 1,2,\ldots,k, \end{aligned} \tag{6}$$

where the functions $\{g_j(x);\ j = 1,2,\ldots,k\}$ are the constraint functions from the original problem that are treated as active the current estimated solution, and where $\{u_j;\ j = 1,2,\ldots,k\}$ are approximations of their Lagrange multipliers. The quantity r is, of course, the penalty parameter that is used in $P(x,r)$. It can now be seen that (6) involves constraints which are both linearized and *perturbed* forms of the original ones. Derivations and discussions of each of these subproblems can be found in the source papers that have been cited already. We are concerned here with the possible advantages of one form against the other. Clearly (6) will usually be a "smaller" problem than (5); further, the solution of (6) can be written down immediately whereas (5) will generally be solved by an iterative process. We must remember, however, that the formulation of (6) involves the preliminary identification of active constraints, as well as the choice of r and u to ensure convergence. A further point, which is not so obvious, is that (6) can be posed so that it always has a well-defined solution; but in problem (5) it may happen that the linearizations of nonlinear constraints do not admit any feasible point, or that locally the normals of the binding constraints are not linearly independent. Specifically, Biggs (1975) shows that, if A denotes the matrix whose rows are $\{\nabla g_j^T;\ j = 1,2,\ldots,k\}$, and if u is defined by

$$(\tfrac{1}{2}rI + AB^{-1}A^T)u = AB^{-1}\nabla F - g, \tag{7}$$

then the solution of (6) is

$$p = B^{-1}(A^T u - \nabla F). \tag{8}$$

Thus, if the matrix B is positive definite, then, for $r > 0$, u exists to satisfy (7), even if the active constraint normals are linearly dependent. Because the existence of u implies the existence of p, we have a well-defined search direction. The expressions (7) and (8) are the basis of a particular minimization algorithm called REQP.

Clearly one would like to resolve some of the questions that arise concerning problems (5) and (6). The relative merits of IQP and EQP subproblems are discussed well by Murray and Wright (1980). One needs to consider the following kinds of questions. When all factors are taken into account, which subproblem involves more work? Which subproblem produces better search directions? Apart from rather special situations where constraint normals are (nearly) linearly dependent, is the presence of the penalty parameter r an advantage or a disadvantage? Should we attempt to satisfy the linearized constraints exactly (thus approaching the solution in the manner of a projection method)? Or should we use r to hold the solution estimates "off" the constraints until close to x* (thus approaching the solution along a "penalty trajectory")?

The recent work of Schittkowski (1980a) has thrown some light on the first two questions, but his extensive comparisons must also reflect other incidental differences between the algorithms tested. A more uniform evaluation has recently been attempted by the author (Bartholomew-Biggs, 1981b). If the original problem (1) has only equality constraints, then the only difference between the subproblems (5) and (6) is due to the r-term. Therefore, in order to give an idea of the usefulness (or otherwise) of this penalty parameter, a set of equality constrained problems was solved using the algorithm REQP, not only in the usual way (with a decreasing sequence of positive r values) but also with r set to zero throughout. The results of this experiment are quite interesting. Out of 18 examples, only 3 were solved significantly faster by the algorithm with $r = 0$. For about half the problems the differences in efficiency between the two approaches were very small. However, on 4 examples at least, it proved noticeably better to make use of nonzero r values.

These results suggest that a new IQP subproblem should be considered, namely

$$\text{minimize } p^T\nabla F + \tfrac{1}{2}p^TBp$$
$$\text{subject to} \begin{cases} \text{either } c_i + p^T\nabla c_i \geq -\tfrac{1}{2}ru_i \\ \text{or } \quad c_i + p^T\nabla c_i = -\tfrac{1}{2}ru_i \end{cases}, \quad i = 1,2,\ldots,m. \tag{9}$$

This subproblem, rather than (5), can be regarded as the true IQP analogue of (6). Subproblem (9) can be formed and solved in a way that is similar to using expressions (7) and (8) to solve (6). Initial computational experiments have identified occasions when (9) is superior to (5), and, just as for the equality constrained examples, there is seldom any serious disadvantage in including the r-term. It seems, therefore, that a good QP algorithm might include the penalty term at points far from the solution, switching to $r = 0$ at a suitable stage to obtain the most rapid ultimate convergence.

It must be pointed out, however, that numerical experience with the EQP subproblem (6) has also been good. Sometimes the IQP methods will solve a problem in fewer iterations than are needed using an EQP. But almost invariably the *time* required by an EQP algorithm is less.

These questions of subproblem formulation are important for the recursive QP methods, which is why we have devoted attention to them so far. However, a successful algorithm will depend upon the existence of good procedures for certain other important calculations, and we shall look briefly at some of these and the work that is being done.

The QP subproblem produces a search *direction*: the minimization algorithm must determine a suitable step size in this direction, which is normally done by means of a line search with respect to some function LS(x). A number of different forms for LS(x) have been suggested, all of which aim to combine the objective function and constraints in a balanced way. For example, some algorithms take LS(x) as P(x,r). Others use Q(x,r) or else some approximation to the Lagrangian function. Most of the choices that have been tried work well in practice, but two goals, namely stronger convergence results and greater confidence in convergence from bad starting points, have led to several recent proposals for line search strategies. We mention the work of Heath (1978), Maratos (1978), Chamberlain, Lemaréchal, Pedersen and Powell (1980) — who suggest the "watchdog" technique which takes account of *two* line search functions — and Bartholomew-Biggs (1981a). This is an area in which further important practical work will probably be done in the next few years.

Another important part of the calculation is the updating of the matrix B at each iteration. It has already been stated that convergence is not necessarily impaired if we make a positive definite approximation to $\nabla^2 L$, even when $\nabla^2 L$ is an indefinite matrix. In practice we need to consider which of the many low-rank correction formulae will be most suitable for building up a good estimate of $\nabla^2 L$, while avoiding the danger of B becoming singular or unbounded. Powell (1978a) suggests an updating strategy with these points in mind. Van der Hoek (1980) gives an experimental evaluation of a number of different updates, mostly in the class known as "self-scaling" formulae. Interestingly his conclusions are relevant to the use of an "active set" strategy for defining the QP subproblem. His observations suggest that, when Powell's update is used, it is possible for "wrong" constraints to be retained too long in the active set. The reasons are not clear, but the remark does illustrate the complex interdependence of the various parts of the algorithms.

Finally we will say a word about computational methods for solving IQP subproblems. Because, in many cases, the major part of the work is devoted to the calculation of p, it is worth considering how to make this phase more efficient. We may take advantage of the positive definiteness of B. Moreover, we would like to exploit the fact that B differs only by a low-rank correction from the matrix in the QP that was solved on the previous iteration. Finally we can expect, near the solution, to have available a good estimate of the active constraints of (5) from the previous QP in the sequence. In this vein, Schittkowski (1981a) has tested a modification of Powell's (1977) Fortran subroutine; the modification replaces the solution of (5) by an algebraically equivalent linear least squares problem. He reports significant reductions in computer times.

In this short paper we have considered some current research areas relating to the solution of constrained minimization problems by recursive quadratic programming. The debate concerning the relative merits of IQP and EQP subproblems does not yet seem to have been resolved. Both ideas have their merits (and advocates!). It does seem, however, that there is growing interest in exploring the relationships between QP methods and various penalty functions. This interest is, at least in part, connected with the need to choose a merit function for the line search procedure. The line search is important, both as a means for promoting convergence from bad starting points, and also because the superlinear convergence of QP methods depends upon the use of the *undamped* correction $x := x + p$ near the solution. Longer term research in recursive quadratic programming techniques could well be devoted to extending their range of applicability beyond small to medium size problems. For certain practical calculations the fact that the methods typically generate *infeasible* points is also regarded as a serious disadvantage, so there may be further interest in feasible point variants for particular applications.

4.3

Augmented Lagrangian and Differentiable Exact Penalty Methods*

Dimitri P. Bertsekas

Department of Electrical Engineering and Computer Science,
Massachusetts Institute of Technology,
Cambridge, Massachusetts 02139, USA.

1. Introduction

The original proposal of an *augmented Lagrangian* method by Hestenes (1969) and Powell (1969) may be viewed as a significant milestone in the recent history of constrained optimization. Augmented Lagrangian methods are not only of practical importance in their own right, but have also served as the starting point for a chain of research developments centering around the use of penalty functions, Lagrange multiplier iterations, and Newton's method for solving the system of necessary optimality conditions.

Augmented Lagrangian methods became quite popular in the early seventies, but then yielded ground to the algorithms based on Newton's method for solving the system of necessary optimality conditions, that are usually referred to as *recursive quadratic programming* (RQP) techniques. The author believes, however, that augmented Lagrangian methods will probably maintain for a long time a significant position within the arsenal of computational methodology for constrained optimization. In fact their value

*This research was conducted in the M.I.T. Laboratory for Information and Decision Systems with partial support provided by National Science Foundation Grant No. NSF/ECS 79-20834.

may increase further as interest shifts more towards large problems. I will try to outline some of the reasons for this assessment, and I will survey briefly the state of the art of Augmented Lagrangian methods in the next section.

On the other hand there is extensive evidence that for many problems, particularly those of relatively small dimension, RQP techniques are considerably more efficient than augmented Lagrangian methods. Locally convergent variants of RQP methods have been known for many years and have seen considerable use in control theory and economics. Their broad acceptance in mathematical programming practice became possible, however, only after a methodology was developed that allowed global convergence based on the monotonic reduction of exact penalty functions. The use of a nondifferentiable exact penalty function for this purpose was originally proposed by Pschenichny (1970; Pschenichny and Danilin, 1975). His work became widely known in the Soviet Union, but went largely unnoticed in the West where nondifferentiable exact penalty functions were introduced independently by Han (1977a), in connection with iterations based on RQP. The work of Powell (1978b) showed how to use effectively quasi-Newton approximations within the nondifferentiable exact penalty/RQP framework, and contributed significantly to the popularization of the overall approach. There are many significant contributions in this area and they are covered extensively in other papers in this volume. It is interesting to note that the RQP direction together with a stepsize of one does not necessarily lead to a decrease in the value of the nondifferentiable exact penalty function even arbitrarily close to a solution (Maratos, 1978). This shortcoming is potentially serious since it may prevent superlinear convergence in situations where it otherwise might be expected. To bypass this difficulty it is necessary to introduce modifications in the algorithm such as those suggested by Mayne and Polak (1978) and Chamberlain, Lemaréchal, Pedersen and Powell (1980).

Recently there has been some interest in the use of differentiable exact penalty functions in connection with RQP. A class of such functions has been proposed by Di Pillo and Grippo (1979). There is an interesting connection between the Newton direction for minimizing any function in the Di Pillo—Grippo class and the Newton direction for solving the system of necessary optimality conditions, which has been noted independently by Bertsekas (1980a) (in connection with second derivative algorithms) and by Dixon (1980) (in connection with quasi-Newton methods). It is also interesting that the class of exact penalty functions proposed by Fletcher (1970b) can be derived (and indeed expanded) via the Di Pillo—Grippo class (Bertsekas, 1980a). A further link in the chain of these developments is established in Bertsekas (1980b), where it is shown that the RQP direction, that is derived from a positive definite approximation to the Hessian of the Lagrangian (see, for example, Powell, 1978b), is a descent direction for any function in Fletcher's class arbitrarily far from a solution, provided that the penalty parameter is sufficiently large. Furthermore, a stepsize of one near the solution decreases the value of the penalty function, so the difficulty noted by Maratos (1978) in connection with nondifferentiable exact penalty functions does not arise. These results, which will be described in Section 3, have placed differentiable exact penalty functions on an equal footing with nondifferentiable ones in terms of desirable descent properties. More research should be expected in this area, as evidenced by the recent work of Boggs and Tolle (1981) and Han and Mangasarian (1981) reported during the meeting. We mention also the two parameter differentiable exact penalty proposed by Boggs and Tolle (1980), which is related to Fletcher's class of penalty functions and to Newton's method for solving the system of necessary optimality conditions.

In this paper we will restrict ourselves exclusively to the equality constrained problem

$$\left.\begin{array}{ll}\text{minimize} & f(x)\\ \text{subject to} & h(x)=0\end{array}\right\},\qquad \text{(ECP)}$$

where $f\colon \mathbf{R}^n \to \mathbf{R}$ and $h\colon \mathbf{R}^n \to \mathbf{R}^m$ are assumed to be three times continuously differentiable.

We focus our attention on local minima/Lagrange multiplier pairs (x^*,λ^*) that satisfy the following second order sufficiency assumptions for optimality:

$$\left.\begin{array}{ll}\nabla_x L_0(x^*,\lambda^*) = 0, & h(x^*) = 0\\ z^T\nabla^2_{xx}L_0(x^*,\lambda^*)z > 0 & \end{array}\right\},\qquad \text{(S)}$$

where $L_0\colon \mathbf{R}^{n+m} \to \mathbf{R}$ is the (ordinary) Lagrangian function

$$L_0(x,\lambda) = f(x) + \lambda^T h(x),$$

where x^* is in the set

$$X^* = \{x\colon \nabla h(x) \text{ has rank } m\},$$

and where z is any nonzero vector such that $\nabla h(x^*)^T z = 0$. In our notation all vectors are considered to be column vectors, and the superscript T denotes transposition. The usual norm on the Euclidean space $\mathbf{R}^n$ is denoted by $\|\cdot\|$ (i.e., $\|x\| = (x^T x)^{\frac{1}{2}}$ for all $x \in \mathbf{R}^n$). For a mapping $h\colon \mathbf{R}^n \to \mathbf{R}^m$, $\{h_i;\ i = 1,2,\ldots,m\}$ are the coordinates of h and $\nabla h(x)$ denotes the $n\times m$ matrix whose columns are the gradients $\{\nabla h_i(x);\ i = 1,2,\ldots,m\}$. Whenever there is a danger of confusion we indicate explicitly the arguments of differentiation.

For the most part we make no attempt to state results precisely, and to give complete references to individual contributions. A detailed analysis of each point made in the paper together with references may be found in the author's book (Bertsekas, 1982), which will be published by Academic Press. For surveys of analytical and computational properties of augmented Lagrangian methods the papers by Bertsekas (1976) and Rockafellar (1976) are recommended.

2. Augmented Lagrangian Methods

The basic form of the augmented Lagrangian method requires the solution of a sequence of problems of the form

$$\left.\begin{array}{ll} \text{minimize} & L_{c_k}(x,\lambda_k) \\ \text{subject to} & x\in\mathbf{R}^n \end{array}\right\}, \tag{1}$$

where for $c \geq 0$, $L_c\colon \mathbf{R}^{n+m} \to \mathbf{R}$ is the augmented Lagrangian function

$$L_c(x,\lambda) = f(x) + \lambda^T h(x) + \tfrac{1}{2}c\,\|h(x)\|^2, \tag{2}$$

and where the sequence of penalty parameters $\{c_k\}$ satisfies $0 < c_k \leq c_{k+1}$ for all k. The initial multiplier vector λ_0 is given and subsequent multiplier vectors λ_k, $k \geq 1$ are generated by some updating formula, such as the first order iteration

$$\lambda_{k+1} = \lambda_k + c_k\, h(x_k), \tag{3}$$

where x_k solves (perhaps approximately) problem (1). There is also a second order iteration

$$\lambda_{k+1} = \bar{\lambda}_k + \Delta\lambda_k, \tag{4}$$

where $\bar{\lambda}_k = \lambda_k + c_k h(x_k)$ is the first order iterate, and where $\Delta\lambda_k$ together with some vector Δx_k solves the system

$$\begin{pmatrix} H_k & \nabla h(x_k) \\ \nabla h(x_k)^T & 0 \end{pmatrix} \begin{pmatrix} \Delta x_k \\ \Delta\lambda_k \end{pmatrix} = - \begin{pmatrix} \nabla_x L_0(x_k,\bar{\lambda}_k) \\ h(x_k) \end{pmatrix}. \tag{5}$$

Here H_k is either the Hessian $\nabla^2_{xx} L_0(x_k,\bar{\lambda}_k)$ of the ordinary Lagrangian function L_0 evaluated at $(x_k,\bar{\lambda}_k)$, or some quasi-Newton approximation thereof. Note that the system (5) is also the basis of RQP methods, which points to the significant relations between augmented Lagrangian methods and RQP.

The convergence properties of the method are quite well understood. There are several results in the literature which state roughly that under second order sufficiency assumptions one can expect convergence of (3) or (4) from an initial multiplier λ_0 which is arbitrarily far from a solution, provided the penalty parameter c_k becomes sufficiently high eventually. The rate of convergence of $\{x_k, \lambda_k\}$ is typically linear if the simple first order iteration (3) is used and $\{c_k\}$ remains bounded, while (4) gives a superlinear rate of convergence.

There are a large number of variations and extensions of the idea of the augmented Lagrangian method. For example extensions are available to handle inequality constraints and nondifferentiable terms in the objective and constraint functions. It is possible to use quadratic penalty functions for these purposes, although they introduce second derivative discontinuities into the augmented Lagrangian function. An alternative that the author has found preferable on several occasions, which does not suffer from second derivative discontinuities, is the use of one of several other kinds of penalty functions — for example an exponential function. Other variations include different stepsize choices in the first order iteration (3), and methods based on a partial elimination of constraints. For example, if (ECP) is augmented by nonnegativity constraints on x, i.e. if the problem has the form

$$\begin{aligned} &\text{minimize} \quad f(x) \\ &\text{subject to } h(x) = 0, \quad \text{and } x \geq 0, \end{aligned}$$

then it may be more convenient to eliminate only the (presumably more difficult) constraints $h(x) = 0$ via a penalty. Minimization of $L_c(x, \lambda_k)$ over x should then be carried out subject to the restriction $x \geq 0$. This points to an important advantage of the augmented Lagrangian method, namely the flexibility it affords in changing the structure of a given problem to one that is more favourable. This flexibility can prove decisive in the solution of large problems, because much depends there on being able to exploit the existing structure. Finally, there is a rich theory associated with augmented Lagrangian methods, revolving around duality, convexification of nonconvex problems, the proximal point algorithm, and related subjects, which can play an important role in the analysis of specific problems, and can provide the basis for the development of new algorithms.

Typical advantages cited in favour of the augmented Lagrangian approach are its robustness, and its ease in programming and tuning for a given calculation. Furthermore the method is broadly applicable since it is capable of solving problems in which the second order sufficiency conditions are not satisfied (although the efficiency usually deteriorates in these cases). Its disadvantages versus other competing methods are primarily in two areas. First feasibility of the generated iterates is not maintained, so, if the algorithm is prematurely terminated, it will not provide a feasible solution. For some problems this can be an important or even a decisive drawback. The second disadvantage manifests itself primarily in small problems and is based on a comparison of the relative efficiency of the method versus RQP techniques. A substantial amount of computational evidence points to the fact that (well tuned) RQP methods require considerably fewer function evaluations to converge than augmented Lagrangian methods. On the other hand, each iteration of the augmented Lagrangian method requires less calculation overhead, particularly for problems of large dimension. Therefore a precise comparison of computational effort for a given problem depends on the relative costs of function and derivative evaluations, and calculation overhead per iteration. We conclude that usually for a given type of problem there is a critical size (or dimension) above which either a first order or a second order augmented Lagrangian method is computationally more efficient than RQP methods, and below which the situation is reversed.

3. Differentiable Exact Penalty Methods

An interesting class of differentiable exact penalty functions for (ECP) was recently introduced by Di Pillo and Grippo (1979). Its basic form is

$$P_0(x,\lambda;c) = L_0(x,\lambda) + \tfrac{1}{2}c\|h(x)\|^2 + \tfrac{1}{2}\|M(x)\ \nabla_x L_0(x,\lambda)\|^2. \quad (6)$$

A more general version, which is essentially the same as one proposed in Di Pillo, Grippo and Lampariello (1980), is given by

$$P_\tau(x,\lambda;c) = L_0(x,\lambda) + \tfrac{1}{2}(c + \tau\|\lambda\|^2)\ \|h(x)\|^2 + \tfrac{1}{2}\|M(x)\ \nabla_x L_0(x,\lambda)\|^2. \quad (7)$$

In (6) and (7) it is assumed that $c > 0$, $\tau \geq 0$, and $M(x)$ is an $m{\times}n$ twice continuously differentiable matrix function on the set X^* (defined in Section 1), such that $M(x)\nabla h(x)$ is invertible for all $x \in X^*$. For example one may choose $M(x) \equiv \nabla h(x)^T$ or $M(x) \equiv [\nabla h(x)^T \nabla h(x)]^{-1} \nabla h(x)^T$. When $\tau = 0$ the function (7) is identical to expression (6), but it seems that the presence of a positive value of τ can have a substantial beneficial effect in algorithmic applications.

The main fact concerning the function (7) is that, roughly speaking, for any value of $\tau \geq 0$, local minima/Lagrange multiplier pairs (x^*,λ^*) of (ECP) can be identified with local minima (with respect to *both* x and λ) of $P_\tau(\cdot,\cdot;c)$ provided c exceeds a certain threshold value. There is an extensive analysis that clarifies the "equivalence" just stated, and that quantifies the threshold level for c, but in view of space limitations we cannot go into details. It is worth noting, however, that this threshold level depends on eigenvalues of certain second derivative matrices, and is largely unrelated to the magnitudes of Lagrange multipliers, even though these magnitudes determine the corresponding threshold level for nondifferentiable exact penalty functions.

There is an interesting connection between Newton-like methods for minimizing $P_\tau(.,.;c)$ and Newton's method for solving the (n+m)-dimensional system of necessary conditions $\nabla L_0(x,\lambda) = 0$. Let (x^*,λ^*) be any local minimum/Lagrange multiplier pair satisfying the local sufficiency conditions (S). It can be shown that the Newton (or second order RQP) direction

$$d_N = \nabla^2 L_0(x,\lambda)^{-1} \nabla L_0(x,\lambda)$$

can be expressed as

$$d_N = B_\tau(x,\lambda;c) \ \nabla P_\tau(x,\lambda;c),$$

where B (.,.;c) is a continuous (n+m) × (n+m) matrix satisfying

$$B_\tau(x^*,\lambda^*;c) = [\nabla^2 P_\tau(x^*,\lambda^*;c)]^{-1}.$$

In other words near (x^*,λ^*) the *RQP direction approaches asymptotically the Newton direction for minimizing* $P_\tau(.,.;c)$. An interesting corollary is that the Di Pillo—Grippo penalty function can serve within a neighbourhood of (x^*,λ^*) as a *descent function* for RQP methods. However, a result that is more interesting from a practical point of view is that a descent property of this type holds *globally* within an arbitrarily large compact subset of X^*. A version of this result, that can be applied to the exact penalty function that was introduced by Fletcher (1970b), is given below.

For $x \in X^*$ consider the function

$$\hat{P}_\tau(x;c) = \min_\lambda P_\tau(x,\lambda;c).$$

Since P_τ is, for each x, a positive definite quadratic function of λ, one can carry out the minimization with respect to λ explicitly. A straightforward calculation yields the minimizing vector

$$\hat{\lambda}(x) = -[\nabla h(x)^T M(x)^T M(x) \ \nabla h(x) + \tau \| h(x) \|^2 I]^{-1} [h(x) + \nabla h(x)^T M(x)^T M(x) \ \nabla f(x)]$$

and the equation

$$\hat{P}_\tau(x;c) = P_\tau[x,\hat{\lambda}(x);c].$$

For specific choices of M and τ, $\hat{P}_\tau(x;c)$ is a penalty function in the class of Fletcher (1970b). For example, if $\tau = 0$ and if $M(x) = [\nabla h(x)^T \nabla h(x)]^{-1}\nabla h(x)^T$, we obtain $\hat{P}_\tau(x;c) = L_0[x,\lambda(x)] + \frac{1}{2}(c-1)\|h(x)\|^2$, where $\lambda(x) = -[\nabla h(x)^T\nabla h(x)]^{-1}\nabla h(x)^T\nabla f(x)$.

The penalty function $\hat{P}_\tau(x;c)$ also has nice (and global) descent properties in connection with directions generated by RQP techniques, as shown in the following proposition due to Bertsekas (1980b, 1982):

Proposition 1: Let X be a compact subset of X*, and let $\mathbf{H}$ be a bounded set of symmetric n×n matrices, such that, if $H \in \mathbf{H}$, if $x \in X$, and if z is any vector in $\mathbf{R}^n$ satisfying $\nabla h(x)^T z = 0$, then

$$b_\ell \|z\|^2 \leq z^T H z \leq b_u \|z\|^2 ,$$

where b_ℓ and b_u are positive constants. Then, for every $x \in X$ and $H \in \mathbf{H}$, the solution $(\Delta x, \lambda)$ of the system

$$\begin{pmatrix} H & \nabla h(x) \\ \nabla h(x)^T & 0 \end{pmatrix} \begin{pmatrix} \Delta x \\ \lambda \end{pmatrix} = -\begin{pmatrix} \nabla f(x) \\ h(x) \end{pmatrix}$$

exists and is unique. Further, there exist positive constants $\bar{c}$ and w such that, for all $c \geq \bar{c}$, $x \in X$ and $H \in \mathbf{H}$, the direction Δx satisfies the inequality

$$\nabla \hat{P}_\tau(x;c)^T \Delta x \leq -w\|\nabla \hat{P}_\tau(x;c)\|^2.$$

Proposition 1 shows that the algorithm

$$x_{k+1} = x_k + \alpha_k \Delta x_k, \tag{8}$$

where Δx_k together with some vector λ_{k+1} is obtained by solving, for a positive definite matrix H_k, a system of the form

$$\begin{pmatrix} H_k & \nabla h(x_k) \\ \nabla h(x_k)^T & 0 \end{pmatrix} \begin{pmatrix} \Delta x \\ \lambda \end{pmatrix} = -\begin{pmatrix} \nabla f(x_k) \\ h(x_k) \end{pmatrix}, \tag{9}$$

has global convergence properties provided c is chosen sufficiently large. The following proposition shows its superlinear rate of convergence properties. We consider the case where α_k is chosen by the Armijo rule with an initial stepsize of one, i.e. $\alpha_k = \beta^{m(i)}$, where m(i) is the least nonnegative integer m that satisfies

$$\hat{P}_\tau(x_k;c) - \hat{P}_\tau(x_k+\beta^m \Delta x_k;c) \geq -\sigma\beta^m \nabla\hat{P}_\tau(x_k;c)^T \Delta x_k, \tag{10}$$

for some constants β, $\sigma \in (0,\frac{1}{2})$.

<u>Proposition 2</u>: Let x* be a local minimum of (ECP) which, together with a Lagrange multiplier λ*, satisfies the sufficiency assumptions (S). Assume that the algorithm (8)—(10) generates a sequence $\{x_k\}$ converging to x* and that the sequence $\{H_k\}$ in (9) is bounded and satisfies

$$\frac{\Delta x_k^T [H_k - \nabla^2_{xx} L_0(x^*,\lambda^*)] Z^*}{\|\Delta x_k\|} \to 0$$

where Z* is any constant $n \times (n-m)$ matrix whose columns are a basis of the tangent space $\{z: \nabla h(x^*)^T z = 0\}$. Then:

a) There exists an index $\bar{k}$ such that, for all $k \geq \bar{k}$, the stepsize α_k is one.

b) The rate of convergence of the sequence $\{\|x_k - x^*\|: k = 1,2,3,\ldots\}$ is q-superlinear.

The conditions of Proposition 2 are always obtained if $H_k = \nabla^2_{xx} L_0(x_k,\lambda_k)$, and usually they hold in practice if H_k is generated by the variable metric formula of Powell (1978b).

It is not possible at present to provide a comparison between RQP techniques that use differentiable and nondifferentiable exact penalty functions for descent. Both types of methods behave identically when x_k is so close to a solution that the superlinear convergence property takes effect. Far from a solution their behaviour can be quite different, and furthermore the threshold values for the penalty parameter in both methods can differ greatly on a given problem (these values can have a substantial influence on efficiency when far from a solution). Methods based on differentiable exact penalty functions require more overhead per iteration because they involve more complex expressions (although not as much overhead as may appear at first sight — see Bertsekas, 1982), and present extensions to deal with inequality constraints are not very "clean". On the other hand they have the theoretical advantage (which may be of value in practice) that they do not require modifications to induce superlinear convergence.

4.4

Penalty Function Methods*

A.R. Conn

Department of Computer Science, University of Waterloo,
Waterloo, Ontario, Canada N2L 3G1.

Historically (around 1960!), there were no effective numerical algorithms available for nonlinear, constrained optimization, but good techniques for unconstrained optimization were starting to be developed. For example, Rosen's (1960) gradient projection method (perhaps the first effective technique for *linear* constraints) was published then, and Davidon's (1959) variable metric method was formulated in 1959, was re-presented in 1963 (Fletcher and Powell, 1963), and is still the basis of many current unconstrained algorithms. Therefore the original motivation for penalty functions was that they changed the extremely difficult nonlinear, constrained optimization problem into a sequence of (it was hoped) easier unconstrained problems for which reasonable algorithms existed. The questions I wish to pursue here are : why are we still interested in penalty functions, should we continue to be interested in them, and (finally) what is their future?

The major immediate advantage of penalty functions, as we have already remarked, is that they represent the constrained problem in terms of unconstrained problems. However, this is of more significance than just the historical point concerning the availability of superior algorithms for the unconstrained calculation.

*This work was supported in part by NSERC grant number A8639.

Typically, the reformulation of the problem enables one to move away from nonlinear constraint boundaries, especially when one is not in the vicinity of a stationary point. In addition, to my mind, a considerable advantage is the intuitive global interpretation of the meaning of both the penalty function and the corresponding iterates of any related algorithm. Any particular penalty function, in effect, incorporates two aims, that of minimizing the objective function and that of satisfying the constraints. Consequently, if one is so far from a solution that one can make large gains in the objective function value, then it is of less importance to satisfy the constraints accurately. However, when one is close to a solution, then feasibility is likely to become a more dominant issue. Penalty functions are merely a consistent method of measuring progress in these two, often naturally conflicting, aims.

A major disadvantage is the choice of penalty parameters. In other words, the choice to give to the weights that fix the relative importance of the objective function and the various constraint functions.

It is, perhaps, worth pointing out that several aspects of penalty function methods can be closely related to techniques that are used in augmented Lagrangian, successive quadratic programming, and reduced gradient methods. In some respects penalty functions provide a more natural interpretation of these techniques (see for example Coleman and Conn, 1980b; Powell, 1978b; Gill, Murray and Wright, 1981; Han, 1977a; Tapia, 1977). With these relationships in mind, it is certainly legitimate to state that some of the best algorithms available are, in effect, penalty techniques.

A particular subclass of penalty functions are the so-called exact penalty functions, that is those that give rise to a finite sequence of unconstrained problems, which may be differentiable (see Paper 4.3 by Bertsekas in this volume) or non-differentiable. The non-differentiable forms are typified by the penalty function of Zangwill (1967a) and Pietrzykowski (1969), and have recently enjoyed a broad popularity. Essentially, in spite of their piecewise nature, these functions are simpler and more closely related to the original problem and its Lagrangian function than either the non-exact penalty functions or the differentiable exact penalty functions. Therefore they are useful even in the special cases of linear programming (Conn, 1976) and quadratic programming (Conn and Sinclair, 1975). I predict that, with the possible exception of trajectory methods (Murray and Wright, 1978), exact penalty functions will continue to hold more promise than the non-exact penalty functions, as long as one is not unduly concerned with the intermediate generation of infeasible points. However, in practice, one often requires even intermediate points to be feasible (see Gutterman's discussion contribution 3.32). For such problems those penalty methods referred to as barrier methods are the only viable penalty techniques, and the trajectory barrier methods are then particularly promising.

Trajectory analysis is certainly an approach that deserves more attention.

One consequence of this interest in exact penalty functions is that I expect there to be much more effort than there is at present on special line searches, trust regions, and other step size selecting mechanisms for piecewise differentiable functions. Some initial work in this area has been done by Murray and Overton (1979) and Charalambous and Conn (1978). It has also been suggested at this meeting (Davidon, discussion contribution 1.20), that conic approximations may be particularly useful for line searches because of their applicability to functions with poles (this remark also applies to differentiable penalty functions). However, conic approximations appear to me to be more useful when the positions of the poles are not known, which is not the case here. In particular, some of the methods mentioned above rely strongly for their success on an *a priori* knowledge of the location of the neighbourhoods of the non-differentiable points.

It is difficult to foresee any significant advances in the

local convergence properties of current algorithms applied to small dense problems that does not presuppose equivalent advances in unconstrained optimization. The reason for this comment is that, in using the term local, I am assuming that one is in the neighbourhood of a solution, and, in particular, that one has identified correctly the active constraints. Therefore the essential problem at hand is to find a stationary point of a Lagrangian, or, equivalently, to solve an unconstrained problem in a reduced space. In fact, the current state of local convergence of penalty function methods is analogous to the corresponding state of unconstrained optimization with two exceptions. There is still no published quasi-Newton method that updates only a projected Hessian matrix and that maintains a superlinear convergence rate; such methods are considered by Coleman and Conn (1981). Secondly, whenever we have preferred to choose our search directions from a model that is distinct from the line search merit function (as is common in most of the successive quadratic programming methods; see Paper 4.2 by Bartholomew-Biggs), there has been difficulty in guaranteeing that a stepsize of unity is eventually acceptable (see, for example, Chamberlain, Lemaréchal, Pedersen and Powell, 1980). At present a unit stepsize is an essential ingredient in any algorithm that claims to mimic efficient unconstrained algorithms asymptotically.

Thus, one current difficulty is to identify when the region of asymptotic convergence has been reached. I suggest that more effort should be spent on studying the earlier stages of iterative methods. An isolated example of genuine non-asymptotic analysis is given by Overton (1981).

A related question concerns the type of penalty transformation that is used. In Paper 2.3 on non-differentiable optimization, Lemaréchal considered two distinct extreme kinds of algorithm, namely a) when the set of all generalised gradients is used explicitly, and b) when only one element of the set is used. He then refers to the method of bundles, which can be regarded as a compromise of b) that is closer to a). There is also the possibility of a compromise of a) closer to b), which suggests a penalty transformation that involves only a few constraint functions.

For example, consider the problem

$$\text{minimize} \quad f(x), \quad x \in \mathbf{R}^n$$

$$\text{subject to } c_i(x) \geq 0, \quad i \in I,$$

and consider replacing the usual exact penalty function

$$p(x,\mu) = f(x) - \frac{1}{\mu} \sum_{i \in I} \min[0, c_i(x)]$$

by the exact penalty function

$$\hat{p}(x,\mu) = f(x) - \frac{1}{\mu} \min[0, \min_i c_i(x)] ,$$

i.e. we replace the sum of the infeasibilities in the penalty term by the greatest infeasibility. We have in mind the possibility of "guessing" j such that $c_j(x) = \min\{c_i(x)\}$, and then defining

$$\tilde{p}(x,\mu) = f(x) - \frac{1}{\mu} c_j(x).$$

From time to time we expect to update our guess. Assuming one uses a projection-like method for determining a descent direction for $\tilde{p}$, the optimal number of active constraints to include in the projection is an open question. It is also uncertain how often one should update j.

The main advantage of this approach would be economies of computation far from the solution when the original problem contains many constraints. A disadvantage is that one has to correct for an inappropriate choice of j. It is of interest to note that, assuming j always corresponds to a violated constraint, there does exist a positively weighted exact penalty function of the form

$$\bar{p}(x,\mu) = f(x) - \frac{1}{\mu} \sum_{i \in I} w_i(x) \min[0, c_i(x)],$$

such that the sequence generated by $\tilde{p}$ produces a monotonically decreasing sequence of values of $\bar{p}$.

With reference to the identification of the correct active set, there are some known difficulties that have been largely ignored. For example, serious problems can be caused by degeneracy (due to redundant active constraints), near-degeneracy, spurious constraints that are nearly active (all these difficulties occur often in nonlinear ℓ_1 data fitting), and near-zero multipliers. Near-zero multipliers are discussed by Gill and Murray (1977b).

Furthermore these problems are not necessarily confined to neighbourhoods of local optima. In fact, a related problem that requires closer examination is false stationary points; an example is given in Coleman and Conn (1980c).

Another algorithmic detail that has been largely ignored is the question of the choice of penalty parameters. Although, in the case of exact penalty functions, their threshold values are known in terms of the optimal Lagrange multipliers (see Luenberger, 1970, for example), there is little understanding of which values are ideal. At present, there is no global algorithm with an exact penalty merit function for which sophisticated updating techniques are applied to the parameters, and yet the resulting algorithm is robust and superlinearly convergent.

As in all areas of optimization, scaling is a significant problem for which few useful results are known.

Since some recent advances can be related to minimizing a quadratic function subject to quadratic constraints (Coleman and Conn, 1980b), perhaps such models will be more generally applied. This is the simplest approach that allows curved constraints. Further, it would enable, for example, the practising model builder to replace 2n upper and lower bounds by one elliptical constraint, which is clearly more reasonable if one is using a penalty function algorithm, instead of a feasible direction algorithm.

It is heartening to note that the experience of nonlinear programmers with piecewise problems, that are motivated by exact penalty functions, has influenced the algorithmic and theoretical development of many other piecewise differentiable problems, incuding linear ones. (For examples, other than those already mentioned, see Bartels, 1980; Bartels and Conn, 1981; Bartels, Conn and Charalambous, 1978; Bartels, Conn and Sinclair, 1978; Calamai and Conn, 1980, 1981; Fletcher and Watson, 1980; Murray and Overton, 1980, 1981; Watson, 1981.)

I anticipate that these fruitful offshoots will continue to grow, and, hopefully, will include broader aspects of non-smooth optimization.

I also feel that advances in large sparse nonlinearly constrained optimization will involve penalty function techniques. For example, MINOS-augmented (Murtagh and Saunders, 1980) is essentially a method based on Robinson's (1972) algorithm and the method of MINOS (Murtagh and Saunders, 1978). However, it is possible to replace the objective function of Robinson's algorithm by any exact penalty function, which poses some interesting challenges. For example, the usual dense techniques involve precise projections, orthogonal matrix decompositions, careful multiplier estimates, and the precise determination of quadratic minima, but sparse techniques may have to be less precise. In addition, for many large scale problems, the work of the matrix calculations of an iteration is very much greater than the cost of evaluating the objective function. For a fuller discussion of large sparse optimization problems see Gill, Murray and Wright (1981).

I am sure that much has been omitted in this brief introduction. One topic that comes to mind is the impact of computer hardware on the subject; for example, parallel processing may alter entirely the choice of a suitable penalty function. Another topic is whether penalty functions have any significant contributions to make to discrete optimization or global optimization.

I would like to emphasize the possibility of using penalty functions for purely linear problems, quadratic problems, and any other problems with special structure. It can be very helpful, provided the penalty function does not destroy the advantages attainable from the original structure.

Thus, I hope that in this paper I have made it clear that we should, indeed, still be interested in penalty functions, that they have certain inherent desirable properties, and that there is still research to be done.

4.5

Reduced Gradient Methods

L.S. Lasdon

Department of General Business,
School of Business Administration,
University of Texas, Austin, Texas 78712, USA.

1. Summary of Current Status and Performance

Reduced gradient (RG) methods use the active constraints of a problem to express certain variables (called basic) in terms of others (nonbasic). The objective then becomes a function of the nonbasics alone, called the reduced objective, and its gradient is the reduced gradient. This gradient is used to determine a search direction for the nonbasics, and a line search determines the step along this direction. During the line search a basic variable may equal or violate one of its bounds. If so, a new partitioning of the variables is defined, and the process is repeated on the new reduced objective.

More precisely, let the nonlinear program be

$$\text{minimize} \quad f(x),$$

$$\text{subject to } c(x) = 0 \tag{1}$$

and

$$\ell_i \leq x_i \leq u_i, \quad i = 1,2,\ldots,n, \tag{2}$$

where $x = (x_1,x_2,\ldots,x_n)$ and $c(x) = (c_1(x),c_2(x),\ldots,c_m(x))$. All constraints have been rephrased as equalities by using logical (slack or surplus) variables, which are included in x, and the ℓ_i

and u_i are given lower and upper bounds respectively.

Denoting the basic and nonbasic variables as x_b and x_{nb} respectively, the constraint equations become

$$c(x_b, x_{nb}) = 0.$$

The Jacobian matrix of c may be similarly partitioned as

$$\frac{\partial c}{\partial x} = \left(\frac{\partial c}{\partial x_b}, \frac{\partial c}{\partial x_{nb}} \right) = (B, B_{nb}),$$

where, for simplicity, the variables are assumed to be renumbered so that the basics are the first m components of x.

Let $\bar{x}$ be the current values of the variables. Then the specific variables to be chosen as the basic variables must be selected so that B, evaluated at $\bar{x}$, is nonsingular. In this case the constraints (1) can be solved (at least conceptually) for x_b in terms of x_{nb} to yield the basics as a function of the nonbasics, $x_b(x_{nb})$. This representation is valid for all x_{nb} sufficiently near $\bar{x}_{nb}$. The objective function is then reduced to a function of x_{nb} only,

$$f(x_b(x_{nb}), x_{nb}) = F(x_{nb}), \tag{3}$$

and the original problem (1)—(2) (at least in the neighbourhood of $\bar{x}$) is transformed to a simpler *reduced problem*:

$$\text{minimize } F(x_{nb}),$$

subject to the bounds on x_{nb}. The function F is called the *reduced objective* and its gradient ∇F is called the *reduced gradient*.

GRG algorithms solve the original problem by solving (perhaps only partially) a sequence of reduced problems. These are usually solved by a method which uses ∇F. At a given iteration, with nonbasic variables $\bar{x}_{nb}$ and basic variables $\bar{x}_b$, the reduced gradient is computed as follows:

a) Solve the square system

$$B^T\lambda = \partial f/\partial x_b \tag{4}$$

for the simplex multiplier vector $\lambda \in \mathbf{R}^m$.

b) Compute the reduced gradient

$$\frac{\partial F}{\partial x_{nb}} = \frac{\partial f}{\partial x_{nb}} - B_{nb}^T\lambda . \tag{5}$$

Note that all partial derivatives are evaluated at the current point $\bar{x}$.

Following Murtagh and Saunders (1978), the nonbasic variables are further partitioned into s superbasic variables, x_s, which are to be varied in the current iteration, and n-m-s remaining nonbasic variables, x_n, which are at one of their bounds and will remain there. In most RG algorithms, the reduced gradient with respect to the nonbasic variables, $\partial F/\partial x_n$, is used only to determine if one of the components of x_n should be released from a bound to join the superbasic set. This decision can be made at each iteration, as in Lasdon, Waren, Jain and Ratner (1978), or after an optimization over the current superbasics is completed to within a loose tolerance, as in Murtagh and Saunders (1978). In either case, the reduced gradient with respect to the current superbasics, $\partial F/\partial x_s$, is used to form a search direction, $\bar{d}$. Both conjugate gradient and variable metric methods have been used to determine $\bar{d}$. Then a one dimensional search is initiated, whose goal is to solve the problem

$$\underset{\alpha \geq 0}{\text{minimize}}\ F(\bar{x}_{nb} + \alpha\bar{d}). \tag{6}$$

In (6), $\bar{d}$ has been extended to include zero components for the nonbasics at bounds. This minimization is done only approximately, and is accomplished by choosing a sequence of positive values $\{\alpha_1, \alpha_2, \ldots\}$ for α. For each value α_i, $F(\bar{x}_{nb} + \alpha_i\bar{d})$ must be evaluated. By (3), this is equal to

$$f(\ x_b(\bar{x}_{nb} + \alpha_i\bar{d}),\ \bar{x}_{nb} + \alpha_i\bar{d}\ )$$

so the basic variables $x_b(\bar{x}_{nb} + \alpha_i\bar{d})$ must be determined. These satisfy the system of equations

$$c(x_b, \bar{x}_{nb} + \alpha_i\bar{d}) = 0,$$

where $\bar{x}_{nb}$, α_i, and $\bar{d}$ are known and x_b is to be found. If x_b appears nonlinearly in any constraint, then this system must be solved by an iterative procedure. Usually, a variant of Newton's method is used.

If the initial point x_0 provided by the user is not feasible, RG algorithms use a 2 phase approach, as does the simplex algorithm for linear programming. Phase 1 attempts to determine a feasible point or show there is none, while phase 2 optimizes the original objective if a feasible point is found. The phase 1 procedure of the algorithm implemented in the code GRG2 (see Lasdon, Waren, Jain and Ratner, 1978) minimizes the sum of absolute values of the constraint violations. This minimization is done subject to the currently active constraints. As in most phase 1 procedures for linear programming, once an infeasible constraint becomes satisfied, it remains satisfied.

Many variants of RG algorithms have been proposed and implemented. Wolfe (1967) first proposed the idea for problems with linear constraints. Extensions for this class of problems have been developed by Murtagh and Saunders (1978) and Shanno and Marsten (1979). Both procedures can solve large sparse problems, using LP-based sparse matrix methods to handle the basis matrix. If a conjugate gradient method is used to determine the search direction, problems with hundreds or thousands of nonbasics can be solved. Recently Beck (Beck, Lasdon and Engquist, 1981) has specialized this idea to nonlinear network problems, where the linear constraints express flow conservation in a network. The convex simplex method, which varies one nonbasic at a time, has also been used in this context.

The case of nonlinear constraints was pioneered by Abadie (Abadie and Carpentier, 1969), who called the procedure generalized reduced gradients (GRG). Later variants were developed by Lasdon and Waren (1978), Gabriele and Ragsdell (1977), and Heltne and Littschwager (1975). Both Gabriele (1980) and Lasdon and Waren (1978) have implemented versions for large sparse systems. A closely related procedure for large nonlinearly constrained problems has been described by Beale (1974). Each "function evaluation" requires the solution of a linearized version of the problem by an LP calculation.

By now, several carefully implemented GRG codes exist for small to medium size problems, mostly due to the authors mentioned above. These codes have performed very well in comparative tests. Only sequential quadratic programming (SQP) algorithms have been competitive or superior. A summary of these results is given by Lasdon (1980). Several GRG codes have been distributed by their authors, and are now widely used in practice.

2. Advantages and Disadvantages of Reduced Gradient Algorithms

2.1 Advantages

(a) By eliminating some variables, GRG algorithms operate in a space of reduced dimension. Often, at least toward the end of the iterative process, there are few superbasic variables and quasi-Newton algorithms locate the optimum quickly.

(b) Extensions to large sparse problems are conceptually simple. One may utilize existing sparse matrix and conjugate gradient technology.

(c) Excursions outside the feasible region are small relative to competing algorithms. This is important in solving problems where some functions are not defined for certain values of their arguments (e.g. logs and fractional powers), which is a common occurrence in practice.

(d) The concepts underlying RG methods are simple and direct. As straightforward extensions of the simplex algorithm, RG methods are easy to understand. It is likely that the direct way they solve NLP problems is responsible for their robustness.

2.2 Disadvantages For nonlinear constraints, most of the computational effort in GRG is devoted to satisfying the currently binding constraints at each step. Hence algorithms which do not attain feasibility until termination (e.g. SQP) often require fewer function evaluations. However, these methods also require more time per evaluation, since an optimization subproblem must be solved at each step. For linear constraints, this disadvantage of reduced gradient methods disappears, since it is straightforward to satisfy active constraints without iteration.

3. Future Trends

RG algorithms are and will continue to be the dominant procedures for linearly constrained problems. Their unique ability to handle large problems overcomes the theoretical disadvantage that some reduced problems may become badly conditioned (Gill and Murray, 1974b). This implies that they will also play an important role in algorithms which solve sequences of linearly constrained subproblems. In Murtagh and Saunders (1980) the MINOS RG code is used to solve the subproblems arising in a variant of Robinson's method. RG codes may also be applied to solve the quadratic programming subproblems in Powell's (1978b) SQP algorithm. In any such application, RG methods permit one to use advantageously the optimal solution of one subproblem as a starting point for the solution of the next subproblem. More research is needed to ascertain the relative advantages of RG versus other procedures in such applications, especially quadratic programming.

GRG algorithms also have strong advantages for certain classes of nonlinearly constrained problems. In discrete time optimal control, the nonlinear constraints are a system of difference equations, which can be solved for the endogenous or state variables given the independent or control variables. Here there is a natural partitioning of the variables into basic and nonbasic, and most implicitly defined models are easily solved for the basics given the nonbasics. Not surprisingly, GRG methods have been predominant here for many years. In problems involving econometric models (Lasdon and Waren, 1980), with 10 to 20 time periods and hundreds of basic variables but only 2 or 3 controls per period, a problem with thousands of nonlinear constraints is reduced to one with fewer than a hundred nonbasics, plus simple bounds. Similar comments apply to problems arising in electrical power systems (Lasdon and Waren, 1980). Here the hundreds of currents and voltages in a network may be found very efficiently given a relatively small number of independent variables, e.g. generator voltages and phases, and transformer and capacitor settings.

For nonlinearly constrained problems of general structure, GRG algorithms will be increasingly challenged by SQP methods. Better implementations of SQP procedures will be developed, and the algorithms themselves will be improved. In some cases, such methods will be considerably faster than GRG. This will be true especially in problems where the constraints are hard to satisfy, for example optimal control calculations with several terminal conditions.

4. Ideas on Useful Future Research

For sparse nonlinearly constrained problems, more work is needed to develop, test, and compare GRG implementations. Important issues include the selection of the basic variables at each step, basis updating, and how often reinversions are done. Few codes exist for solving large nonlinearly constrained NLP's and practically no comparisons have been made.

As mentioned earlier, many practical problems have "hard" constraints — constraints which must be strictly satisfied in order for some portion of the model to be defined. In Sarma and Reklaitis (1979), exterior penalty and augmented Lagrangian algorithms fail to solve such a problem arising in chemical process design, while a GRG algorithm is successful. However, all algorithms except interior penalty procedures can run into trouble here. More work is needed to develop efficient algorithms which can deal with "hard" constraints.

There are a myriad of algorithmic options available within the GRG framework. Some which seem worthwhile have yet to be investigated. Issues which appear to be important are: (1) the line search strategy, (2) the choice of basic variables at each step, (3) determination of the point at which a basis change is made, (4) the use of improved equation solving procedures to determine the basic variables, (5) tactics used when the problem becomes degenerate, (6) scaling of variables and constraints, and (7) proper tolerance setting and dynamic tolerance adjustment.

Since nonlinear programs have such variety, it is unlikely that any algorithm will be best for all problems. We need more information on which methods are best for which kinds of problems and why. Testing to shed light on this issue should include both "made-up" and generated problems and real-world problems of various types. Such testing would be facilitated by the availability of a library of portable well implemented routines with compatible input formats for the leading NLP algorithms.

Discussion on "Nonlinear Constraints"

4.6 Steihaug

What are the benefits of 1-norms and ∞-norms over 2-norms for penalty terms and trust regions?

4.7 Fletcher

Various considerations affect the decision as to which norm to use with an exact penalty function. Regarding the function as $\nu f(x) + \|c(x)\|$, then the 2-norm is smooth except at a feasible point, but it does not give rise to QP-like subproblems in a trust region type of method. Both the 1- and ∞-norms do give rise to QP-like subproblems, but the discontinuities in the 1-norm lie on the constraint boundaries and their positions are not influenced by constraint scaling, whereas those in the ∞-norm lie along the surfaces $|c_i(x)| = \max_t |c_t(x)|$ and so are affected. I am not sure to what extent this is significant in practice. The choice of norm in defining the trust region is also of importance. The 1- and ∞-norms both lead to linear constraints which preserve QP-like subproblems but the ∞-norm is more convenient. The only objection might be that, when the radius of the trust region is small, the direction of the correction is to a corner of the 'box' and is not the steepest descent direction. A 2-norm trust region corrects this property but no longer gives QP-like subproblems. It might be used in a Hebden–Moré type of approach (Fletcher, 1981a, Algorithm 2), at the expense of solving a sequence of subproblems.

4.8 Moré

In theory it is necessary to find the global solution of the local model in some methods of Paper 4.1 in order to guarantee convergence. This seems to be an objection against the use of the maximum norm in the local model. How serious is this objection?

4.9 Fletcher

In the trust region method with an ℓ_∞ trust region, and when exact second derivatives are used, then there is a disparity between the practical computation of a local solution and the requirement of the theory to guarantee a global solution of the subproblem. However, the use of pseudo-constraints (see Fletcher, 1981b) to solve the ℓ_1QP subproblem does ensure that an objective function reduction is always predicted, so circumstances under which the method could break down are extremely unlikely. When a positive semi-definite Hessian approximation to $W^{(k)}$ is used then the difficulty does not arise.

4.10 Moré

What is meant by a "suitable global convergence result"?

4.11 Fletcher

My global result for the trust region method (and also when second order corrections are used) requires only that the calculated vectors of variables are bounded and that the smooth functions are twice continuously differentiable. The result is that there exists an accumulation point which satisfies first order conditions for a minimizer of the composite objective function (or the ℓ_1 penalty function), that is there is no strict direction of descent at the accumulation point. I am hopeful of being able to prove stronger results about second order conditions when second order corrections are used.

4.12 Griewank

Under what assumptions can one ensure that this minimizer corresponds to a feasible point of the underlying constrained problem.

4.13 Fletcher

It is quite possible that a local minimizer of the exact ℓ_1 penalty function is not a feasible point of the corresponding nonlinear programming problem, as is true for most penalty functions. It is also possible that a uniformly infeasible sequence $\{x^{(k)}\}$ is determined such that $\phi(x^{(k)}) \to -\infty$. There is no easy guaranteed answer to this problem, and I regard it as an inevitable consequence of using a penalty function. In practice it does not arise frequently, and the usual cure is to increase the weighting of the constraint terms. In the context of an exact ℓ_1 penalty function it is worth trying $\nu = 0$ in an attempt to locate a feasible point. One is successful if the objective function is reduced to zero, but, if a local minimizer with positive value is found, then either no feasible point exists or global minimization techniques (usually expensive) are required to locate the feasible point.

If the nonlinear programming problem has a solution at which second order sufficient conditions hold, then this point is also a local minimizer of the exact ℓ_1 penalty function.

4.14 Powell

Should not the problem of balancing the magnitudes of h and f be viewed as a disadvantage of the ℓ_1 exact penalty function methods of Paper 4.1?

4.15 Fletcher

I agree that choosing the relative scaling of the objective function and penalty terms in an ℓ_1 penalty function may not always be straightforward. One has to avoid having the parameter ν of my Paper 4.1 too small, because then steep sided curved grooves may occur. However, following my work on the use of second order corrections, this difficulty seems to me to be less serious, and it can be avoided if necessary by re-solving the problem with a more suitable ν. The opposite aspect is to choose ν to be smaller than the threshold value and also to avoid $\phi(\underline{x}^{(k)}) \to -\infty$ for infeasible points. Currently I do not see how best to choose or adjust ν automatically, although a good method would be valuable. In any case it is usually not difficult for the user to repeat the calculation with different ν in order to settle on a suitable value. I do not believe that the difficulty is any more severe than with other penalty methods which also require the user to balance the objective function and penalty terms.

4.16 Lemaréchal

I have two comments. Firstly, I cannot believe that the ℓ_1 exact penalty function method of Paper 4.1 is significantly different from existing ones. It has to be a variant of one of the given iterated quadratic programming methods. Actually, its main advantage, in my opinion, is that it looks at the same object from another side, so it brings new possibilities for understanding, and this may give improvements for both classes of methods.

My second comment concerns EQP versus IQP. There exists a principle from common sense: separate the difficulties. Here there are two difficulties: nonlinearity and combinatorics (due to $\leq$). Since the direction finding problem (QP) avoids the first difficulty, common sense suggests that one should add to it the second difficulty, thus leading to IQP. Of course this is not a definitive answer since common sense requires sometimes more thought than is apparent at first glance.

4.17 Wright

Certain fundamental questions remain unresolved concerning the relative merits of an EQP or IQP formulation of the subproblem; many of these issues are discussed in detail in Murray and Wright (1980). The same considerations apply in all projected Lagrangian algorithms, including those in which the subproblem involves a general objective function and linearized constraints (such as the methods of Robinson, 1972; Rosen and Kreuser, 1972; Rosen, 1978; Murtagh and Saunders, 1980; and Best, Braüniger, Ritter and Robinson, 1981). I should like to consider three main points.

Firstly, in my view it is misleading to state that the IQP formulation avoids the need for an active set strategy. With either form of QP subproblem, a prediction is made of the active set. Using the active set (and/or multiplier estimates) of an IQP subproblem is simply *another form* of active set strategy.

Secondly, the impression is sometimes given that by doing the extra calculations associated with an IQP subproblem, a consistently better prediction of the active set will be obtained. If this were true, solving an IQP subproblem would be analogous to performing a very accurate line search in unconstrained optimization, in that more work would be required at each iteration, but fewer iterations would be needed for convergence. However, it is not true that the prediction produced by an IQP subproblem is always more reliable than that resulting from some other strategy for predicting the active set; it is not even known whether the IQP prediction is better in general.

Results are known (e.g. Robinson, 1974) that guarantee a correct prediction by an IQP subproblem in some neighbourhood of the solution, under certain conditions on the Hessian of the QP subproblem. However, it is not obvious that a meaningful prediction of the active set will be made by an IQP subproblem at an arbitrary point. In fact, it is easy to construct examples in which the IQP prediction of the active set is completely wrong. Therefore, much further study and research is required in order to investigate the theoretical properties of active set strategies as well as their performance in practice.

Finally, I should like to mention one suggestion (from the Murray and Wright paper) that is intended to improve the reliability of the IQP prediction of the active set and the corresponding Lagrange multiplier estimates. It is helpful to compare the IQP multipliers with (say) a first-order estimate computed at the new point with the predicted active set treated as equalities. If these estimates do not display some agreement, the prediction of the active set (and the multipliers) may be unreliable.

4.18 Fletcher

I would like to point out that I personally have researched both EQP and IQP types of algorithm for nonlinear programming and feel that both have their uses in certain situations. However, for general use where function evaluation costs are significant, my results indicate that on larger problems substantially more iterations (and hence function evaluations) are required by an EQP method to determine the correct active set, so that the IQP method performs better. Regarding the correct determination of the active set, I would also like to point out that the trust region method can play an important role. Of course, when started near the solution, the sequential QP approach gives a good estimate of the true active set directly, but this may not be true remote from the solution. The trust region method, however, only attempts to find a set of locally active inequality constraints, and it is very likely that this will be done correctly.

4.19 Overton

One of the advantages of solving an EQP rather than an IQP is that one needs to approximate only the projected Hessian. For example, if gradient differences are used, they are required only in the null space of the active constraint Jacobian.

4.20 Gay

I think it may be best for low dimensional inequality constrained problems to update an approximation to the full Hessian.

4.21 Rosen

Most problems with nonlinear constraints also contain linear constraints. Often, only a few of the constraints are nonlinear. Therefore, any good method should take full advantage of this possibility, and should reduce to a good linearly constrained algorithm when all constraints are linear. "Almost linear" constraints should also be handled efficiently. This is true of projected Lagrangian methods (Rosen and Kreuser, 1972; Robinson, 1972; Rosen, 1978; Murtagh and Saunders, 1980), but is not necessarily true of many penalty or augmented Lagrangian methods.

Both projected Lagrangian methods and sequential methods are similar in that they use a quadratic approximation to the objective function, and linearize some, or all, of the nonlinear constraints. The key difference seems to be the relative frequency with which these two approximations are carried out.

In the projected Lagrangian method the constraint linearization is fixed during a major iteration, while the quadratic approximation to the Lagrangian is updated. In sequential QP the quadratic approximation is fixed during each QP major iteration.

The choice between these two approaches seems to depend on the relative nonlinearity of the objective and constraint functions. Perhaps they can be merged so as to modify automatically the relative frequency of approximation.

4.22 Fletcher

It is possible, but not always straightforward, to take account of linear constraints when using an IQP type of method (e.g. sequential QP or my trust region method). It is necessary for the software for QP or ℓ_1QP to recognise that the same linear constraints are active from one call to the next, and hence avoid refactorizing (partially) the matrix of active constraints. For example this may be done in the factorization $A = QR$ by having the linear constraint normals in the leading columns of A. I believe this consideration is important, and it clearly has implications for the design of QP or ℓ_1QP software.

4.23 Lootsma

What further gains in efficiency and robustness are to be expected from the new algorithms of Paper 4.1? To be more precise, does one expect improvements of 20%, 30%, 40% or improvements which are an order of magnitude better?

4.24 Fletcher

Substantial progress in nonlinear programming has already been made, for example the number of iterations for solving Colville's second test problem (Colville, 1968) has come down from about 500 to less than 10 over the years, and it is difficult to envisage any further improvement on results of this nature. However, it is possible to look for more reliability in standard methods. Without being clairvoyant it is difficult to predict in what area the next substantial jump in progress might be — perhaps it will occur in large scale nonlinear programming.

4.25 Powell

In my opinion there are good reasons for reserving the term "variable metric" for quadratic programming methods that preserve positive definite second derivative matrices, even when there are nonlinear constraints. In unconstrained optimization one can obtain search directions for variable metric methods by applying the method of steepest descents to transformed variables, where the transformation makes the current second derivative matrix equal to the identity. If one makes the equivalent transformation in the constrained case, then the required search direction becomes a projected steepest descent direction. This is the main characteristic of variable metric methods.

4.26 Sargent

May I suggest a slight amendment to Contribution 4.25. Dennis and Walker (1980) have been studying least-change secant and inverse-secant methods and divide them into two classes: "fixed scale" and "iteratively rescaled", depending on whether the metric measuring the change is fixed or changes at each iteration. May I suggest the use of the term "variable metric" in place of "iteratively rescaled"? It has the advantage that it makes the metric referred to precise, and the class includes the DFP and BFGS formulae, which were the original variable metric methods.

4.27 Tanabe

The function of the penalty term in the method of Paper 4.2 is to overcome difficulties due to a singular Jacobian matrix. I would like to suggest that the penalty term be introduced only when the iteration encounters a singularity.

4.28 Bartholomew-Biggs

The penalty parameter term in the EQP subproblem *does* provide a remedy for linear dependence among constraint normals far from the solution. This was not the original motivation for including it, however. The subproblem is based on the idea of finding an efficient way of approximating the minima of a sequence of penalty functions. Since the exterior point penalty function may have a well defined minimum, even in the presence of "redundant" constraints, it is to be expected that the EQP subproblem will be fairly insensitive to such situations. Near the solution as $r \to 0$ there may still be numerical difficulties.

4.29 Conn

I like to handle degeneracy by perturbing the right hand sides of the relevant active constraints. However, unlike the linear case where one may perturb a vertex of a linear polytope, in the nonlinear case one can no longer expect the perturbed optimal solution to be at a vertex.

This is especially distressing when one has reached a degenerate optimal solution, since one may now spend a great deal of time solving the perturbed nonlinear problem to determine that the solution has been found. It is possible, for example, in the extreme case when the solution to the perturbed problem is actually interior to the perturbed feasible region, that one has to drop all the current activities after perturbation.

4.30 Dembo

As I see it, the penalty parameter r in Paper 4.2 has many important features, most of which are related to the global behaviour of the given algorithm. However, locally it is not sufficient for r to simply go to zero in order to get superlinear convergence. It must tend to zero sufficiently rapidly. The inexact Newton rate of convergence characterization, given by Dembo, Eisenstat and Steihaug (1980), provides an easy way to decide just how fast r should go to zero in order to obtain superlinear or quadratic convergence.

4.31 Polak

In the context of engineering design, the time needed to solve a moderate sized quadratic program is negligible relative to the time needed for function evaluations via system simulation. Hence we can contemplate methods which solve more than one QP per iteration in order to achieve better behaviour.

4.32 Bartholomew-Biggs

We have had experience of optimization problems where the function evaluations are very expensive. Indeed there are cases where a "function evaluation" involves the actual operation of plant for a time with the current values of the control variables. In such cases the overhead costs of an optimization iteration are negligible — and we should probably want to do as much work as possible to squeeze out the best solution estimate from the information we already have.

In these terms it may be argued, *a priori*, that IQP will probably require rather fewer function evaluations than EQP. We need to distinguish, however, whether the objective function is expensive because it involves a lot of calculation or simply because it has many variables. The remarks of Professor Bertsekas suggest that QP approaches may not compare well with augmented

Lagrangian techniques for very large problems in optimal control with certain structures.

4.33 Schnabel

There are some optimization applications where the user cannot afford to find the minimum, but rather is interested in which algorithm will provide the "best" solution in a fixed amount of computing time. I wonder whether other people have encountered such problems, and whether we know which algorithms we should use in this case. One such experience of mine was with an unconstrained minimization application. Even here, where feasibility is no problem, is it clear that the normal "best" algorithms are still best?

4.34 Gutterman

Schnabel's comment extends some remarks previously made. In the simplest case, an algorithm which maintains feasibility may, in a sense, be "better" than a more rapidly convergent algorithm that does not maintain feasibility since the information available when the budget is exhausted is of value. More complicated situations would require knowledge of the relationship between accumulated cost and accuracy for each algorithm to be compared.

4.35 Lootsma

Barrier function methods have the attractive property that they maintain feasibility during the computations, so that one does not run into trouble if the objective function and/or some of the constraints are undefined outside the constraint set (think of terms like $\sqrt{(1-x_1^2-x_2^2)}$ or $\ln(1-x_1^2-x_2^2)$). Do other methods like exterior penalty functions, augmented Lagrangians, reduced gradients or quadratic approximations, have precautions to handle these difficulties? A useful technique is given in Section 1 of Fletcher's paper 4.1 — ed.

4.36 Mifflin

There are nonbarrier ways to stay feasible that require no objective function evaluations outside the feasible set and that do not use constraint function values at feasible points — see Mifflin (1980, 1981). The latter paper, presented in a research seminar here, introduces a new type of penalty approach, with an automatically determined parameter, which is not ill-conditioned in degenerate situations. These ideas also give the user more freedom in modelling a problem and in writing the function evaluation subroutine.

4.37 Meyer

In the case of convex feasible sets, piecewise linear inner approximation of the feasible set ensures that the resulting set is a subset of the original feasible set. Thus, this type of approximation has the property that iterates will be feasible.

4.38 Wright

There is a feasible point algorithm for nonlinear constraints (Wright, 1976; Murray and Wright, 1978) that is based on computing the search direction from a quadratic programming subproblem, and then executing a linear search with respect to the logarithmic barrier function (which thereby serves as a merit function). The iterates from this algorithm tend to converge nontangentially to the active constraints, rather than along constraint boundaries (as with a GRG method). The subproblems do not become ill-conditioned as the iterates converge, and feasibility is retained throughout.

4.39 Sargent

There is no real competition between obtaining feasibility and making progress towards the solution, since there is no reason why one should not try to do both together, using the Hessian of the Lagrangian as the metric in which to measure constraint violations. More specifically, the direction given by solving the QP in recursive QP methods is a descent direction for reducing constraint violations, so, if feasibility is important at an early stage, one can use the steplength to ensure violation reduction until feasibility is achieved, and thereafter revert to the use of the descent function to measure the progress towards the solution. I shall be describing such a method in my research seminar.

4.40 Polak

It seems to make sense to use hybrid methods for engineering design problems, composed of two phases, where phase I is a method of feasible directions (see, e.g., Polak, Trahan and Mayne, 1979) for obtaining a good starting point for phase II, which is a local method such as successive quadratic programming. The phase I — phase II methods of feasible directions are very fast when far from a solution and very reliable. The cross-over is to take place when the Newton or quasi-Newton method starts exhibiting at least linear convergence. For details of a general theory that enables one to construct such cross-over tests see Polak (1976). For example, a cross-over from the Armijo gradient method to the Newton method for the unconstrained optimization problem:

$$\min_{x \in \mathbf{R}^n} f(x), \quad g(x) \equiv \nabla f(x), \quad H(x) \equiv \frac{\partial^2 f(x)}{\partial x^2}$$

is given by the following algorithm:

Parameters: γ, α, $\beta \in (0,1)$, $K_1 >> 1$, $K_2 \geq 1$.

Step 0: Set $i = 0$, $j = 0$, $\ell = 0$.

Step 1: Compute $h(x_i) = H(x_i)^{-1} g(x_i)$.

Step 2: If $\| h(x_i) \| \leq K_1 \gamma^j$, set $j = j+1$, set $x_{i+1} = x_i - h(x_i)$, set $i = i+1$ and go to step 1. Else proceed.

Step 3: Compute the least positive integer k that satisfies the condition

$$f(x_i - \beta^k g(x_i)) - f(x_i) \leq -\alpha\beta^k \| g(x_i) \|^2,$$

set $x_{i+1} = x_i - \beta^k g(x_i)$, and set $i = i+1$.
If $\| g(x_i) \| \leq K_2 \gamma^\ell$, set $\ell = \ell+1$ and go to step 1.
Else go to step 3. □

4.41 Tapia

I would like to mention that Boggs, Tolle and Wang (1979) have proposed an exact penalty function that is similar to the one of Di Pillo and Grippo that is mentioned in Paper 4.3. They have been working with it for several years and have obtained some interesting results.

Moreover the version of Fletcher's (1970b) exact penalty function, mentioned in Paper 4.3, breaks down when the penalty constant is zero. It is possible to overcome this flaw by using a slightly different expression for the multipliers. This alternative choice, which was also used by Fletcher, includes the constraint functions. Specifically I suggest that one uses the multipliers

$$\lambda(x) = [\nabla h(x)^T \nabla h(x)]^{-1}[h(x) - \nabla h(x)^T \nabla f(x)]$$

instead of

$$\lambda(x) = -[\nabla h(x)^T \nabla h(x)]^{-1} \nabla h(x)^T \nabla f(x).$$

4.42 Bertsekas

I became fully aware of the work of Boggs and Tolle in the last few days. I think that there are some substantial differences between their work and that of Di Pillo and Grippo but there are certainly some strong similarities.

Regarding your point on the formula best suited to approximate Lagrange multipliers, I agree that the one I used (for simplicity) in my presentation is not the best. Several better formulae are available, including the one you mention.

4.43 Coope

Although augmented Lagrangian methods are easily programmed, my own experience is that they are not so easily "tuned". Optimal tuning seems to me to be much more problem dependent for augmented Lagrangian methods than for variable metric methods.

When comparing overheads between variable metric and augmented Lagrangian methods it is important to consider the number of constraints present. For example, if there is only one constraint, then the QP subproblem of a variable metric method can be solved in $O(n^2)$ operations. When there are many constraints this figure approaches $O(n^3)$, but then so does the updating of the λ parameters of an augmented Lagrangian method if second order information is used.

4.44 Bertsekas

I agree that it is extremely difficult to implement a mechanism that will automatically provide a starting value of the penalty parameter in an augmented Lagrangian method, and that will work well for all or even most problems. This is probably typical of all methods that use penalty functions.

Your point regarding computational savings in recursive quadratic programming algorithms for problems involving few constraints is well taken. Nevertheless, the required overhead per iteration is substantial and, in my view, for almost every type of problem there will be a threshold value of size beyond which this overhead dominates the computation and makes recursive quadratic programming methods inferior to augmented Lagrangian methods. This is particularly so since augmented Lagrangian minimizations can be solved by conjugate gradient methods, which have only an $O(n)$ computational overhead per iteration.

4.45 Lootsma

It is well known that quadratic exterior penalty functions or ℓ_2 penalty functions are dangerous, because, due to a poor choice of penalty parameter, there may be no finite minimum, and then the run will be stopped prematurely. The non-differentiable ℓ_1 penalty functions obviously have the same difficulties. There is an attractive variant of exterior quadratic penalty functions which avoids these pitfalls. It is the method of moving truncations (Lootsma, 1974), and is based on ideas in Huard's (1967) method of centres and Morrison's (1968) minimization of least squares. Moreover, it is well known how the moving trucations must be controlled to achieve linear or superlinear convergence to the optimum value. May these ideas provide useful extensions to the algorithms of Paper 4.4? Does anybody have experience with such a variant?

4.46 Conn

I am fairly certain that it is possible, but I know of no experience with this variant of the ℓ_1 penalty function.

4.47 Lootsma

It is not easy to achieve certain widely proclaimed advantages of exterior penalty functions. Obviously, one only has to evaluate the *violated* constraints and their gradients, but if the code is organised in such a way that all the problem functions and their gradients are evaluated in one call (which makes it simpler for the user to supply his problem), then the advantage is lost. One always has this conflict between certain computational advantages on one hand, and simplicity of use on the other.

4.48 Conn

I do not see the need to evaluate only the *violated* constraints as a major motivation for the exterior penalty functions, especially if one is considering the exact ℓ_1 penalty function method.

However, I do agree that computational advantages and simplicity of use are often conflicting aims.

4.49 Mangasarian

I wish to point out that, besides the nondifferentiable ℓ_1 exact penalty function, strongly advocated in Paper 4.4, there exist exact differentiable exact penalty functions (Han and Mangasarian, 1981) which when used with special algorithms, such as SOR methods, can handle very large problems.

4.50 Lasdon

Current differentiable exact penalty functions require knowledge of first derivatives of the problem functions in order to evaluate the penalty function. This is a strong disadvantage. Minimizing them by a first derivative method requires second derivatives of the problem functions. Even using them in a line search will involve extra gradient evaluations, which is particularly bad when gradients are approximated by differencing.

4.51 Mangasarian

Differentiable exact penalty functions do need extra derivatives. However, for special problems such as quadratic programs this poses no difficulty. For general nonlinear programs, this additional cost is hopefully compensated by the global differentiability of the penalty function and the finiteness of the penalty parameter.

4.52 Dixon

Although most differentiable exact penalty functions need extra derivatives, it was shown in Dixon (1980) that the Di Pillo—Grippo exact penalty function can be combined with the recursive quadratic programming algorithm, in such a way that global convergence and superlinear convergence properties are retained without using second derivatives.

4.53 Polak

A general method for adjusting exact penalties automatically is described in Polak (1976). It postulates a test function $t_c(.)$ that satisfies the following three conditions: (i) $t_c(.)$ is lower semi-continuous, (ii) if $t_c(x) \leq 0$ and if x satisfies the optimality condition for the unconstrained (penalized) problem, then x satisfies the optimality condition for the original problem, (iii) for every x in $\mathbf{R}^n$ there exist $\hat{\rho} > 0$ and $\hat{c} \geq 0$ such that $c \geq \hat{c}$ and $\|x - \hat{x}\| < \hat{\rho}$ imply $t_c(x) \leq 0$.

For the problem

$$\min \{f(x) \,|\, g(x) = 0\} ,$$

the Fletcher penalty function is

$$f_c(x) = f(x) + \psi(x)^T g(x) + \tfrac{1}{2}c \|g(x)\|^2$$

with $\psi(x) = -\left[\frac{\partial g(x)}{\partial x} \frac{\partial g(x)}{\partial x}^T\right]^{-1} \frac{\partial g(x)}{\partial x} \nabla f(x)$.

The function $t_c(.)$ can be taken to be

$$t_c(x) = -\|\nabla f_c(x)\|^2 + \frac{1}{c}\|g(x)\|^2,$$

and c is increased if $t_c(x) > 0$, i.e. if infeasibility "exceeds nearness to optimality" for min $f_c(x)$. Thus, if we use the Armijo gradient method for min $f_c(x)$, then a suitable algorithm is as follows:

Parameters: α, $\beta \in (0,1)$ and $\{c_j;\ j=0,1,2,\ldots\}$ is any increasing divergent sequence of positive numbers.

Step 0: Set $i = 0$, $j = 0$, $c = c_0$.

Step 1: If $t_c(x_i) > 0$, increase c to the first c_j such that $t_{c_j}(x_i) \le 0$.

Step 2: Set $x_{i+1} = x_i - \beta^k \nabla f_c(x_i)$, with β^k computed by the Armijo rule (which depends on α as in Contribution 4.40). Set $i = i+1$ and go to Step 1. □

It can be shown that if c becomes unbounded then $\{x_i\}$ is also unbounded. Hence, if $\{x_i\}$ remains bounded and if $\tilde{x}$ is an accumulation point of $\{x_i\}$, then $\nabla f_c(\tilde{x})$ and $g(\tilde{x})$ are both zero, so $\tilde{x}$ satisfies the first order optimality condition for the original problem. Of course we need to assume that $\partial g(\tilde{x})/\partial x$ has maximum rank.

4.54 Conn

I suggested in Paper 4.4 that there is no robust, global and superlinearly convergent algorithm with an exact penalty merit function for which sophisticated updating techniques adjust the parameters automatically. The closest to this ideal at present is probably the work that has just been mentioned, but I still believe that we have only a rudimentary idea of suitable parameters and their automatic adjustment.

4.55 Todd

The early subgradient method of Shor, with step sizes of the form $\lambda_k = 1/k$ (Poljak, 1967), while leading to very slow convergence, does allow estimation of dual variables. Unfortunately, step sizes that give faster convergence seem not to yield enough information for such estimates.

4.56 Wolfe

While it is correct that multipliers are not explicitly available when subgradient optimization is used, they are in fact close at hand. As Todd indicated, near convergence one finds a large number of steps whose net displacement is nearly zero; sums of the step length multipliers associated with individual constraints thus yield approximations to the appropriate Lagrange multipliers. More about the determination of multipliers and verification of optimality is discussed in the final section of Held, Wolfe and Crowder (1974).

4.57 Byrd

Exactly how is an initial feasible point computed in a phase 1 iteration of Paper 4.5.

4.58 Lasdon

The phase 1 of GRG2 uses the ℓ_1 norm of the constraint infeasibilities as its objectives. The line search terminates when one or more constraints become satisfied. The algorithm ensures that once a constraint becomes satisfied it stays satisfied, which is all very similar to the "standard" LP phase 1.

4.59 Shanno

GRG methods have been developed in a way that allows for easy and natural handling of linear variables in the objective function. Much of the reported work on other algorithms ignores this possibility, and modifications of current published algorithms may be necessary for many large problems that have a great deal of linearity in the objective function.

4.60 Gill

Such modifications may not be needed in null space methods for constrained optimization, because the size of the reduced Hessian is independent of the way in which the independent variables are defined.

4.61 Dembo

Please comment on strategies for changing the superbasic variables in GRG algorithms.

4.62 Lasdon

The computational experience reported by Shanno and Marsten (1979) for large linearly constrained problems implies that it is usually best to add several promising nonbasics to the superbasic set, at least when conjugate gradient methods are being used.

4.63 Mangasarian

What conditions imply superlinear convergence in GRG?

4.64 Lasdon

Consider the sequence $\{x_k\}$ produced by GRG, where each x_k is the final point in a line search. Assume that (a) $\{x_k\} \to x^*$ where x^* is a solution, and (b) there exists K such that, for all $k \geq K$, the same partitioning of variables in basic, superbasic, and nonbasic is used. Then the original problem is equivalent to that of minimizing the reduced objective function determined by the partitioning in (b). This is an unconstrained problem, and the rate of convergence of GRG will be that of the underlying unconstrained minimizer used. There are several minimizers (e.g. BFGS) which are known to be Q-superlinearly convergent.

4.65 Sargent

If one uses a descent test on the objective function in the GRG algorithm, the questions of whether one has the right active set and whether $\{H_k\}$ converges to $\nabla_{xx}L(\hat{x},\hat{\lambda})$ are irrelevant to superlinear convergence, since the steplength does not necessarily tend to unity, and convergence is then at most linear (Sargent, 1981).

4.66 Powell

Due to the procedure for satisfying the active nonlinear constraints in the GRG method, there may be more noise in the objective function than usual. Is this a serious difficulty in practice?

4.67 Lasdon

The reduced objective in GRG does have more noise than usual since evaluating it requires the numerical solution of a system of nonlinear equations. This implies that care must be taken to ensure that the equations are solved to within a tolerance that is neither too loose nor too tight.

4.68 Dembo

It appears that GRG methods are a good example of algorithms that have to work with noisy functions, one never "solves" the constraints exactly, and hence the reduced objective function value is only known to a limited accuracy which depends on the constraint tolerances. With reference to John Dennis's research seminar on noisy functions, has this caused any difficulties in practice when a quasi-Newton method is used on the reduced problem?

4.69 Lasdon

In our computations we have been able to solve some problems only by tightening the constraint tolerance near the solution. In particular, we set the constraint feasibility tolerance to its default value of 10^{-4} and, when the algorithm terminated, we reduced the tolerance to 10^{-6} and started again. The termination point with tolerance of 10^{-4} was not very close to the solution, while the final point using 10^{-6} was. This suggests that the noise level from 10^{-4} was too large.

4.70 Polak

In Polak (1971) there is a theory for constructing implementable algorithms, which takes into account errors in function as well as other evaluations. In such algorithms the cost does not necessarily decrease monotonically. The precision of approximation must be increased according to some simple tests if convergence is to be ensured. There are several applications of this theory in the literature (see, e.g., Klessig and Polak, 1973; Mukai and Polak, 1976, 1978).

4.71 Dixon

Yamashita (1979) recently proposed an algorithm for nonlinearly constrained approximation that uses a sequence of feasibility tolerances. In his algorithm he introduces two parameters δ_1 and δ_2, both of which decrease in a geometric progression. At each stage the constraints are satisfied within δ_1 and then the function is approximately minimized within δ_2 in this region. One of my students Osman-Hashim (1981) implemented the method, and it seems to be very reliable and efficient.

PART 5

Large Nonlinear Problems

5.1

Algorithms for Very Large Nonlinear Optimization Problems

E.M.L. Beale

Scicon Computer Services Ltd.,
Milton Keynes MK11 3EJ, England.

1. Introduction

We can both formulate and solve optimization problems with several thousand bounded variables and a few thousand other constraints, provided that the variables are continuous, the objective function is linear, and the constraints are both sparse and linear. This is useful, because linear programming models often become large, particularly when they are used to study operations that extend over many locations and time periods.

All is not lost when these simplifying assumptions are relaxed. But for problems with many constraints it seems safe to assume that in practice the constraints will always be expressible in some compact way. Otherwise there must be serious doubts as to whether the mathematical problem being solved does represent the situation that the user has in mind. So it is appropriate to start a discussion of algorithms for large problems by considering formulations that may be useful to model builders. This is done in Section 2. Different algorithmic approaches are considered in the remaining sections.

2. Problem Formulation

The simplest modification to a linear programming problem is to specify that some variables must be integers. This is one of the reasons why integer programming is widely used, even though some formulations cannot be solved economically. (Of course another reason is that integer programming provides a way of modelling the task of finding global optima of certain problems, including ones that involve economies of scale.) Nearly as simple is the inclusion of nonlinear functions of scalar arguments in an otherwise linear, or integer, programming problem; so these *separable* problems will continue to deserve algorithmic attention.

In more general linear programming formulations, it is useful to divide the variables into *linear variables* x and *nonlinear variables* y, such that the problem is linear once the values of the nonlinear variables have been chosen. We let n and s be the number of components of x and y respectively. Therefore the canonical formulation is:

$$\text{maximize} \quad x_0$$

$$\text{subject to } x_0 + \sum_{j=1}^{n} a_{0j}(y)\, x_j = b_0(y),$$

$$\sum_{j=1}^{n} a_{ij}(y)\, x_j = b_i(y), \qquad i = 1,2,\ldots,m,$$

$$0 \le x_j \le \beta_j, \qquad j = 1,2,\ldots,n,$$

$$\ell_k \le y_k \le u_k, \qquad k = 1,2,\ldots,s.$$

Allowing the coefficients a_{ij} to depend on y often greatly reduces the number of variables that have to be classified as nonlinear.

It is useful to distinguish between linear variables with at least one variable coefficient and linear variables whose coefficients are all constant. This is because trial values of both the nonlinear variables and the linear variables with variable coefficients are relevant to the computation of local linear or quadratic approximations. These two sets of variables may therefore be called the *relevant variables*.

Within this general formulation, there are two alternative approaches for specifying the optimization calculation. The choice between them is fundamental to the design of optimization software, and it also influences the choice of algorithm. These approaches can be described as *local problem definition* and *global problem definition*.

Local definition means that the user supplies a routine to compute the values, and perhaps the derivatives, of the objective and constraint functions for any given set of trial values of the relevant variables. The standard optimization software then interacts with this routine as often as necessary. This is how hill-climbing software usually works.

Global definition means that the user supplies a complete specification of the problem, so that the standard optimization software can continue without interruption until the problem is solved. This is how linear programming software works.

Both approaches are useful. Local definition is more flexible. Global definition tends to produce more efficient software, and in particular is more attractive when seeking a global optimum in the presence of integer variables and other nonconvexities.

The distinction between these approaches is not clear-cut when applied to algorithms. For a restricted class of calculations, one can build what would otherwise be a user routine into the standard software, leaving the user to supply only the data for this routine with the rest of the problem specification; but the interface may be cumbersome. Algorithms that are developed as extensions of hill-climbing algorithms tend to be suited to local definition, while algorithms developed as extensions of the simplex method for linear programming tend to be suited to global definition.

In algorithms for software using local definition, the precise meaning of an iteration is important. When developing general algorithms for unconstrained optimization, we usually think of function and derivative calculations as potentially much more laborious than any computing done within the optimization routine itself. This seems to be a reasonable assumption when developing general purpose software for constrained optimization. Therefore one may define the number of iterations to be the number of sets of function evaluations, and it may be of little consequence if the routine takes 3000 simplex steps between iterations.

However, on most test examples, and on many real problems, 3000 simplex steps take much more effort than one set of function and derivative evaluations. Therefore we certainly need algorithms that take more than the minimum number of iterations to solve problems, in order to avoid excessive work within each iteration. If function evaluations are really simple, it may be best to integrate them into the optimization software, or in other words to use global problem definition.

3. Algorithms for Globally Defined Problems

As noted earlier, global definition provides the possibility of finding global optima to nonconvex problems. In Paper 1.1, Schnabel refers to random search methods and to tunnelling methods for such problems. These methods can be applied to large constrained problems, but a *branch and bound* approach seems more systematic. Such an approach is discussed by Beale and Forrest (1976) and by McCormick (1976). McCormick coins the term *factorable* for problems whose nonlinear functions can be expressed in terms of constraints that are linear except for sums of nonlinear functions of single arguments. Beale and Forrest show how such functions can be handled within a production mathematical programming system, using *special ordered sets* (due originally to Beale and Tomlin, 1970) and *automatic interpolation*. *Chains of linked ordered sets* (Beale, 1980), which are a useful extension of special ordered sets, deal more directly with product terms; an earlier version of chains of linked ordered sets is described by Beale (1978) and by Beale and Forrest (1978).

Separable programming, special ordered sets, and their extensions need only minor enhancements to IBM's standard *MPS Input Formats* for linear programming. This is because the optimization calculations have the formal structure of linear programming problems, perhaps using collective definitions for sets of variables that represent all possible members of separable sets, and, for nonconvex problems, there are restrictions on the combinations of variables that may be nonzero. The user may need to do some algebra to express his class of problems in this form, and he must then write a *matrix generator* to generate any specific problem from his input data. If specialists in unconstrained optimization regard this as an unreasonable imposition on the user, they should note that similar work is needed with linear programming models. Large problems cannot be solved efficiently without some effort on the part of the user.

Such separable problems are normally solved without considering the derivatives of the Lagrangian explicitly. This is a simplifying feature; but it limits the attainable accuracy, even with automatic interpolation, unless one regains the accuracy by using multiple precision arithmetic. Therefore, if all the nonlinear functions can be differentiated analytically, it might be expedient to add the constraints that the Kuhn–Tucker conditions be satisfied, even though these conditions are technically redundant.

For example, if we have the problem:

$$\text{maximize} \quad x_0$$

$$\text{subject to } x_0 + \sum_{k=1}^{s} f_{0k}(z_k) + \sum_{j=1}^{n} a_{0j}\, x_j = b_0,$$

$$\sum_{k=1}^{s} f_{ik}(z_k) + \sum_{j=1}^{n} a_{ij} x_j = b_i, \quad i = 1,2,\ldots,m,$$

$$0 \leq x_j \leq \beta_j, \quad j = 1,2,\ldots,n,$$

$$\ell_k \leq z_k \leq u_k, \quad k = 1,2,\ldots,s,$$

then, to avoid what is otherwise a near dual degeneracy, one might add a constraint of the form

$$\nabla f_{0k}(z_k) + \sum_{i=1}^{m} \pi_i \nabla f_{ik}(z_k) = 0,$$

for each k such that z_k is an interior point of $[\ell_k, u_k]$. To apply separable programming to this calculation, for each k one makes a partition of the interval $[\ell_k, u_k]$ of the form

$$\ell_k = Z_{k0} < Z_{k1} < \ldots < Z_{kt} = u_k,$$

and one introduces non-negative variables $\{p_{k\ell}: k = 1,2,\ldots,s; \ell = 0,1,\ldots,t\}$ representing the "probability" that $z_k = Z_{k\ell}$. These probabilities are sometimes denoted by $\lambda_{k\ell}$. The standard formulation is then to maximize x_0 subject to the constraints

$$x_0 + \sum_{k=1}^{s} \sum_{\ell=0}^{t} p_{k\ell} f_{0k}(Z_{k\ell}) + \sum_{j=1}^{n} a_{0j} x_j = b_0,$$

$$\sum_{k=1}^{s} \sum_{\ell=0}^{t} p_{k\ell} f_{ik}(Z_{k\ell}) + \sum_{j=1}^{n} a_{ij} x_j = b_i, \quad i = 1,2,\ldots,m,$$

$$\sum_{\ell=0}^{t} p_{k\ell} = 1, \quad k = 1,2,\ldots,s,$$

$$\sum_{\ell=0}^{t} p_{k\ell} Z_{k\ell} - z_k = 0, \quad k = 1,2,\ldots,s,$$

$$0 \leq x_j \leq \beta_j, \quad j = 1,2,\ldots,n.$$

The proposed additional constraints then have the form

$$\sum_{\ell=0}^{t} p_{k\ell} \{\nabla f_{0k}(Z_{k\ell}) + \sum_{i=1}^{m} \pi_i \nabla f_{ik}(Z_{k\ell})\} = 0,$$

but the best way to introduce them is not clear. Do we treat the multipliers $\{\pi_i; i = 1,2,...,m\}$ as constants, determined from the rest of the problem, or do we make them decision variables, defined by further constraints specifying that the reduced costs of all basic variables must vanish?

Note that, for each k, we must in any case avoid giving non-zero values to nonadjacent probabilities in the set $\{p_{k\ell};\ \ell = 0,1,...,t\}$, in order to ensure that $\Sigma_\ell\, p_{k\ell}\, f_{ik}(Z_{k\ell})$ is an adequate approximation to $f_{ik}(z_k)$. Special ordered sets use branch and bound methods to enforce this.

4. Algorithms for Locally Defined Problems

Most of the nonlinear programming problems that I have had to solve were nonlinear because they included economies of scale. Such problems often have many local optima, and also they often contain integer variables. Therefore I have used the methods outlined in the previous section more often than hill-climbing methods based on local definition.

However, hill-climbing methods are important. Many problems are naturally formulated as linear problems because there are no precise data on any nonlinear effects, but some formulations include product terms. In particular we may need to determine both the composition and the distribution of a stream of material. Specifically, if x_i denotes the total quantity of material i in the stream, and if y_d denotes the proportion of the stream assigned to destination d, then the amount of material i reaching destination d is $x_i y_d$. In this case, the x_i are normally treated as linear variables and the y_d as nonlinear variables.

Some multi-time period problems are such that the coefficients in the equations for one time period depend on the decisions made in earlier time periods. In oil production problems, for example, the well productivities in a reservoir depend on how much oil has already been produced from the reservoir, and on how much water or gas has been injected to maintain pressure. Here the number of wells operating in different modes may be linear variables, but their coefficients in the expressions for production capacity are functions of the nonlinear variables.

In any primal approach to the nonlinear programming problem, the variables, including the slack variables of any inequality constraints, can be divided into *basic*, *independent* (or *superbasic*) and *other nonbasic* (i.e. at their lower or upper bounds). The problem then reduces to an embedded unconstrained problem in the space of the independent variables, but we must be prepared to change the set of independent variables if necessary. Note that most algorithms can be organized so that no linear variable is ever independent.

There are four main approaches to this problem:

1. Some form of *successive quadratic programming* (SQP).

2. *Successive linear programming* (SLP). This approach was pioneered by Griffith and Stewart (1961) under the title *method of approximation programming* (MAP).

3. A *reduced gradient* (RG) approach.

4. *Reduced gradient approximation programming* (RGAP), which is a hybrid between SLP and RG.

SQP is theoretically attractive. The need to store and manipulate information about the Hessian of the Lagrangian is an obvious difficulty on large problems, but, as Dembo shows in Paper 6.3, it is not necessarily insuperable.

SLP is more generally applicable. It is based on local linear approximations to the objective and constraint functions. Tolerances are imposed on the permitted deviations of the nonlinear variables from their current trial values. Palacios-Gomez, Lasdon and Engquist (1981) have found that with nonlinear constraints one must work with a composite objective function (an exact ℓ_1 penalty function), which is a weighted sum of the objective function and the constraint violations. If violations of the linearized constraints are regarded as overwhelmingly more significant than improvements in the objective function, then SLP is forced to take very small steps. The algorithm then converges slowly, and may fail to reach a true local optimum. However, with a composite objective function, Palacios-Gomez, Lasdon and Engquist have achieved good results that are similar to those reported by industrial practitioners.

Reduced gradient methods have proved very effective, as indicated in Lasdon's Paper 4.5. The following two aspects of these methods deserve further study.

One is the exploitation of the presence of linear variables, which could make reduced gradient methods both more efficient and more like other methods for large problems. Each iteration could be divided into two phases. In the first phase one adjusts the linear variables to their optimal values for the current trial values of the nonlinear variables, which is done by taking simplex steps. The only unusual feature is that all nonlinear variables, whether basic, independent or other nonbasic, must be regarded as having temporary lower and upper bounds equal to their current trial values. So any basic nonlinear variable must be made independent, rather than have its trial value changed, even though this requires the introduction of a basic linear variable at its lower or upper bound. The second phase of the iteration seeks better trial values of the independent variables. Having chosen a search direction, one may have to make several simplex steps to remove linear variables at their bounds from the basis, but eventually one should be able to make a genuine improvement. This suggestion may sound complicated, but in practice one could expect the first phase to be vacuous most of the time, and the fact that no linear variable need ever be independent should be a real advantage in many calculations.

The other aspect of reduced gradient methods that deserves further study is the treatment of nonlinear constraints. The original proposal of Abadie and Carpentier (1969) involves another type of two phase iteration, in which one first moves in the tangent plane of the active constraints, and one then moves in an orthogonal direction to satisfy these constraints again. This is awkward, at least theoretically, and the approach pioneered by Murtagh and Saunders (1980) in their code MINOS/AUGMENTED may be better. This code performs major iterations, each consisting of several minor iterations. Each major iteration starts with trial values of both the decision variables and the Lagrange multipliers. The problem is then approximated by linearizing the constraints, and one works with an objective function that is the sum of three parts: (1) the original objective function, (2) the deviations of the true constraint functions from their linear approximations, each multiplied by the trial values of their Lagrange multipliers, and (3) the sum of the squares of these deviations, multiplied by a suitable (Levenberg) parameter. This problem is then solved using MINOS (a reduced gradient program for linear constraints), and its steps are regarded as minor iterations. Because the third term in the MINOS/AUGMENTED objective function makes the subproblem nonlinear in any variable with a nonconstant coefficient in any constraint, a code that fully exploits the presence of linear variables might require a modification of this term.

RGAP is an approach that concentrates on the embedded unconstrained problem. In principle, a function evaluation in this problem requires the solution of a conditional nonlinear programming subproblem, but in practice, since this subproblem has no independent variables, one can hope to solve it very rapidly by SLP; in current implementations its solution is approximated by making only one iteration of SLP. The gradient components for the embedded problem are derived from the shadow prices and the derivatives of the objective and constraint functions with respect to the independent variables. More details are given in Batchelor and Beale (1980) and in Beale (1978).

Note that monotonic progress with a "merit function" is not needed in SLP or RGAP. If the approximate objective function does not improve, one restarts with smaller tolerances.

The local convergence rates of these methods are that SQP should be quadratically convergent, reduced gradient methods and RGAP can be made superlinear, while SLP is at best R-linear unless there are no independent variables. However, for large practical problems, dynamic scaling of the variables may be more important than good local convergence properties. SLP lends itself very naturally to this dynamic scaling, for the scaling is an integral part of the method. Dynamic scaling can also be applied to reduced gradient and RGAP methods, but not so obviously to SQP methods.

The importance of dynamic scaling arises from the fact that most large problems are "weakly connected" in the following sense. The decision variables can in principle be divided into s sets such that, if one fixes the variables in (s-1) of the sets at sensible feasible values, then the conditionally optimal values of the variables in the remaining set will not depend "much" on which values are chosen for the fixed variables. If these s sets could be identified, then we could solve just s small subproblems instead of one large one.

In practice we cannot do this, but we do want our algorithms to solve the s subproblems in parallel and not in series. How do we ensure this?

Part of the answer is to scale the variables properly. It seems that this must be done dynamically, unless we use a scale invariant algorithm such as SQP with computed Hessians. Present dynamic scaling algorithms are much more primitive than Davidon's variable metric method, but they can be far more efficient when the number of variables is very large. It is suitable to apply dynamic scaling when the algorithm restarts its quasi-Newton or conjugate gradient procedure, and a good practical technique is to scale up variables whose gradient components have had a constant sign over many iterations, and to scale down variables whose gradient components have been oscillating.

5. Algorithms Using Orthogonal Rotations

Gill and Murray (1977a) advocate the use of methods based on orthogonal rotations for linear and quadratic programming. The extent to which such methods should be used is debatable, but there is no doubt that orthogonal rotations can be helpful on some very large optimization problems.

A particular instance arises if one has a minimization problem with linear constraints and a convex objective function whose Hessian is exactly or approximately diagonal and computable. We can now scale the variables so that the Hessian becomes (exactly or approximately) the unit matrix. An advantage of applying orthogonal rotations to the variables when the Hessian is the unit matrix is that this nice property is maintained. Specifically, if the objective function is $F(x) = c^T x + \frac{1}{2}x^T x$, if the active constraints are $Ax = b$, and if we write $Qx = y$, where Q is any orthogonal matrix, then the objective function and active constraints become $F(x) = F(Q^T y) = c^T Q^T y + \frac{1}{2}y^T y$ and $AQ^T y = b$ respectively. Now Q can be chosen so that AQ^T is lower triangular. In this case it is convenient to use the constraints to fix the first m components of y, where m is the number of active constraints. The unconstrained problem in the remaining variables can be solved very easily by a gradient method, or by a conjugate gradient method if the objective function is not truly quadratic. Adding constraints to the active set is easy. The best way to remove an active constraint may be to restart the process of forming Q.

5.2

The Implications of Modelling Systems on Large Scale Nonlinear Optimization Codes *

Arne Drud

Development Research Center, The World Bank,
Washington, DC 20433, USA.

Research in large scale nonlinear optimization is directed towards creating more and more efficient software for large nonlinear optimization problems. A reasonable starting point for a discussion of large scale nonlinear optimization is therefore a discussion of costs and efficiency and their relations to the overall modelling work. A modelling exercise will usually consist of data collection, model formulation, generation of input to the solution algorithm, optimization of the model, and generation of various reports. When we consider costs we must think of both manpower and computer costs of all these steps of the modelling exercise.

When discussing large scale optimization, we must first realize that almost all large models have some sort of structure. A 1000-equation model will almost certainly have less than 20 "types" of equations or macro-equations. Also, the data used in a large model will usually consist of relatively small compact tables. Thus, using conventional algebraic notation, it is often possible to define a large model using little space. However,

*The views and interpretations in this document are those of the author and should not be attributed to the World Bank, to its affiliated organizations or to any individual acting in their behalf.

difficulties in the practical world arise when the optimization algorithms require the model description in a different format, usually dependent on the algorithm and the internal data structure used in the optimizer.

It is generally possible to save computer time in the optimization phase by supplying more detailed information, e.g., first and/or second partial derivatives, sparsity patterns, etc. However, there are costs associated with the generation of this detailed information. In small scale nonlinear programming, it has usually been assumed that the user punches the input to the solution algorithm and, if we use the criterion that a piece of input is useful only when the costs of generating and handling it are smaller than the savings in the optimization phase, we often arrive at the conclusion that we should cut down on the input to the optimization algorithm. In large scale optimization the trade-off is different. Because of storage limitations, it may be essential to use and therefore supply some type of structural information, and the question is more how detailed should the information be? No matter what degree of detail we choose, it is simply impossible to punch all input by hand, so the input to the optimizer must be generated by a computer program that transforms the compact model and raw data into the data structure used by the optimizer. The computer program can either be a specially written program, or it can be a general purpose program that uses an easy description of the model and the data as input. In large scale LP, several general purpose systems (called matrix generators) exist. Therefore model dependent matrix generators are now only used under special circumstances. Unfortunately, most existing matrix generators are designed for the data structures of commercial LP systems, and they cannot be changed to handle nonlinear models. However, the idea of general systems that can generate input to the optimization algorithms automatically is very appealing, and the feasibility of such "modelling systems" is currently being proven by some experimental studies.

A modelling system can be defined as a computer system that stands between the modeller and one or more solution algorithms. From the users' point of view, the modelling system makes it easy to formulate the model using conventional algebraic notation. Thus the model is entered into the computer in a compact format that is natural to the modeller, and the modelling system takes care of all communication with the solution algorithms. To make model formulation easy, a powerful modelling system should have good facilities for handling sets and set operations, mappings and partitionings, it should have a module that handles input and preprocessing of the input tables, and it should have some report writing facilities.

GAMS (General Algebraic Modelling System) is currently being developed by Alexander Meeraus in the World Bank along these lines. An important feature of the system is that the model formulation is a general algebraic formulation, so it is not specialized towards any particular solution algorithm. Transformations of a model into a particular format required by a solution algorithm are made by the system. Currently, GAMS can interface with various LP systems, and an interface with Lasdon's GRG2 code is being prepared.

The main implication of using a modelling system is that the costs of generating input, and especially of generating additional detailed input, will become much smaller. We will therefore have to rethink our definitions of useful input to large scale nonlinear optimization codes. The research areas I will suggest are therefore all of the form: for a given piece of information about the optimization model, how can we use this information in a particular optimization method; can we revise the method or develop new methods that use the information more efficiently; what are good ways of representing the information; what are the potential savings from using it; and what are the extra costs of generating, storing, retrieving and manipulating the information? To make the issues clearer, let me with a few sample examples describe how different pieces of information can be analyzed in the context of a large scale generalized reduced gradient (GRG) algorithm.

Sparsity information In a GRG algorithm, because all the matrix operations on the Jacobian are similar to operations performed in a linear programming code, the same sparse matrix algorithms can be used directly. However, we can use the sparse matrix information more efficiently, by changing the algorithm slightly in the part where we choose the basic variables, in order that the basis matrix becomes closer to being lower triangular. The savings in the optimization code from a sparse matrix implementation are evident. It will allow us to work with problems of a size we could not handle in a fully allocated code, and the computer costs will be decreased dramatically for large problems. The generation of the sparsity pattern is a cheap and straightforward matter for any modelling system, and we already know from linear programming how to manipulate the information.

Linearity information It is always valuable to know if a constraint is linear, since a first order approximation to a linear function is equal to the function itself. In the GRG algorithm, we maintain feasibility during the search by performing Newton iterations. Thus linear constraints will always be satisfied, so we need not re-evaluate their functions. But when is a function linear? If it contains some linear terms plus a nonlinear term, but all variables in the nonlinear term are temporarily fixed, the function is in practice linear. Also a term like $x \cdot g(y)$ is linear in x as long as y does not change. If we define linearity of a function as a dynamic property that depends on which variables are temporarily fixed, then most models will be much more linear than with a static definition. We could even think of using a feedback from the linearity information in the decision on which variables to keep temporarily fixed, in order to make the number of linearities large. We can certainly generate the necessary information, both on static and on dynamic linearity, quite cheaply. The static linearity information is easy to handle and utilize. But can the dynamic linearity information be handled efficiently, and, if so, are the extra savings larger than the extra administrative costs?

Exact derivatives It is straightforward to automate the generation of partial derivatives. We can either use point derivatives, whose numerical values are computed by using the chain rule directly on the expressions that occur, or we can generate the actual analytic expressions for the derivatives. Since derivatives can be made available easily, we should consider if they are useful in a GRG context. Of course first partial derivatives are very helpful, but can we utilize the second derivatives of the constraint functions, how can they be stored, and what are the potential savings? Further, will third order information ever be useful?

Information on the actual functional form In the GRG algorithm, we reduce the problem in the same way as in the simplex algorithm: m basic variables are adjusted to satisfy m equality constraints, and local changes to the remaining variables are subject only to simple lower and upper bounds. Each time the nonbasic variables are changed, the values of the basic variables are computed, so that they can be used in the evaluation of the objective function and so that their bounds can be tested. In linear programming we have analytic expressions for the basic variables, but in the GRG algorithm we compute the values of the basic variables numerically with a Newton-type method. As long as the nonlinearities are represented by a subroutine that works like a black box from the point of view of the optimization code, one cannot do better. However, in most large models, if the values of s(<m) basic variables, called "spike variables", are fixed, then the remaining m-s basic variables can be determined analytically from m-s of the constraint equations. Thus m-s basic variables are eliminated from the calculation, and s constraint equations remain to define the spike variables. In order to use such an approach, the optimization code has to know the variables that can be eliminated analytically by each constraint, and it has to be able to perform the elimination. The transformations must be done rapidly as changes of basis will require some equations to be solved with respect to other variables. This type of optimization approach opens some interesting joint research with researchers in computer science, especially in compiler construction and in the manipulation of symbolic and algebraic formulae.

The advantages of available information are clear in the first two examples, but in the last two they are doubtful, mainly because we do not know how the information can be handled and utilized. But the potential benefits seem to be large, and these are certainly interesting research topics.

So far, I have written about modelling systems as if they can already do all we want them to. This is not the case. Research in the design of modelling systems will be an important area in the next decade. I will describe a few selected research areas in order to show that this research should be of interest to optimization code developers.

Selection of optimization algorithms In a modelling system, we can easily distinguish between linear programming problems, quadratic programming problems, problems with linear constraints and a general nonlinear objective function, and general nonlinear programming problems, and, within each of these groups, we can distinguish between small and large problems. Based on this information, we can choose an optimization code. As the number of reliable optimization codes increases, we will probably find that even within the groups mentioned above some codes are better for some types of problems and other codes are better for other types of problems. Since most users will not know how to choose the best solver, the modelling system should be able to choose one automatically. Therefore we need to build a set of decision rules that are based on easily recognized characteristics of the model. This seems to be an interesting area for the use of adaptive learning procedures.

Automatic reformulation of models Most optimization problems can be formulated in different ways, and the ease with which they can be solved will often depend on the formulation. We should investigate the characterization of good formulations (depending on the optimization code) and, if we are successful, we should try to develop automatic procedures for reformulating the models. A simple example we worked with was a model where both x_i and $\log(x_i)$ appeared many times. Almost all equations were nonlinear due to the logarithms. After $\log(x_i)$ was replaced by y_i, and after the equations $y_i = \log(x_i)$ were added, almost all of the old constraints became linear, and the nonlinearities became very simple. The model grew in size but it was easier to solve.

Utilization of macro structure The sparse matrix techniques mentioned above utilize the micro structure of the model. But in a modelling system it is easy to detect the macro structure, and we should consider ways of using it, e.g. in different types of decomposition algorithms. Little work has been done in this area, but combined research on structure detection and decomposition may prove useful for very large models.

Automatic generation of special purpose algorithms If we get a good handle on how to characterize models, and on identifying the algorithm types that are good for various problem types, we can extend the idea of the automatic selection of an algorithm and start thinking about tailoring a special purpose algorithm to each model, i.e. we build an algorithm from a set of building blocks selected to suit the exact characteristics of the model. This route is very different from the main route today where we try to develop general purpose algorithms that can solve almost any problem. But special purpose codes may often be much faster, because the large overheads of a general purpose code can be omitted.

Concluding, the comments and examples above are intended to show that one of the very important research areas in large scale nonlinear optimization is the model formulation and modelling system area, and its implications for the type of input the optimization codes are able to accept. Research is necessary both on how to use the more detailed model information more efficiently and on how to generate suitable model information automatically.

5.3

On the Unconstrained Optimization of Partially Separable Functions

A. Griewank* and Ph.L. Toint[†]

Department of Applied Mathematics and Theoretical Physics,
University of Cambridge,
Silver Street, Cambridge CB3 9EW, UK.

[†]Department of Mathematics,
Facultés Universitaires de Namur,
Namur, Belgium.

1. Introduction

We consider the problem of minimizing a smooth objective function f of n real variables. For $n > 200$ we can only hope to locate a local minimum of f within the usual limitations on storage and computing time by using a minimization algorithm that exploits some special structure of f. One such possibility is that the Hessian $G(x)$ of $f(x)$ has clustered eigenvalues at a minimizer x^*, in which case conjugate gradient and limited memory variable metric methods were found to work quite well (Gill and Murray, 1979). However, in general, the performance of these methods is rather unpredictable since, except for certain test functions, the eigenvalue structure of G at or near x^* is usually not known. Therefore we pursue the traditional approach of approximating f by local quadratic models, which is computationally feasible even for large n if f has a certain separability

*The work of A. Griewank was fully supported by a research grant of the Deutsche Forschungsgemeinschaft, Bonn.

structure. This structure is always implied by sparsity of G, and depends only on the way in which the components of x enter into f, and not on the numerical values of f or its derivatives.

2. Partial Separability of $f \in C^n(\mathbf{R}^n)$

Let E_0 denote the vertex set of the unit cube in $\mathbf{R}^n$ and, for any $e = (\varepsilon_1, \varepsilon_2, \ldots, \varepsilon_n)^T \in E_0$, define

$$|e| \equiv \sum_{i=1}^{n} \varepsilon_i,$$

which is an integer in [0,n]. For any $x = (\xi_1, \xi_2, \ldots, \xi_n)^T \in \mathbf{R}^n$ and for any $e \in E_0$, let x_e be the vector

$$x_e = (\varepsilon_1 \xi_1, \ \varepsilon_2 \xi_2, \ldots, \varepsilon_n \xi_n). \tag{1}$$

Thus the points $\{x_e; \ e \in E_0\}$ are the vertices of the rectangular cube

$$Q(x) = \{(\eta_1 \xi_1, \ \eta_2 \xi_2, \ldots, \eta_n \xi_n)^T : 0 \leq \eta_i \leq 1, \ i = 1,2,\ldots,n\}.$$

As a consequence of Taylor's theorem it can be shown that the identity

$$\frac{\partial^n f(x)}{\partial \xi_1 \ \partial \xi_2 \ \cdots \ \partial \xi_n} = 0, \quad \text{all } x \in \mathbf{R}^n, \tag{2}$$

holds if and only if

$$\sum_{e \in E_0} (-1)^{|e|} f(x_e) = 0, \quad \text{all } x \in \mathbf{R}^n. \tag{3}$$

Any function for which these equivalent conditions hold will be called partially separable, because (3) can be rewritten as

$$f(x) = (-1)^{n+1} \sum_{k=0}^{n-1} (-1)^k \sum_{e \in E_0, |e|=k} f(x_e). \tag{4}$$

Thus we have expressed the value of the function f at x in terms of its values at the $2^n - 1$ other vertices x_e of $Q(x)$, all of which have at least one zero component.

In order to refine this additive decomposition of f, we consider all mixed partial derivatives of the form

$$D^e f(x) = \frac{\partial^{|e|} f(x)}{\partial^{\varepsilon_1}\xi_1 \, \partial^{\varepsilon_2}\xi_2 \, \ldots \, \partial^{\varepsilon_n}\xi_n}, \quad \text{where } e \in E_0.$$

For each e such that $D^e f \equiv 0$, the term $f(x_e)$ can be expressed as a linear combination of the terms $\{f(x_{\hat{e}});\ \hat{e} \in E_e\}$ where E_e is the set $\{\hat{e} : \hat{e} \leq e$ componentwise and $\hat{e} \neq e\}$. Further, if $\hat{e} \leq e$ componentwise, then the function $f(x_e) \pm f(x_{\hat{e}})$ can be expressed as a function of x_e. Thus we may reduce equation (4) to an identity of the form

$$f(x) = \sum_{e \in E} f_e(x), \tag{5}$$

where the following conditions are satisfied. The set E is the "subset of maximal elements of E_0", which means that $D^e f \not\equiv 0$ for all $e \in E$, but, if $\bar{e} \in E_0$, if $\bar{e} \geq e$, and if $\bar{e} \neq e$, then $D^{\bar{e}} f \equiv 0$. Further, the functions $f_e(x)$ are in $C^n(\mathbf{R}^n)$, and they satisfy the conditions

$$\partial f_e / \partial \xi_i \equiv 0 \Leftrightarrow \varepsilon_i = 0 \tag{6}$$

and

$$D^e f_e \equiv D^e f, \tag{7}$$

because $f_e(x)$ depends just on the nonzero components of x_e. The decomposition (5) is not unique, but any two functions f whose sets E are identical will be said to have the same separability. Assuming that no component of the gradient $g = \nabla f$ vanishes identically, we find $E \neq \emptyset$ and, for any $v \in \mathbf{R}^n$,

$$v_e = 0 \text{ for all } e \in E \Rightarrow v \equiv 0, \tag{8}$$

where v_e is defined by analogy to (1), and where E is still the subset of maximal elements.

In many applications a decomposition of the form (5) arises naturally, for example through the use of finite elements or other discretizations of variational problems. In particular we note that a user can only know that the Hessian G is sparse, if f is given in the form (5), where certain pairs of variables do not occur nontrivially in the same element function. As a measure of separability we use the depth

$$d = \max\{|e| : e \in E\} \in [1,n],$$

which equals the highest order of any nonvanishing mixed partial derivative of f. If d = 1 we have separability in the traditional sense, and if d = n the function f is not partially separable at all. The depth d is always less than or equal to the maximal number of nonzero entries in any row or column of G, which is often used as a measure of sparsity.

3. Restricted Transformation Invariance

Let $f \in C(\mathbf{R}^n)$ be fixed, and let $t : \mathbf{R}^n \to \mathbf{R}^n$ be a nonsingular affine transformation (i.e. the sum of a translation and a nonsingular linear transformation). A minimization algorithm is said to be invariant with respect to the transformation t on the domain of f if, when applied to f and to the composition $f \circ t$, it generates iterates $x^{(j)}$ and $y^{(j)}$ respectively such that

$$x^{(j)} = t(y^{(j)}) \quad \text{for all } j \geq 0,$$

provided that $x^{(0)} = t(y^{(0)})$ and that all other initial values are suitably adjusted. Then the corresponding sequences of function values are identical, so, as far as the particular algorithm is concerned, the problem of minimizing the transformed objective function $f \circ t$ is essentially the same as that of minimizing the original f.

Newton's method and the quasi-Newton schemes in the Broyden family are, for any f, invariant with respect to all t (see e.g. Powell, 1971). Such uniformly affine invariant methods can be expected to perform reasonably well on f if at least one of the compositions $f \circ t$ is a "nice", well-scaled function. On the other hand, it is quite clear that any method which is able to exploit special structure of f can only be invariant with respect to those t for which $f \circ t$ has the same structure as f.

If f is partially separable, a general affine transformation will immediately destroy the identity (2), unless f happens to be a polynomial of degree less than n. However, nothing is lost if a method that can exploit partial separability reduces in the general case $d = n$ to a good uniformly affine invariant method. Then the former method can be expected to perform always at least as well as the latter. Therefore a method that exploits partial separability cannot and should not be uniformly affine invariant. It can be easily seen that f and $f \circ t$ have the same separability (i.e. the same E), if the linear part of t is block-diagonal, where the i-th and j-th variables belong to the same block only if the i-th and j-th components of e are identical for all $e \in E$. In other words, t is only allowed to mix variables which occur either together or not at all in the element functions f_e. Hence we will attempt to find partial separability exploiting methods, which are invariant with respect to this restricted class of transformations, and which reduce to a uniformly affine invariant scheme in the general case $d = n$.

4. Algorithmic Consequences of Partial Separability

Because the decomposition (5) arises naturally, we can ask the user to supply, for each $e \in E$, the nonzero entries of the gradient $g_e = \nabla f_e$, and possibly also the Hessian $G_e = \nabla^2 f_e$, as a function of the $|e|$ variables on which f_e depends nontrivially. Often several f_e have the same algebraic form, and differ only in the values of certain parameters and the assignment between the internal variables and the components of x. This is certainly the case in finite element applications where the number of structurally different elements is usually quite small. Then the user may only have to code a few subroutines for evaluating $\{g_e;\ e \in E\}$, and possibly $\{G_e;\ e \in E\}$, and to supply the index vectors $e \in E$ in a suitable form. The detection of sparsity in G and the corresponding storage allocation, as well as the accumulation of g and G from the values of the g_e and G_e, can be left to the minimization routine.

If there are constants $\{c_e > 0;\ e \in E\}$, such that g_e and G_e can be evaluated in $|e|c_e$ and $\frac{1}{2}|e|(|e| + 1)c_e$ arithmetic operations respectively, then the ratio between the work w(G) and w(g) to compute G and g is bounded above by $w(G)/w(g) \leq \frac{1}{2}(d+1)$, where d is still the depth of the separability. Hence Newton's method is always an attractive proposition if d is small, which may be the case even if G is nonsparse. If some G_e are not explicitly available, they can be approximated by differencing using $|e|$ additional gradient evaluations, so $w(G)/w(g)$ is still not greater than d. In general, this is the cheapest way of computing G by finite differences.

Quasi-Newton methods calculate a step s from the linear system

$$Gs = (\sum_{e \in E} G_e)s = -g. \tag{9}$$

The system can be solved approximately by iterative schemes like SSOR or CG without the need to sum the G_e to obtain G. This is also true if the G_e are replaced by approximations B_e.

Another important consequence of partial separability is that steps that alter only a few components of x affect only some of the element functions f_e, so that the values of f and its derivatives at the new point can be obtained comparatively cheaply. Such sparse steps are particularly useful far from any minimizer, when the local quadratic model is indefinite or otherwise unreliable. By tearing or nested dissection (George, 1977), one might also construct a hierarchy of smaller weakly connected subproblems, and treat them to a certain extent separately. However, if R-superlinear convergence is to be achieved, one occasionally has to take a linking step that alters all variables simultaneously. Apparently this approach has so far not been developed in the case of nonquadratic optimization.

The well known methods in the Broyden class, namely BFGS and DFP, converge Q-superlinearly after some $O(n)$ steps, require $[\frac{1}{2}n^2 + O(n)]$ storage locations, and involve $O(n^2)$ arithmetic operations per step. Some of these three characteristics are likely to be unacceptable if n is large. So far all attempts to overcome these disadvantages by sparsity exploiting quasi-Newton methods (e.g. Toint, 1977; Shanno, 1980) have resulted in the loss of three other nice features of BFGS and DFP with suitable line searches. These are: (1) finite termination on quadratics, (2) guaranteed positive definiteness of the matrix B that approximates $G = \nabla^2 f$, and (3) uniform affine invariance as discussed in Section 3. The following approach to exploit partial separability promises some improvement over earlier methods, but it does not recover all the nice properties of BFGS and DFP.

5. Partitioned Quasi-Newton Updating

Instead of approximating G directly, we suggest storing approximations $B_e \approx G_e$ for all $e \in E$. Because of (6) and (7), each B_e is required to have the same sparsity as ee^T so that the nonzero entries of B_e, like those of G_e, form a principal minor of order $|e|$. After an arbitrary step from x to $x^+ = x + s$, the

B_e are updated to new approximations B_e^+ such that, for all $e \in E$,

$$B_e^+ s = B_e^+ s_e = y_e = g_e(x^+) - g_e(x), \tag{10}$$

which implies the overall secant condition

$$B^+ s = (\sum_{e \in E} B_e^+) s = y = g(x^+) - g(x). \tag{11}$$

Here the projected steps s_e are defined by analogy to (1). If $s_e = 0$ then we may set $B_e^+ = B_e$, but by (8) some of the vectors $\{s_e;\ e \in E\}$ are nonzero. Clearly the collection of equations (10) gives much more second order information than the single condition (11), which forms the basis of published standard and sparse updating formulae for quasi-Newton methods.

Since the nontrivial $|e| \times |e|$ minors of the matrices B_e are dense and symmetric, one can in principal apply any conventional rank two update to satisfy (10). However, in contrast to the usual situation, the projections s_e of the quasi-Newton step $s = -B^{-1}g$ usually depend not only on g_e, B_e and f_e, but also on all other element functions. This has the consequence that, even if f is quadratic and exact line searches are performed, no algebraic relations like conjugacy are likely to hold between consecutive directions s_e. In fact these projected steps may be linearly dependent for certain $e \in E$, so that rank one or Davidon (1975) updates (see also Schnabel, 1977), which could possibly recover finite termination, are likely to be highly unstable in the non-quadratic case. Therefore there seems little point in performing accurate line searches, which in any case cannot ensure that all the inner products $y_e^T s_e$ are positive. This is so because, even at an isolated minimizer x^* where $G(x^*) > 0$, some of the element Hessians $G_e(x^*)$ may have negative eigenvalues.

Therefore we can only apply BFGS or DFP throughout, if it is known beforehand that all f_e are globally convex, which implies

$$G_e(x) \geq 0 \quad \text{for all } x \in \mathbf{R}^n \text{ and } e \in E.$$

In this case we say that the decomposition (5) is convex. While certainly strong, this assumption is not unrealistic as it is satisfied in many finite element applications, but typically the matrices $G_e \geq 0$ have null vectors v with $v_e \neq 0$. Under such conditions the DFP update seems to be unstable, but good numerical results can be achieved with the BFGS formula as reported by Griewank and Toint (1981a). The resulting partitioned variable metric method maintains $B = \Sigma B_e > 0$, it converges locally and Q-superlinearly as shown in Griewank and Toint (1981b), and it has the restricted invariance properties that are formulated at the end of Section 3, for it reduces to BFGS if $d = n$.

If only a few element functions f_e are nonconvex, and if the corresponding values of $|e|$ are comparatively small, one may consider computing their Hessians G_e either explicitly or by finite differences; the approximations $B_e \approx G_e$ for the remaining $e \in E$ are updated as above. However, because the resulting $B = \Sigma B_e$ is not necessarily positive definite, trust regions (see Sorensen's paper, this volume) or steps of negative curvature (Moré and Sorensen, 1979) have to be implemented. The same complication arises if some of the G_e are approximated using fixed scale least change updates like PSB, for which Q-superlinear convergence can be easily established using the techniques of Powell (1975). Furthermore, because such methods have very limited invariance properties, and because numerical experiments are not encouraging, the following approach to convexify a given decomposition seems more promising.

If $G(x) > 0$ at the current iterate, there are symmetric matrices C_e with the sparsity of ee^T such that

$$C \equiv \sum_{e \in E} C_e \geq 0, \tag{12}$$

and such that, for each $e \in E$, the matrix $\tilde{G}_e(x) \equiv G_e(x) - C_e$ satisfies

$$v^T \tilde{G}_e(x)\, v > 0, \quad \text{if } v_e \neq 0; \tag{13}$$

thus the matrices $\{\tilde{G}_e;\ e\in E\}$ are positive semi-definite, and C has the sparsity of G. We note that sometimes condition (13) requires $C_e \not\equiv 0$, and that it implies that the modified element functions

$$\tilde{f}_e(x) \equiv f_e(x) - \tfrac{1}{2}(x-x^*)^T C_e\ (x-x^*)$$

are convex in a neighbourhood of the current iterate. Therefore, if a step s from x remains in the neighbourhood, and if $s_e \not\equiv 0$, we have

$$y_e^T s_e > s_e^T C_e s_e . \tag{14}$$

It follows that, if $\tilde{B}_e$ is an approximation to $\tilde{G}_e$ that satisfies the positive definiteness condition that is analogous to (13), it can be updated by the BFGS formula

$$\tilde{B}_e^+ = \tilde{B}_e - \frac{\tilde{B}_e s_e s_e^T \tilde{B}_e}{s_e^T \tilde{B}_e s_e} + \frac{(y_e - C_e s_e)(y_e - C_e s_e)^T}{y_e^T s_e - s_e^T C_e s_e} .$$

Thus the matrices $B_e^+ \equiv \tilde{B}_e^+ + C_e$ satisfy the secant conditions (10), and we have $v^T \tilde{B}_e^+ v > 0$, except for null vectors v with $v_e = 0$. Consequently (8) and (12) imply

$$B^+ = \sum_{e\in E} B_e^+ = C + \sum_{e\in E} \tilde{B}_e^+ > 0,$$

which shows that the positive definiteness of $B = C + \Sigma\, \tilde{B}_e$ is maintained.

Whenever the inequality (14) is violated for some $e\in E$, at least the corresponding matrix C_e must be reset. Provided $G(x) > 0$, the collections $\{C_e;\ e\in E\}$ for which (12) and (13) hold form a convex set K(x) with nonempty relative interior which varies continuously in x. Therefore one can expect to reach after finitely many iterations and resets a collection $\{C_e;\ e\in E\}$ in $K(x^*)$, such that the method proceeds from then on as if f had a convex decomposition; thus Q-superlinear convergence would be obtained. However, no practical procedure for updating $\{C_e;\ e\in E\}$ with the desired transformation invariance properties has yet been developed. Another open question is what should be done far from x^* when $K(x)$ is empty because $G(x)$ is indefinite. By a suitable line search we can ensure $y^T s > 0$, so conditions (12) and (14) can always be compatible, even though (13) may not be attainable. Therefore one might hope to develop a workable scheme of updating the C_e such that (13) is violated only finitely often and (12) is always satisfied. We may not know whether (13) is ever actually satisfied, but Q-superlinear convergence seems likely.

6. Tentative Conclusions and Future Research

If f is partially separable with a small depth $d < n$, then second derivatives may be obtainable analytically or by differencing without much additional coding or computing, in which case Newton's method is an attractive proposition. Otherwise partial separability can be exploited by sparse steps or partitioned updating. The latter approach yields good theoretical and practical results if the decomposition of f is convex, but it is not yet clear how nonconvex element functions should be treated. So far the convergence results are purely local, and suitable safeguards must be developed to ensure global convergence at least to a stationary point of f. The structure of the decomposition (5) suggests that sometimes a direct solution of the linear system (9) based on a dissection ordering (George and Liu, 1979) may be more efficient than iterative schemes.

Since many problems are in fact discretizations of variational problems, one can use mesh refinement or multigrid techniques (Brandt, 1977) to limit the number of expensive iterations on the finest grid. In some applications f has a decomposition which is not strictly additive, but involves products or compositions of simpler functions whose second derivatives could also be approximated in a partitioned fashion. Finally, we note that the concept of partial separability may be extended to constrained optimization problems.

7. Acknowledgements

It is a pleasure to acknowledge here the numerous and helpful suggestions that were made by L.C.W. Dixon and by a number of nice people present at the 1981 meeting on Mathematical Programming in Oberwolfach.

5.4

Local Piecewise Linear Approximation Methods*

R.R. Meyer

Computer Sciences Department,
University of Wisconsin,
Madison, Wisconsin 53706, USA.

1. Introduction

Perhaps the best known piecewise linear approximation method is separable programming, which in its traditional form goes back at least as far as Charnes and Lemke (1954). Two major disadvantages of the basic separable programming approach are that (1) it requires unacceptably large increases in the size of the approximating problem in order to achieve high accuracy, and (2) the global approximation concept employed does not extend in a reasonable manner to nonseparable problems. Both of these shortcomings may be circumvented by replacing global approximations by approximations over suitably chosen neighbourhoods. Moreover, the use of piecewise linear approximations determined by function values leads to a number of advantages over approaches based upon first and second order information. These advantages include (1) guaranteed improvement in the objective value without the need for a line search, (2) the ability to include information obtained from Lagrangian relaxations (thereby allowing data from both feasible and infeasible solutions to be combined in the approximations), (3) ease of calculation of tight lower bounds on the optimal

*Research supported by National Science Foundation Grant MCS-7901066.

value, and (4) the ability to deal with very large problems via iterative use of existing high-quality network and LP software.

2. The Separable Case

2.1. Linear constraints The basic ideas of the local piecewise linear approximation approach are most easily stated for the problem

$$\left.\begin{array}{ll} \underset{x}{\text{minimize}} & F(x) = \sum_{i=1}^{n} f_i(x_i), \\ \text{subject to } Ax = b, & \ell \le x \le u \end{array}\right\},$$

where $x = (x_1\ x_2\ \dots\ x_n)^T$, where each f_i is a continuous (differentiability is not required) convex function on the non-degenerate interval $[\ell_i, u_i]$, and where $Ax = b$ is a system of m equations. At the j-th iteration of the method, the original problem is replaced by an approximating problem of the form:

$$\left.\begin{array}{ll} \underset{x}{\text{minimize}} & F^{(j)}(x) = \sum_{i=1}^{n} f_i^{(j)}(x), \\ \text{subject to } Ax = b, & \ell^{(j)} \le x \le u^{(j)} \end{array}\right\}, \qquad (P^{(j)})$$

where $f_i^{(j)}$ is a piecewise linear approximation to f_i over an interval $[\ell_i^{(j)}, u_i^{(j)}]$ that contains $x_i^{(j-1)}$, the optimal value of x_i from the *preceding* iteration. (In the initial iteration, however, $\ell_i^{(0)} = \ell_i$, $u_i^{(0)} = u_i$, and *no* feasible starting point is required.) The function $f_i^{(j)}$ agrees with f_i at $x_i^{(j-1)}$, $\ell_i^{(j)}$, and $u_i^{(j)}$ (and, depending on the variant of the algorithm employed, may also agree with f_i at some additional points in $[\ell_i^{(j)}, u_i^{(j)}]$); also $f_i^{(j)}$ *dominates* f_i on $[\ell_i^{(j)}, u_i^{(j)}]$. It is shown by Kao and Meyer (1981) that the intervals $[\ell_i^{(j)}, u_i^{(j)}]$ may be constructed in such a way that the values $\{F(x^{(j)});\ j = 1,2,3,\dots\}$ of the objective function are non-increasing, and such that every accumulation point of the sequence of iterates $\{x^{(j)};\ j = 1,2,3,\dots\}$ is an optimal solution of the original problem. The problem $(P^{(j)})$ is easily transformed into a linear programming calculation, and may then be solved using high-quality LP software. Further, if the original problem is a network problem, then each $(P^{(j)})$ may be transformed into a network problem, to allow the use of the extremely fast techniques of network optimization.

Observe that the solution of $(P^{(j)})$ yields not only a feasible solution and an *upper bound* for the optimal value of $F(x)$, but also a value $\lambda^{(j)}$ for the dual variables corresponding to the constraints $Ax = b$. These dual values (or various refinements of these values) may then be used to construct the Lagrangian relaxation problem:

$$\left.\begin{array}{l} \underset{x}{\text{minimize}} \quad \sum_{i=1}^{n} f_i(x_i) - \lambda^{(j)T}(Ax - b), \\ \\ \text{subject to } \ell \leq x \leq u. \end{array}\right\} \qquad (D^{(j)})$$

Since the objective function of $(D^{(j)})$ is separable, and since there are no coupling constraints, $(D^{(j)})$ may be decomposed into n one dimensional problems. The solution of $(D^{(j)})$ thus provides both a *lower bound* on the optimal value and a corresponding vector $\bar{x}^{(j)}$ that may be used (subject to some tolerance safeguards) in the construction of the piecewise linear approximations $\{f_i^{(j+1)};\ i = 1,2,\ldots,n\}$ for the next iteration. In most practical problems the solution of the one dimensional problems in $(D^{(j)})$ is available from a closed form expression for a solution of $f_i'(x_i) = c$, where c is a constant. Alternatively, approximate solutions may be obtained by any suitable method for one dimensional optimization. Moreover, "primal" techniques, which do not require derivatives, and which depend on error estimates of the approximations to the functions $\{f_i;\ i = 1,2,\ldots,n\}$, can also be used to calculate lower bounds and corresponding vectors $\bar{x}^{(j)}$ (Kao and Meyer, 1981).

Computational experience with this approach on a variety of problems (including network problems with hundreds of constraints and variables) has shown that both the upper and lower bounds are rapidly convergent, with six figure agreement of these bounds being attained in 10 — 20 iterations.

2.2. Nonlinear constraints Each separable nonlinear constraint of the form

$$\sum_{i=1}^{n} g_i(x_i) \le \beta,$$

where g_i is continuous and convex on $[\ell_i, u_i]$, may be approximated in the same manner as the objective function. That is, a local piecewise linear approximation about the current iterate is substituted for each $g_i(x_i)$, leading to the replacement of each nonlinear constraint by a linear constraint. With this technique of constraint approximation, the feasible set of the approximating problem will be a subset of the original feasible set, guaranteeing feasibility of the solutions of the subproblems. Lower bounds on the optimal value may be obtained by the Lagrangian approach that has been described.

3. The Nonseparable Case

The technique of the previous section is easily generalized to the nonseparable convex problem:

$$\left.\begin{array}{ll} \underset{x}{\text{minimize}} & F(x), \\ \text{subject to } g(x) \le 0, & \ell \le x \le u, \end{array}\right\}$$

where $g(x) = (g_1(x)\ g_2(x)\ \dots\ g_m(x))^T$, and where F and the g_i are continuous convex functions. Given a feasible point $x^{(j)}$ (which may be determined initially by applying local piecewise linear approximation techniques to the first phase of a phase I — phase II procedure, or from a penalty function version of the given problem), the objective function may be replaced by a piecewise linear function $F^{(j)}$ defined on a closed neighbourhood

$N(x^{(j)})$, which is the convex hull of a collection of points that are near $x^{(j)}$. By considering a collection of points of the form $\{x^{(j)} \pm \delta^{(j)} e_i; \ i = 1,2,\ldots,n\}$, where $\delta^{(j)} e_i$ is a multiple of the i-th co-ordinate vector, it may be seen that this approach is a straightforward generalization of the technique for the separable case. However, neighbourhoods $N(x^{(j)})$ that are determined by other collections of points are also allowable, including simplices with n+1 vertices. Similar approximations may be used for the constraint functions $\{g_i(x); \ i = 1,2,\ldots,m\}$, and the resulting problem may be shown to be equivalent to a linear program, whose feasible set is a subset of the intersection of $N(x^{(j)})$ and the original feasible set. If $x^{(j)}$ is not an optimal solution of the original problem, then $N(x^{(j)})$ may be constructed so that the optimal solution $x^{(j+1)}$ of the local approximating problem is feasible, and so that the relation $F(x^{(j+1)}) < F(x^{(j)})$ is obtained in the objective function. Therefore line searches are unnecessary, but they may be used to accelerate convergence since $(x^{(j+1)} - x^{(j)})$ will be a descent direction. Lower bounds may still be obtained via the Lagrangian relaxation, but now, in general, there is no decomposition into one dimensional problems, so other lower bound techniques (such as linearization) may be more efficient.

Preliminary computational experience with this approach on relatively small, nonseparable problems has been promising. It should be noted that, in contrast to the separable case, differentiability is required in the proof that the method converges.

4. Directions for Future Research

One particularly promising direction for further research is an implicit grid approach, in which additional segments of a piecewise linear approximation are calculated as needed when the boundary of the current approximation interval is reached. Another promising idea is to expend less effort on the optimization of subproblems, since improvement of the original objective function is assured once the objective of the approximating subproblem has been improved. Both of these devices have yielded encouraging results in preliminary tests. Other techniques that offer promise for accelerating convergence in the neighbourhood of an optimal solution are extrapolation methods for selecting grid points, and the use of projected SOR (Mangasarian, 1981b) for solving quadratic subproblems. Finally, from the dual point of view, rather than simply taking the dual variables $\lambda^{(j)}$ from the solution of the j-th approximating problem, the duals could be calculated by using first derivatives at the new point $x^{(j+1)}$, or by using subgradient techniques.

Discussion on "Large Nonlinear Problems"

5.5 Gutterman

It is not always clear what "successive function evaluations" means in Paper 5.1. I presume it should mean evaluation of the nonlinear portion of the objective and constraint functions.

5.6 Beale

By "function evaluations" I mean returns to the user routine for new information. Therefore a normal LP code, for example, makes no function evaluations after the initial input.

5.7 Lootsma

What performance is expected from decomposition methods such as the Dantzig-Wolfe decomposition for large linear problems, or Benders decomposition for mixed-integer programming?

5.8 Beale

I have no recent experience of using decomposition techniques, so I am reluctant to pronounce on their potential. I believe, however, that, in general terms, automatic column generation (as used for example for interpolation in separable programming), and automatic row generation, will become increasingly useful in the solution of large optimization problems.

5.9 Shanno

Research at Stanford University on decomposition for large linear problems (Perold and Dantzig, 1978), has led to interesting new algorithms, which are currently not widely known. Recent work on generalizing these algorithms by Roy Marsten (some of it is described in Marsten and Shepardson, 1978), looks extremely promising, as numerical instabilities encountered in early studies appear to be overcome. This work should generalize in a fairly straightforward manner to nonlinear problems with special structure.

5.10 Beale

This work should indeed be noted, although it is not clear to me that even modern decomposition techniques always work well on practical problems. Beale, Hughes and Small (1965) found that problems with more than about 50 common rows were difficult, and that problems with more than about 100 common rows were impossible. I understand that James Ho observed that a more precise description of this difficulty is that good progress depends on having not more than about 6 proposals in the basis from any one subproblem. Otherwise the basis is nearly singular. Note that in the mid 1960's we were not forming proper triangular factors of our bases, and this increased our numerical difficulties. On the other hand, the triangular factorization procedures that have been in production LP codes since the early 1970's make these codes much better for large sparse problems than earlier codes, even without exploiting special structure.

5.11 Lasdon

In solving large sparse quadratic programming problems by reduced gradient methods, the number of superbasic variables may become so large that Newton's method cannot be used. Because all variable metric and conjugate gradient methods generate the same sequence of points in such cases (assuming exact line searches), one may as well use the simplest conjugate gradient method.

5.12 Goldfarb

I would like to point out that, when one is using a conjugate direction algorithm, a restart is necessary whenever a constraint is added to the active set. This means that all of the second order information accumulated up to that point is essentially lost, unless one follows the type of approach advocated by Best and Ritter (1976), which is quite expensive.

5.13 Wolfe

The function of nonlinear variables that is obtained by optimizing with respect to the linear variables of a linearly constrained calculation is not smooth. Is this of concern to the procedures of Paper 5.1, or would Lasdon be concerned about it in GRG?

5.14 Beale

As you say, the objective function is nonsmooth when treated as a function of the independent variables only. It seems reasonable to assume that its global behaviour on large problems will not be very different from a smooth function. We hope in the final stages to have identified the active constraints correctly, for then the behaviour will be smooth.

But we do need a way of landing on a point where the gradient is discontinuous, or in other words of reducing the number of independent variables by introducing a new active constraint without making a currently active constraint inactive. We meet this need by making an exploratory step from time to time, that is essentially an ordinary SLP (sequential linear programming) step.

5.15 Powell

It is mentioned in Paper 5.1, that diagonal scaling techniques are more suitable than general variable metric methods for very large problems, but for small problems variable metric algorithms are more successful. Are there any ideas on how these two approaches could be combined into a single algorithm that is suitable for a wide range of cases?

5.16 Beale

The possibility of combining dynamic variable scaling with other variable metric algorithms without waiting for a restart could well be a fruitful field for research.

5.17 Boggs

The idea in Paper 5.2 of providing extensive software tools to help in all phases of the optimization process is excellent. In another context, that of solving certain classes of partial differential equations, a similar system is being developed (Engquist and Smedsaas, 1980). The designers in this case view the initial output of the system as a "first draft" of the computer code. If only one problem is to be solved, then nothing is changed. If a production code is desired, then modifications are made. I think that the system just described could be viewed in this way. The system automatically takes care of much of the tedious detail, but leaves room for modifications which the experienced analyst could provide.

5.18 Drud

The approach in the PDE-solver will only work if the computer code makes sense to the user. In GAMS we store much of the information in a sparse format, so the automatically generated subroutine will often contain information in a form that is not meaningful to an ordinary user. We have therefore given attention to the automatic generation of efficient problem subroutines.

5.19 Lootsma

What is the approximate maximum size of the problems that can generally be solved by the programs of Paper 5.2? I ask this, because I do not know what we mean by large or very large problems, and I would like to have a more precise definition.

5.20 Drud

The GAMS-modelling system contains a data base module, modules for handling general equations, and interfaces to various solution algorithms. All parts of the system are implemented using sparse matrix techniques, and it should be straightforward to work with several thousand constraints and variables. Currently, the limit in size seems to be due to the solution algorithms.

5.21 Ragsdell

What is a large problem? Do the number of variables and constraints determine the size? I think we need to keep nonlinearity in mind when discussing these questions. A good example of my point is structured optimization problems as a class. Here the objective is a linear function of the design variables (cross sectional areas), and there are linear and nonlinear equality and inequality constraints. Some of these constraints are so nonlinear that, to my mind, a problem with only 500 variables and 500 constraints is large. I sense that other participants would not take this view but I wish to suggest that they do not give enough attention to the nonlinearities.

5.22 Schnabel

I think it is important to accompany the results of an optimization algorithm with some estimate or comment on how believable these results are: for example the algorithm may provide confidence intervals on the parameters. By this I mean not only estimates of the sensitivity of the optimization model near the optimal parameters, but also (if possible) the sensitivity of the optimal parameters to the data that were used in forming the model. In the absence of such information, users sometimes draw conclusions from optimal parameters that are actually badly determined by the data. Has this area been studied sufficiently by the statistical community, and are there techniques that optimization software is not using but should be using in its *a posteriori* analyses?

5.23 Beale

The general problem of what to do about uncertainties in the data seems to me to be too confused to deserve serious study. Statisticians sometimes complain that operational research workers produce "point estimates" as the answers to problems where the data are in principle stochastic. But these statisticians do not seem to understand the scale of the problems. Sometimes there are a very small number of input data that have a major influence on the conclusions, but whose true values are uncertain. Then, as has been said, one does sensitivity analyses to explore various possibilities. If one can quantify the definition of an optimum compromise decision, one can go further and use some form of stochastic programming. But this is unusual, and even less can be done with hundreds of more or less equally uncertain data. Even if one could quantify their variances, we cannot hope to quantify their covariances in a convincing way.

A more tractable problem is what to do about the possibility that the objective function may be flat near the optimum, assuming that the data are incorrect. This also is handled by sensitivity analysis. In particular, one may do parametric programming to study the consequences of imposing an extra constraint and changing its constant term so as to move the solution away from the optimum in a particular direction, which is a one-dimensional process. However, we could never hope to make much sense of the solution to multi-dimensional parametric programming problems.

5.24 Jackson

With regard to contribution 5.22, on whether we have overlooked, or have underplayed, the need for work on the problem of "bad" or poorly known data values, I wish to draw attention to the successful efforts of the linear programming community on sensitivity analysis. This is precisely what should be done about the problems that have been raised. Furthermore, Tony Fiacco at George Washington University has been working for quite some time on sensitivity analysis in nonlinear programming.

5.25 Moré

What is a "sparse step of negative curvature"? More generally, how are indefinite Hessians handled in Paper 5.3?

5.26 Griewank

Steps are said to be sparse if they leave all but a few components of x constant and affect therefore only some of the element functions. When the conjugate gradient routine for solving $Gs = -g$ finds a direction of negative curvature, say c, we try to make c sparse by setting some of its components to zero subject to $c^T Gc$ remaining negative. Then a line search using the procedure of Shanno and Phua (1976) is performed along c. This approach works well if G is the exact Hessian but further development is necessary for the quasi-Newton case.

5.27 Steihaug

In the recommended technique of Paper 5.3, what is used as the merit function of the line search?

5.28 Griewank

During all line searches the merit function is the sum of all the element functions that are affected by the step. Therefore the overall objective function is reduced consistently throughout the calculation.

5.29 Buckley

Please clarify how the various secant equations are satisfied. After taking a step s from one point to the next, one computes the component gradient differences y_e. As well, one has a quasi-Newton Hessian approximation B which is given and stored in terms of approximations B_e to the small element Hessians. Thus $B = \Sigma B_e$, and y (the total gradient difference) is equal to Σy_e. I understand how one may update each B_e to a new approximation B_e^+, say, using a standard quasi-Newton update formula to obtain $B_e^+ s_e = y_e$. However, because $s \neq \Sigma s_e$, please explain why $B^+ s = y$, where $B^+ = \Sigma B_e^+$.

5.30 Griewank

The equation $B^+ s = y$ is satisfied for the following reason. By s_e we mean a vector which equals s in those components that correspond to the variables of x which actually occur in f_e. The other components of s_e are zero. Thus we can write $B^+ s = (\Sigma B_e^+) s = \Sigma (B_e^+ s)$, and, since B_e^+ involves only the variables of f_e, we may write $B_e^+ s = B_e^+ s_e$. Hence $B^+ s = \Sigma B_e^+ s_e = \Sigma y_e = y$.

5.31 Mifflin

Sometimes one can apply a decomposition (nested dissection) approach to objective functions of the form that is studied in Paper 5.3. In this method an outer optimization is done over some (hopefully few) variables such that, when they are fixed, the inner problem separates into several independent (hopefully small) optimization problems that can be solved in parallel.

5.32 Drud

The method described in Griewank's paper seems to be very suitable for optimizing Lagrangian and augmented Lagrangian functions in large scale nonlinearly constrained optimization. A natural decomposition would be to let the element functions be the original objective function and the weighted constraint functions. The method is suitable because in many practical applications, each of these element functions depends only on a small subset of the total set of variables.

5.33 Beale

I agree that Griewank's work has a great potential in constrained optimization.

5.34 Polak

At present, most engineering simulation codes do not compute derivatives with respect to design parameters. Hence gradients for optimization are commonly computed by finite differences, which makes the use of secant methods as proposed in Paper 5.3

at best problematic.

5.35 Toint

I would like to stress the point that, if the element gradients are available, then the known structure of the Hessian matrix in Paper 5.3 allows the matrix to be estimated with the minimal number of differences.

5.36 Goldfarb

It seems to me that, for solving large sparse unconstrained problems, Newton's method should be less expensive than variable metric methods in terms of gradient calculations if one fully exploits the sparsity of the Hessian. This observation and the difficulties involved in implementing a sparse variable metric method were the main reasons why I abandoned my research in this area years ago.

5.37 Todd

Martin Beale has stressed the importance of product terms. Does Bob Meyer have any experience with piecewise linear approximations (with a small number of pieces) to functions of two variables?

5.38 Meyer

Our computational experience thus far has been limited to the extreme cases of separable and "completely" non-separable convex problems. The algorithm, however, extends in a straightforward fashion to intermediate cases such as vector-separable problems in which the objective and constraint functions may be written as sums of functions of distinct vector variables. Certainly, the particular case of two variable functions is of considerable practical interest, and we would like to carry out some computational testing with such problems. Our preliminary numerical results with non-separable problems have been most encouraging, and we are optimistic about the computational merits of the approach.

5.39 Byrd

What step bound expansion or contraction factor was used in Paper 5.4 for the runs that give a linear convergence rate near $10^{-\frac{1}{2}}$. Do you expect the step bound factor to have much influence on the convergence rate?

5.40 Meyer

The step bound expansion was limited to a factor of 10 and the step bound contraction to a factor of 0.2 in the results reported. In our experience, the performance of the algorithm has not been particularly sensitive to those parameters. Doubling or halving those limits seems to influence running time by only a few per cent.

5.41 Lemaréchal

If I understand it properly, the algorithms of Paper 5.4 solve, on each iteration, one linear program followed by n one-dimensional minimizations. Have any numerical experiments been tried in cases when the one-dimensional minimizations cannot be carried out explicitly.

Please comment also on the assumption of convexity. Is it essential, or is it likely that the algorithm will behave well in several nonconvex situations?

5.42 Meyer

We have had computational experience with a variant of the method in which lower bounds on the optimal value and "temporary" bounds on the variables were obtained by techniques that did not require one-dimensional optimization. This variant also performed well, but required about 30% more computing time on most test problems. In all of our test problems and, seemingly, in most applications, the solution of the associated one-dimensional problems has been available in closed form; hence the alternative approaches might be more competitive in those less common cases in which this feature is not present.

In response to your other question, it seems that the results for convex problems may be generalized to nonconvex problems by imposing the additional conditions that an iterate will be accepted only if it is feasible and yields some "sufficient" improvement in the true objective function. (Otherwise, a contraction of the approximation region would be performed.) Convergence to a local rather than a global solution would be expected. This is an interesting area for further research.

5.43 Beale

When seeking a global optimum solution (by branch and bound) to nonconvex problems, one needs global solutions to the one-dimensional optimization problems derived from Lagrangian relaxation.

Beale and Forrest (1976) describe a way of finding them (to within ε), using first and second derivatives. This technique requires the second derivatives to satisfy a piecewise monotonicity assumption, that is valid for all applications they have considered.

5.44 Lasdon

Can problems with nonlinear separable convex constraints be treated by the methods of Paper 5.4?

5.45 Meyer

The same approach used to approximate separable objective functions may be applied to convex separable constraints. The only note of caution that needs to be introduced is that the feasible sets for the approximating problems will be subsets of the original feasible set. Thus, the feasible set of the initial approximation may be empty if the approximation is too crude. Of course, once a feasible point has been obtained, such a point will also be feasible for the next approximating problem. Computational experience recently reported by Thakur (1978) for nonlinear constraints appears quite promising.

5.46 Beale

Convexity of nonlinear constraint functions is only relevant if the constraint functions are on the left-hand sides of less-than-or-equal-to inequalities (or if for other reasons one knows that the Lagrange multipliers cannot be negative). It is perhaps worth noting that Beale and Forrest (1976) used the same Lagrangian relaxation to derive both bounds and new interpolation points on the separable curves. The rest of their implementation is different from the one taken in Paper 5.4, and in particular it could not be extended so easily to nonseparable problems.

5.47 Meyer

One approach that we are investigating in the hopes of obtaining rapid convergence in the terminal iterations (without prohibitive computational cost for the subproblems) involves the solution of quadratic approximating problems via the projected SOR methods of Mangasarian. Such methods have been very efficient in large networks with positive definite objectives.

5.48 Final Comments by Introductory Speaker (Beale)

This and other sessions have indicated the accelerating progress being made in the methodology for finding local optimum solutions to very large nonlinear problems. There also seems to be good agreement about promising research directions.

People whose main interest is the development of algorithms can easily underestimate the tasks of formulating appropriate large scale nonlinear programming models, of transmitting these formulations to the computer, and of interpreting the solutions. So it is good that Drud has followed my general discussion of these tasks with an account of a practical system addressed to them. Meyer's contribution on separable programming is welcome, since such problems are easier to formulate and manipulate than more general nonlinear problems. The demonstration that the methods can be extended to nonlinear functions of more than one argument is encouraging: model-builders like to know that they can change the mathematical structure of their models if necessary without changing the solution strategy.

Satisfactory computational results have been reported on nonlinear optimization problems that are very large by the standards of the 1970's. These have been obtained using SLP, reduced gradients and RGAP; and it is good that these techniques are becoming more similar to each other.

Many contributors, notably Griewank in this session and Dembo in the next, have helped to show how we can expect to make increasing use of explicit information about the Hessian of the Lagrangian, even for very large problems. The different variants of SQP that will be appropriate for different types of problem may have much in common with future developments in SLP and reduced gradient methods. Most of pp. 189–196 of Beale (1967) still seems relevant.

Large scale nonlinear optimization methodology is likely to remain untidy for a fundamental reason. A very large problem is one that can only be solved efficiently if advantage is taken of some special simplifying features. (This is the best answer I can give to Lootsma's and Ragsdell's question "What is a large problem?".) In linear programming we can go a long way by just exploiting sparseness, but this does not cover the whole range of nonlinear problems. In particular, the relative amounts of work in computing local approximations to the objective and constraint functions and in solving the resulting QP or LP subproblems must be considered carefully. The remarks of Lasdon and Goldfarb relate to this point.

I am impressed by the number of speakers, at this and other sessions, who have spoken encouragingly about decomposition methods.

Global minimization of nonconvex problems is an area in which solution methods must exploit simplifying features even in one or two dimensions. But branch and bound methods have been applied successfully to some practical problems of this type with hundreds of variables. So global optimization should not be dismissed as an unrealistic aim, although I believe that research should concentrate on exploiting special structures.

Finally, there is the problem of the validity of the results of optimization calculations. This is an important topic, and my remarks in the discussion may be misinterpreted. There is always the danger that an optimization model may give misleading results, either because the model is wrong, or because the data are wrong, or because other solutions may be as good or almost as good. This danger is enhanced if the model is very large. The practical conclusion from this is that the process of formulating and solving the model must be both quick and cheap. The analyst will then have time to think about his solution, and also the opportunity to try a range of alternative assumptions. Algorithmic research can contribute to making the solution process quicker and cheaper, and to efficient parametric nonlinear programming. I remain sceptical, however, about the feasibility of any general solution to the "data problem" that claims to relieve the analyst of the responsibility for thinking hard about his specific model.

PART 6

The Current State of Software

6.1

Notes on Optimization Software*

Jorge J. Moré

Applied Mathematics Division, Argonne National Laboratory,
Argonne, Illinois 60439, USA.

This paper is an attempt to indicate the current state of optimization software and the research directions which should be considered in the near future.

There are two parts to this paper. In the first part I discuss some of the issues that are relevant to the development of general optimization software. I have tried to focus on those issues which do not seem to have received sufficient attention and which would significantly benefit from further research. In addition, I have chosen issues that are particularly relevant to the development of software for optimization libraries.

In the second part I illustrate some of the points raised in the first part by discussing algorithms for unconstrained optimization. Because the discussion in this part is brief, the interested reader may want to consult other papers in this volume for further information.

In both parts my comments are influenced by my involvement in the MINPACK project and by my experiences in the development of MINPACK-1 (Moré, Garbow and Hillstrom, 1980).

---

*This work was supported by the Applied Mathematical Sciences Research Program (KC-04-02) of the Office of Energy Research of the U.S. Department of Energy under Contract W-31-109-Eng-38.

1. Software

The development of quality optimization software is a difficult task which requires research in many areas. Some of the problems involved in this work are discussed by Gill, Murray, Picken and Wright (1979), and by Moré, Garbow and Hillstrom (1980). The following issues are particularly important and in general have not received sufficient attention. Research in these topics is necessary in order to be able to develop better optimization software.

1.1. Noise

Most optimization software is not designed to solve problems in which the computation of the functions is subject to noise. Instead, it is usually assumed that the problem functions can be evaluated to full machine precision. A problem with the optimization of noisy functions is that, if the algorithm requires the estimation of derivatives by differences and if the difference parameter does not depend on the noise level of the functions, then incorrect derivative approximations are usually obtained and this invariably leads to a failure. Another possible problem is that the termination criteria of the algorithm must then recognize when differences in the objective function values are only due to noise.

Noise in the computation of a function can arise in several ways. For example, if a user has programmed a function in double precision but has forgotten to declare one or more of the variables, then the function is essentially computed in single precision and this leads to a noisy function. As another example, consider the computation of a function whose definition involves the solution of a non-trivial system of nonlinear equations. The computation of such a function can only be carried out to within a certain tolerance and again leads to a noisy function.

It is necessary to develop software which estimates the noise level in the problem functions. Hamming (1971) discusses the principles of roundoff estimation and Lyness (1976) mentions some of the problems associated with the optimization of functions subject to noise. A related problem is the development of software which produces suitable estimates for the derivatives of the functions. Lyness (1977) discusses the principles behind the development of semi-automatic numerical differentiation software.

<u>1.2. Reliability</u> There are many aspects to the development of reliable optimization software, that is, software which solves a wide variety of problems.

One aspect requires robust software which, for example, avoids arithmetic exceptions (e.g., destructive underflows and overflows) and minimizes the effects of roundoff errors. Other requirements of robust optimization software are discussed by Moré (1979, 1980), but these two requirements are particularly important on machines with a restricted exponent range, or with short precision calculations. In general, the aim of robust software is to extend the domain of the algorithm without a serious loss in efficiency.

Another aspect requires that the software be based on algorithms with acceptable theoretical properties. This does not mean that a code should not be developed unless the appropriate results have been established but, if the optimization software is developed strictly on an experimental basis, then the resulting software is usually unreliable.

Reliable optimization software should also be scale invariant. Proper attention to this point helps in the solution of badly scaled problems and, in general, improves the design of the algorithms. Scaling is discussed in more detail by Moré, Garbow and Hillstrom (1981a) in the context of testing optimization software. They point out that, if an algorithm is scale invariant, it need not perform well on a problem, but at least its performance is not changed with the scaling of the problem; on the other hand, the performance of a scale-dependent algorithm usually deteriorates when it is applied to badly scaled functions. They illustrate the latter point with numerical results obtained from two optimization library subroutines. On well scaled problems these two subroutines behaved similarly but their behaviour on badly scaled problems was radically different — one of the subroutines succeeded on about 75% of the problems while the other subroutine failed completely.

Reliability of the software also requires that the termination criteria be designed with great care. It should not mislead the user or be overly pessimistic. Moreover, it should be scale invariant, and, if possible, error bounds should be provided.

1.3. Transportability A desirable attribute of optimization software is transportability, that is, the code should perform satisfactorily on different machines with only a small number of changes. For example, MINPACK software only requires changes to the two subprograms which define the single and double precision machine parameters, and the changes consist of activating the appropriate set of parameters.

Testing of the codes with the PFORT verifier (Ryder, 1974; Ryder and Hall, 1979) and the WATFIV compiler is an important element in the process of developing transportable software. It is also helpful to follow the guidelines of Smith (1977). These precautions guarantee that the code executes on a wide range of machines and compilers.

The importance of testing of this type cannot be over-emphasized, and yet it seems to be not appreciated. For example, we have received several codes intended for the Transactions on Mathematical Software (presumably polished software) that have not passed these tests.

1.4. Performance testing In addition to testing for transportability, it is necessary to test the performance of the code. A minimal requirement is that the code be executed on a reasonably large set of test problems. At Argonne, Moré, Garbow and Hillstrom (1981b) have produced an easy-to-use collection of test problems for systems of nonlinear equations, nonlinear least squares, and unconstrained minimization; Moré, Garbow and Hillstrom (1981a) have also designed guidelines for testing unconstrained optimization software. The Argonne collection is available from the ACM Algorithms Distribution Service, but this version does not include Hessians for the minimization problems. However, Burt Garbow has now prepared codes for these Hessians and the codes are available to any interested researcher.

We have been pleased with our experience with this set of test problems. However, an even better test of the performance of the code is usually provided by users. User feedback is an important element in the process of developing software, although obtaining this feedback is not always easy.

The Argonne set of test problems can be used to test for reliability; see, for example, Hiebert (1979, 1980, 1981). It is also possible to use the Argonne set to test codes on functions subject to noise, by computing the functions in a shorter precision or by randomly perturbing the functions; see, for example, Barrera and Dennis (1979) and Hiebert (1980).

The Argonne set of test problems is, however, not appropriate for testing large scale unconstrained optimization software. There are several lists of test problems for large scale unconstrained minimization (e.g., Toint, 1978; Gill and Murray, 1979), and some of these problems are appropriate for nonlinear least squares and systems of nonlinear equations. It would be very helpful to have available a test problem collection for large scale optimization, say a nonlinear version of the test problems of Everstine (1979) and Duff and Reid (1979).

Easy-to-use collections of test problems should be developed for other optimization areas and made available to prospective researchers. For example, Dembo (1976) has a set of geometric programming test problems with information on the accuracy of the solutions. His problems, and many others, are part of the collection available from Hock and Schittkowski (1981).

It is also important to design guidelines for testing the software. At present, guidelines for testing constrained optimization software have not received sufficient attention.

1.5. Reverse communication Optimization software usually requires that the user provide a subroutine which calculates the problem functions. In some applications, however, this requirement cannot be satisfied, and then it is important to have optimization software with a reverse communication interface.

Software with a reverse communication interface returns to the calling program whenever there is a need to evaluate the problem functions. The calling program must then evaluate the problem functions and return control to the optimization algorithm. This process can be repeated at the discretion of the calling program. In this way communication problems with the calling program are simplified and, in addition, the calling program is given strict control over the optimization process. Software written with a reverse communication interface, however, is more difficult to use and therefore its documentation is more involved.

Mallin-Jones (1978), for example, has described an application in which reverse communication is needed. Another example is the following one.

Consider a system of nonlinear equations of the form $G(x,y) = 0$, $H(y) = 0$. To solve this system it is natural to define $F(x) = H(y(x))$ where $y(x)$ satisfies $G(x,y(x)) = 0$, and to attempt to solve $F(x) = 0$. In this approach, however, to evaluate $F(x)$ it is necessary to solve $G(x,y) = 0$ for y, and then we have a situation in which the evaluation of the problem function requires a call to the optimization algorithm. If the optimization software has a standard interface then this cannot be done in Fortran, but it is not difficult to see that a reverse communication interface allows a single nonlinear equation subroutine to be used.

Most optimization software only uses reverse communication in a few special subroutines. For example, Gill, Murray, Picken and Wright (1979) use reverse communication in their line search, and Moré, Garbow and Hillstrom (1980) use reverse communication in their derivative checker. On the other hand, Gay (1980a) uses reverse communication in his core subroutines, but has a standard interface for the higher level subroutines. In this way the demands for several types of software can be satisfied.

1.6. Auxiliary subroutines The solution of optimization problems often requires auxiliary subroutines to help the user in various aspects of the solution process. For example, the solution of a nonlinear least squares problem is usually accompanied by a request for the covariance matrix at the solution.

There are tasks of this type for which adequate software does not exist. I have already mentioned that software for determining the noise level of a function is needed. As another example, consider symbolic differentiation. There are various programs for this task, but most of them either place unreasonable restrictions on the subroutine which defines the function (for example, the subroutine must consist of a sequence of assignment statements) or are inefficient in the calculation of Jacobians. The thesis of Speelpenning (1980) contains an interesting discussion of symbolic differentiation.

Since tasks of the above type are not strictly part of nonlinear optimization, they have not received sufficient attention. This is unfortunate because they make the solution process easier, and in many cases a user will prefer a package which seems to make the optimization process as easy as possible regardless of the quality of the algorithms behind the package.

2. Algorithms

Although many unconstrained optimization algorithms are fairly well understood, there are still important problems. For example, we have already mentioned that most optimization software is not designed to solve problems in which the functions are subject to noise. It is always possible to ask the user to supply the noise level of the functions, but this is not always convenient. Another problem is that most optimization software is not scale invariant and tends to perform poorly on badly scaled problems. The user could be asked to scale his problem, but a more attractive possibility is to make the scaling automatic. For example, if the Levenberg-Marquardt algorithm uses scale factors based on the current approximation to the Jacobian matrix, then this scaling seems to be satisfactory. A third problem is the need to extend unconstrained optimization algorithms to linearly constrained (simple bounds and general constraints) optimization. There is little software in this area and more research is clearly needed.

The discussion below centres about some of the problems with unconstrained optimization software. In particular, we emphasize the need to extend standard unconstrained optimization algorithms to large scale problems. We discuss some of the algorithms which have reached a reasonable steady state in their development. Conjugate gradient methods for nonlinear problems still seem to be changing at a rapid pace and are not included in this discussion.

<u>2.1. Powell's hybrid method</u> There are many algorithms for the solution of systems of nonlinear equations; see, for example, Bus (1980). In my experience, Powell's (1970b) hybrid method is a good basis for an algorithm for small and medium scale problems; software based on this algorithm is available in MINPACK-1.

The dogleg algorithm used in Powell's hybrid method can be extended to large scale problems since it only requires the Newton direction and a scaled gradient direction, but there are other possibilities, for example, a Levenberg-Marquardt method. The cost of obtaining a Newton direction for large problems has led to the proposal that an inexact Newton direction be used instead. Dembo,

Eisenstat and Steihaug (1980) have shown that an inexact Newton method only needs to control the residuals of the linearized problem in order to obtain good local results. These inexact Newton methods are discussed in detail by Steihaug (1980).

Powell's hybrid method requires an approximation to the Jacobian matrix. This approximation can be obtained by differences or by a quasi-Newton update. If the problem functions are subject to noise, then the use of differences may prove to be unreliable, and in this case an analytic Jacobian matrix (even a noisy one) is especially helpful. However, for many small and medium scale problems, a combination of differences and Broyden's (1965) update works well. For large scale problems it is possible to use Schubert's (1970) extension of Broyden's update, but in this case Curtis, Powell and Reid (1974) have shown that a sparse Jacobian matrix can be determined with only a few function evaluations, and this makes an approach based on differences attractive. Coleman and Moré (1981) have improved the techniques of Curtis, Powell and Reid (1974), and present evidence which indicates that the improved algorithms are nearly optimal on practical problems.

There are other approaches to the solution of large systems of nonlinear equations. For example, Eriksson (1976) notes that, if the Jacobian matrix has a non-trivial block triangular form, then it is possible to reduce the solution of $F(x) = 0$ to the solution of smaller systems of nonlinear equations. A disadvantage of these decomposition techniques is that they require that the components of the function corresponding to the subsystems be evaluated separately. For many problems the components of the function have common subexpressions and then this may not be a reasonable requirement.

To obtain global convergence to a solution of the system of nonlinear equations, Powell's hybrid method enforces a decrease in the Euclidean norm of the residuals at each iteration. This technique has some drawbacks. One of the most important drawbacks is that in some cases the algorithm converges to a point at which the Jacobian matrix is singular and the equations are not satisfied. For general problems this is an unavoidable situation because the solution of a system of nonlinear equations is equivalent to a global minimization problem, but for certain problems it is possible to construct globally convergent methods, that is, methods which always converge to a solution of the system of nonlinear equations. For instance, the simplicial and continuation methods are globally convergent under appropriate conditions (see, for example, Allgower and Georg, 1980; Watson, 1980), and then these methods can be effective. However, for general problems the simplicial and continuation methods are not competitive with Powell's hybrid method.

Another drawback to enforcing a decrease in the Euclidean norm of the residuals at each iteration is that the algorithm is then dependent on the scaling of the functions, and with a poor scaling the algorithms may not perform satisfactorily. It is always possible to allow increases in the residuals, but then global convergence is lost. On the other hand, the algorithms can then be invariant with respect to the scaling of the functions. There does not seem to be a satisfactory resolution to the conflict between global convergence and invariance to the scaling of the functions. Deuflhard (1974) suggests enforcing a decrease in a scaled Euclidean norm, but then the resultant algorithm is not invariant to changes in the scale of the variables.

An interesting possibility for dealing with false convergence problems is the tunnelling algorithm of Levy and Montalvo (1979). This algorithm is designed for the global minimization problem and the reported results are impressive. It deserves to be investigated further.

<u>2.2. Levenberg–Marquardt method</u> Most variations of the Levenberg–Marquardt method differ in the way that the Levenberg–Marquardt parameter λ is chosen. This parameter is used to control the length of the step, and since there is a nonlinear relationship between λ

and the length of step, a direct choice of λ does not always have the desired effect. Indirect choices of λ do allow the determination of a step of the desired length and, therefore, are preferable.

The MINPACK-1 version of the Levenberg—Marquardt method chooses the Levenberg—Marquardt parameter by a scheme proposed by Hebden (1973) and refined by Moré (1978). This version of the Levenberg—Marquardt method has performed well in the tests of Hiebert (1979, 1981) and, in addition, Moré (1978) has shown that it is scale invariant and globally convergent.

A disadvantage of the Levenberg—Marquardt method is that it may converge slowly on large residual least squares. There are several algorithms for large residual least squares problems, among them those proposed by Gill and Murray (1978b), Dennis, Gay and Welsch (1981), and Nazareth (1980). At present, it is not clear if these algorithms are an improvement on the Levenberg—Marquardt method. In fact, Hiebert (1980) tested specific implementations of the Levenberg—Marquardt and large residual methods and concluded on the basis of her testing, that one method did not appear to be superior to the other. More research is needed on this topic.

All of the algorithms mentioned above require the storage of an m by n Jacobian matrix, where m is the number of functions and n is the number of variables, but if the Jacobian matrix is supplied one row at a time, then it is easy to modify the Levenberg—Marquardt algorithm so that it only needs to store an n by n matrix. A disadvantage of supplying the Jacobian matrix row by row is that it is then difficult to take into account common subexpressions, and this can lead to inefficiencies in the computation of the Jacobian matrix. On the other hand, this modification allows the solution of nonlinear least squares problems with a reasonable number of variables but a large number of functions. Software for this case is also provided in MINPACK-1.

For large scale nonlinear least squares there is little software. If there is an ordering of the variables for which the normal equations have a sparse Cholesky factor, then the Levenberg—

Marquardt method can be used, but otherwise an inexact Newton method may be preferable. If the Jacobian matrix is not available, then it can be approximated by either Schubert's update or by differences. Schubert's update should be used with care because, although it is superlinearly convergent for systems of nonlinear equations, it may not converge on nonlinear least squares problems. If differences are used to approximate the Jacobian matrix, then the techniques of Curtis, Powell and Reid (1974) or Coleman and Moré (1981) can be used to reduce the required number of function evaluations.

2.3. Newton's method The two major approaches to Newton's method are the line search approach (e.g., Gill and Murray, 1974a), and the trust region approach (e.g., Fletcher 1980; Gay, 1981; Sorensen, 1980b). There are many differences and similarities between the two approaches. They are similar when the Hessian matrix is positive definite, but when the Hessian matrix has a negative eigenvalue the trust region approach makes strong use of directions of negative curvature, while the line search approach replaces the Hessian matrix by a positive definite approximation.

There is no convincing numerical evidence which shows that one approach is better, but the theoretical results of Sorensen (1980b) for the trust region approach are very strong. In addition, there are few *ad hoc* decisions in the trust region approach, but the same cannot be said for the line search approach when the Hessian matrix is indefinite. On the basis of these arguments, I prefer a trust region approach with Newton's method. On the other hand, with a quasi-Newton method I prefer a line search approach. The reasons for the latter preference are discussed in the section on quasi-Newton methods.

The determination of the step in a trust region approach is a critical calculation. The first reasonable scheme for this problem was proposed by Hebden (1973). This scheme is basically sound except for the treatment of one case. This case — the hard case — requires the determination of a direction of negative curvature. Gay (1981) refined Hebden's scheme and proposed using inverse iteration to obtain the appropriate direction. Moré and Sorensen (1981) have refined Gay's work further and, in particular, have proposed a scheme which uses the LINPACK technique for estimating condition numbers to determine directions of negative curvature at an early stage.

The trust region approach based on the above schemes can be extended to large scale minimization problems, but now it is not clear that the cost of these schemes is justified. A trust region method based on a preconditioned conjugate gradient method may be preferable; see, for example, Dembo and Steihaug (1980).

Newton's method requires either the Hessian matrix or a suitable approximation to the Hessian matrix. For large scale problems a difference approximation to the Hessian matrix can be obtained with the techniques of Curtis, Powell and Reid (1974) or Coleman and Moré (1981). These techniques, however, ignore the symmetry of the Hessian. Powell and Toint (1979) have proposed a technique which takes into account symmetry and usually reduces further the number of gradient evaluations needed to estimate the Hessian matrix. For example, a general tridiagonal Jacobian matrix can be estimated with 3 extra function evaluations, but with the technique of Powell and Toint a symmetric tridiagonal Hessian matrix can be estimated with 2 extra gradient evaluations. Since the technique of Powell and Toint is based on the estimation of the lower triangular part of a permutation of the Hessian matrix, the unsymmetric techniques can be used here.

2.4. Quasi-Newton methods For small and medium scale unconstrained minimization problems, a quasi-Newton update with a line search is an excellent method, and there seems to be agreement that the BFGS update should be used in this algorithm. There is, however, disagreement on how to store the update, how to carry out the line search, and how to choose the initial update. In addition, there is still no proof that quasi-Newton methods are convergent for nonconvex functions.

A quasi-Newton update with a trust region approach should be used with care. The most elegant type of trust region method minimizes the approximation to the objective function in the whole of the trust region (for example, the methods of Fletcher, 1980; Gay, 1981; Sorensen, 1980b), and thus with this type of trust region method the directions of negative curvature for the Hessian matrix and its approximation should be similar. However, quasi-Newton updates do not have this property. Other types of trust region methods only minimize the approximation to the objective function in a subset of the trust region (for example, Powell's, 1970b, dogleg method), and then certain quasi-Newton methods are appropriate. For general smooth functions, the only quasi-Newton update which is known to satisfy the growth conditions needed in Powell's (1975) convergence result for trust region methods is Powell's (1970a) symmetrization of Broyden's (1965) update. In spite of the theoretical support of this update, a BFGS update with a line search approach performs better for small and medium scale problems.

The situation is different for large scale problems. Toint (1977) has extended Powell's (1970a) update to large scale minimization problems, but there is no analogue of the BFGS update which preserves sparsity and, in fact, the existence of reasonable sparse, symmetric, positive definite quasi-Newton updates is questionable. An example of Sorensen (1981) shows that these updates may become unbounded under reasonable conditions; the only possible way to salvage the situation is to modify the quasi-Newton equation in some manner.

6.2

The Current State of Optimization Software: A User's Point of View

Thomas E. Baker

Exxon Corporation, Florham Park, NJ 07932, USA.

1. Applications Areas

Nonlinear optimization software seems to defy a generic approach similar to that which has been successful in the area of linear programming. In fact, surveys of nonlinear programming software are generally organized by a characterization of the types of problems which the software can solve. In order to avoid paraphrasing recent software surveys (Waren and Lasdon, 1979; Wright, 1978; and others), I will concentrate on the types of problems encountered in practical applications areas, especially in the following areas which are of importance to Exxon:

Statistics
Design Optimization
Operations Optimization
Process Control
Manufacturing and Logistics Models
Equilibrium Models
Interactive Algorithmic Computing.

From a user's point of view, the optimization software available for statistics is considerably more advanced than that available for any other applications area. In the past many companies developed software packages for their own use. However, these in-house packages are falling into disuse due to the proliferation of powerful commercial packages which seem to be available for almost every conceivable form of computing environment. The major difficulty with applying statistical optimization packages remains, as always, the problem of applying statistics properly.

The area of design optimization is poorly defined by comparison. When possible, recourse is generally made to models based on LP (linear programming). However, it should be noted that linear, continuous design problems seem to be rare. More often a design problem is defined by means of a set of computations or a process simulation model, and then generally a closed form of the objective function does not exist. Further, analytic derivatives do not exist, decision variables are sometimes discrete, and, instead of being directly available to the optimization software, constraints are often implied by the simulation model. Problems of this sort tend to be difficult to solve, and the optimization software seems to be applied on an *ad hoc* basis.

In the area of operations optimization, most of the comments made above for design optimization still apply, since the problem is generally defined by a process simulation model. However, by virtue of the fact that operations models are run routinely (and hence are better understood), the applications of nonlinear optimization have been more successful. Thus overall operating criteria, such as the maximization of throughput or the minimization of unsaleable production, are translated into detailed operating conditions. Again, due to the lack of suitable mathematical functions, the optimization software applied in practice tends to be unsophisticated and *ad hoc*.

For the purposes of the present discussion, the area of process control is separated from that of operations optimization by virtue of the fact that the process control applications make use of measurements of variables from physical processes. These measurements can be used directly as objective function evaluations, or they can be used as data for the identification of off-line models, which in turn provide objective function evaluations. In either case, the optimization problem is complicated by the introduction of uncertainty. The addition of noise to objective function evaluations seems to cause less difficulty in unsophisticated, small-step algorithms. Again, the software that is applied in practice tends to be *ad hoc*, since it must be tailored to the process control computing environment.

Manufacturing and logistics models are 'mostly linear' models with the addition of some nonlinear functions resulting from process yields, nonlinear blends, pooling of quality/volume relationships, economies of scale, etc. Thus, the nonlinear optimization software used for these applications is usually an extension of the user's linear programming software. The nonlinear relationships are represented in closed form in some instances and by simulations in others. Most successful applications have required that the software be tailored to the user's application and to his LP package. Algorithms based on SLP (successive linear programming) have been applied in practice for many years, and they continue to be the subject of research efforts — an example of an implementation of SLP is given in Section 3 below. Moreover, some of the recent gains in computational efficiency can be attributed to new LP software facilities that allow efficient redefinitions of subproblems.

A similar applications area is the use of equilibrium models (or equilibrium seeking models) to study problems in competitive strategy analysis, multi-echelon organizational structures, market analysis, etc. Again, these models are mostly linear, and the optimization algorithms generally employ the solution of linear programming subproblems, in ways that require the software considerations mentioned above.

Interactive algorithmic computing is the centre of several applications areas that stem from the growing number of interactive problem solving environments. These applications range from simple goal seeking financial calculations, such as minimizing a debt/equity ratio, to complex interactive scheduling systems. Assuming that these applications are truly interactive, the optimization algorithms must be robust in the sense that they must be able to withstand nonsympathetic intervention by users. They must also be fast, in order to preserve the benefits of an interactive system. Over the next few years, the largest growth in applications of nonlinear optimization will most likely be in the area of interactive algorithmic computing.

2. Sources of Software

As mentioned above, optimization software for the statistics area seems to be readily available and relatively well-packaged. However, the areas of design optimization, operations optimization, and process control present the potential user with a special set of requirements involving noise, lack of derivatives, expensive function evaluations, and specialized computing environments. Even though many nonlinear optimization software libraries contain a wide variety of implementations, the probability of finding a suitably packaged algorithm which meets all the requirements for a particular problem in the above areas remains relatively small.

From the user's point of view, nonlinear optimization software can be divided into three categories — statistics packages, extensions of linear programming packages, and all the rest. Applications with functions represented in closed form generally fall into the first two categories. For the general user, to delve into the "all the rest" category requires a careful consideration of both the problem characteristics and the computing environment.

3. Industry Implemented Software

Given that an industrial practitioner is unable to find suitably packaged software to solve a particular problem, what types of algorithms might he be tempted to implement? In general, an industrial practitioner will seek the simplest means to a reasonably good solution. Most of the recent academic algorithm research seems to sacrifice simplicity to computational efficiency. Unfortunately, the high cost of software implementation and maintenance precludes the consideration of complex algorithms; marginal gains in efficiency do not justify additional algorithmic complexity. The implementation of complex algorithms can be justified only if they are the only means to solve an unsolved problem, and even this is not always a sufficient reason.

An example of the simplistic approach often taken by industry is the implementation of successive linear programming for 'mostly linear' nonlinear problems. As mentioned above, SLP software would generally be an extension of the user's existing LP system. One of the simplest approaches is as follows. Let the nonlinear problem be:

$$\begin{aligned} &\text{minimize} \quad f(x), \\ &\text{subject to } g_i(x) = 0 \quad \text{for all } i. \end{aligned} \qquad (NLP_1)$$

Guided by the Taylor series expansion about x_0, we let

$$x = x_0 + d,$$

$$f(x) \approx f(x_0) + d^T \nabla f(x_0),$$

$$g_i(x) = g_i(x_0) + d^T \nabla g_i(x_0) + s_i \quad \text{for all } i,$$

where s_i represents the error of the approximation to $g_i(x)$. Thus the linearized form of NLP_1 becomes:

$$\begin{aligned} &\text{minimize} \quad f(x_0) + d^T \nabla f(x_0) + p \sum_i |s_i|, \\ &\text{subject to } g_i(x_0) + d^T \nabla g_i(x_0) + s_i = 0 \quad \text{for all } i, \end{aligned} \qquad (LP_1)$$

where p is a positive constant that multiplies an ℓ_1 penalty term, which controls the approximation error. Note that the value of s_i, which is determined by the solution of LP_1, is equal to the higher order terms of the Taylor series expansion of $g_i(x)$ if $g_i(x) = 0$. The algorithm proceeds in the usual SLP manner with a sequence of solutions of successive updated versions of LP_1. In the algorithm, the elements of d are confined to a box ($-L_j \leq d_j \leq L_j$ for all j), which expands and contracts with the performance of the major SLP iterations. The algorithm terminates when the absolute values of all elements of d are below a specified tolerance.

The form of LP_1 is similar to the composite objective function formulation used by Palacios-Gomez, Lasdon and Engquist (1981), although the terms s_i represent the error in the approximation instead of the error in the original constraints of NLP_1. The use of these terms in LP_1 is similar to the "elastic programming" approach taken by Brown and Graves (1979) and others. The important points to notice about an SLP implementation of LP_1 are that it is:

Robust — LP_1 always has a feasible solution, allowing the user to start from any initial guess,

Reasonably Efficient — SLP is especially efficient on the highly constrained 'mostly linear' problems found in the petrochemical industry, and

Extremely Simple — What could be simpler?

Our development of the above implementation is not atypical of industrial practitioners, for we started with a simple algorithm, we worked towards a more complex one, and we stopped when reasonable results were obtained. Enhancements to the above algorithm will probably be justified when we have several (about twenty) large scale users.

4. Future Software

Because of the high cost of the implementation and the maintenance of optimization software, industrial practitioners, in general, would rather use packaged software and leave software development to others. As has been the case with optimization software for statistical applications, well-packaged commercial software can easily supplant in-house development efforts. The hope for other applications areas is that well-packaged software will be developed for a wide variety of problem requirements and computing environments.

6.3

Large Scale Nonlinear Optimization*

R.S. Dembo

School of Organisation and Management, Yale University,
Box 1A, New Haven, Connecticut 06520, USA.

1. Introduction

The purpose of this paper is to give a personal perspective on algorithms and software development for large scale nonlinear programming. It is not a survey and is biased and largely coloured by my experience in the area. The paper is organised as follows. First, by way of a few examples, I would like to convince the reader that this is an important and challenging area of research. I will then discuss some of the factors that influence the design of large scale programming algorithms. Finally, I will illustrate some new ideas that I believe will play an important role in the future.

2. Large Real-World Problems do Really Exist!

Large scale nonlinear programming is not an area of research in search of applications. Research on this topic is driven by the need to solve some important problems. For example:

---

*This work was supported in part by grant CT-06-0011 from the Department of Transportation and grant ENG-78-21615 from the National Science Foundation.

2.1. Traffic equilibrium models Currently, large amounts of money are spent in non-academic environments on the modelling of flow in communications and road networks (Florian, 1976). At the core of many of these models is a huge linearly constrained NLP which is used to estimate equilibrium flows on the network (Florian, 1976). For reasonably sized cities or communication systems these problems can have $O(10^6)$ variables and $O(10^5)$ constraints. Their special structure, however, makes them tractable even with current technology.

2.2. Computer tomography Many applications of NLP are emerging in the field of medicine. An extremely important one is the process of automated image reconstruction (computer tomography). It is possible to formulate the problem of reconstructing an image from its components as a very large constrained NLP (Herman, 1980). At present the NLP algorithms that are available for such problems are not competitive with other formulations, that are solved using Fourier transforms.

2.3. Hydroelectric power scheduling Nonlinear programming has been used for some time to assist in the management of hydroelectric power systems (Jacoby and Kowalik, 1980). The problems that are encountered here can have thousands of variables and constraints, are not necessarily separable or convex, and are sometimes not even smooth. Yet NLP models and the algorithms that are used are deemed to be reliable enough to be run on-line.

2.4. Pipe network analysis Pipe network analysis (water distribution) models are used worldwide to assist in the design and daily management of water distribution systems in large cities. These models can often be posed as convex, separable programming problems with network flow conservation constraints (Collins, Cooper, Helgason, Kennington and LeBlanc, 1978). For a city the size of Dallas, Texas the corresponding NLP has approximately 1000 variables and 700 constraints. These problems are currently attracting considerable interest, and their structure is such that very efficient, stable software will probably be available soon.

2.5. Entropy models Entropy models are used in many fields, ranging from image reconstruction to socioeconomic research (Herman, 1980; Eriksson, 1980). They can be posed as nonlinear programming problems with thousands of variables and constraints. The problems that I am familiar with in this area are all either separable or are close to being so, and they have very sparse constraint sets.

3. Factors Influencing the Design of Software/Algorithms

3.1. Trade-offs The trade-offs in large problems are often quite different and far more extreme than those which govern the design of algorithms for small problems. For example, consider the following quadratic programming problem which arises in portfolio analysis:

$$\begin{aligned} \underset{x\in\mathbf{R}^n}{\text{minimize}} \quad & \tfrac{1}{2}x^T(V^TV)x, \\ \text{subject to} \quad & \sum_{j=1}^{n} a_j x_j = b \\ & \sum_{j=1}^{n} x_j = 1. \end{aligned}$$

In this problem V is a known (totally dense) matrix and $\{a_j;\ j = 1,2,\ldots,n\}$ and b are given constants. Problems of interest in finance may have V as large as 5000 rows and 1000 columns. Thus the Hessian, V^TV, is available trivially and no special data structures are used to represent it since it is completely dense. However, recently I was able to implement a conjugate gradient algorithm that solved problems with this structure in less time than it takes to evaluate the Hessian once!

Such extreme trade-offs are likely to be the norm and not the exception. This, in my opinion, will by necessity result in a tendency towards more specialized software. To further emphasize this point, consider the problem mentioned earlier of modelling traffic flow which often takes the following form:

$$\begin{aligned} &\underset{y,x^{(i)}\in\mathbf{R}^n\ (i=1,2,\ldots,k)}{\text{minimize}} && f(y), \\ &\text{subject to } A^{(i)}x^{(i)} = b^{(i)}, && i = 1,2,\ldots,k, \\ &\sum_{i=1}^{k} x^{(i)} - y = 0, && \\ &x^{(i)} \geq 0, && \text{all } i. \end{aligned}$$

This is a nonlinear multicommodity flow problem with a convex, separable objective function. For reasonably sized road or computer networks this problem may have millions of variables and perhaps a few hundred thousand constraints. Thus, simply storing the values of all the variables in core can pose difficulties. Specialized algorithms for this problem are therefore essential, not only for reasons of efficiency but also simply to be able to cope with its size.

Finally, one of the most important and often-cited reasons for specialized software is the important role that data structures and sparsity play in obtaining an efficient algorithm. Data structures are important, but there are exceptions as the portfolio example shows. It also shows that not all large problems are necessarily sparse.

3.2. The availability of derivatives For many large problems, such as in the examples cited, first and second derivatives are often readily available and are cheap to compute. Moreover, in the above examples, one can often express the functions, gradients and Hessians compactly. For example, although traffic equilibrium models are huge, the user-supplied subroutine for these problems to compute functions, gradients and Hessians may require only a few lines of FORTRAN and a table of parameter values.

Even in cases where supplying first and second derivatives is not a trivial task, it is likely to require relatively little work compared with the cost and effort that has to be expended to collect and maintain data. Thus, in contrast to NLP research that has centered on algorithms that do not require second derivatives (since it has been assumed that such information places too large a burden on the user), in large scale programming it seems reasonable to consider devising algorithms that rely on the availability of these derivatives, which are often quite cheap to calculate.

3.3. Problem structure and general purpose software There is definitely a role for "general purpose" software for large problems. Every user cannot be expected to code specialized algorithms for each and every application. The cost of manpower and expertise involved would prohibit this.

A good example of "general purpose" software for large scale nonlinear systems is MINOS (Murtagh and Saunders, 1978). It is designed on the assumptions that the problem being solved has relatively few nonlinear variables, and that second derivatives are not available. Using these assumptions Murtagh and Saunders (1978) have created a system that sets a high standard for future developments, and that is capable of handling a wide range of large problems. Unfortunately, in almost all the examples cited above, most of the problem variables are nonlinear and hence MINOS is not well suited to them. However, these problems possess other structures that are easy to identify, such as network flow constraints. Thus, it is likely that in the future a software library for large scale problems will include different codes to handle a variety of structures, for example "mostly linear", "network flow constraints", etc. etc.

A consequence of the above is that we will have to be more innovative in our use of structure than we have been in the past. The balance between "degree of generality" and "the ability to generate robust, efficient software" is an art.

There are classes of nonlinear functions that are ideally suited to large scale programming, because their domain of applicability is wide enough to model a significant portion of the problems we are likely to encounter, and because they possess properties that are extremely convenient for the design of software that is robust, efficient and easy to use. Some examples are factorable functions (Fiacco and McCormick, 1968) and generalized polynomials (Dembo, 1979).

In building a nonlinear programming model one has a large amount of flexibility as regards the functions that are used. If reliable, efficient software were available for large generalized polynomial programming problems, for example, we would certainly start to see more real-world models exhibiting this form; i.e. the supply will create the demand.

4. Directions for Future Research

4.1. Some constructive theory There are some basic properties that nonlinear programming algorithms must possess if they are to exhibit desirable convergence characteristics. One of these is that an algorithm will be rapidly convergent if and only if the search direction it generates is a sufficiently good approximation to the Newton direction. A natural question to ask is just how close (in terms of easily measured parameters) should the approximate direction be to one generated using Newton's method. This is particularly relevant in large scale systems, where there is a marked trade-off between the cost of approximating the Newton direction and the overall speed of convergence.

This question has been answered (in one particular framework) for systems of nonlinear equations in Dembo, Eisenstat and Steihaug (1980). The result has been used in Dembo and Steihaug (1980) to construct a globally convergent algorithm for large, unconstrained nonlinear programming problems in which a user, by varying a single parameter, can choose to trade-off the asymptotic rate of convergence with the cost per iteration.

The real importance of the above theory lies in its constructive nature. The same philosophy can be applied to constrained problems where the need for controlling such trade-offs is even more essential. Consider the following equality constrained NLP:

$$\text{(ENLP)} \quad \underset{x \in \mathbf{R}^n}{\text{Minimize}} \quad f_0(x) \tag{1a}$$

$$\text{subject to } f_i(x) = 0, \quad i = 1,2,\ldots,m, \tag{1b}$$

where $f_i : \mathbf{R}^n \to \mathbf{R}$. Let $\lambda \in \mathbf{R}^m$ be a vector of multipliers associated with the constraints (1b). Also, let $J(x)$ denote the $n \times m$ Jacobian matrix of the constraint functions $f_i(x)$ in (1b). The columns of $J(x)$ are the gradient vectors $\nabla f_i(x)$.

A feasible point, x^*, is said to satisfy first order optimality conditions if there exist multipliers λ^* such that

$$g_L(x^*,\lambda^*) \equiv g_0(x^*) + \sum_{i=1}^{m} \lambda_i^* \, g_i(x^*) = 0, \tag{2}$$

where $g_i : \mathbf{R}^n \to \mathbf{R}^n$ is used to denote the gradient of f_i. It is well known that, if x^* is the solution of (ENLP), and if $J(x^*)$ is of full rank, then such multipliers do exist. Thus (2) and (1b) form a system of $n+m$ equations in $n+m$ unknowns that may be solved to yield stationary points x^*. The "Inexact Newton Method" of Dembo, Eisenstat and Steihaug (1980) for solving this system consists of the following steps.

For $k = 0,1,2,\ldots$ until convergence do (3a)—(3c) as follows:
Find *some* step (p_k, w_k) such that the norm of the residual vector

$$\begin{pmatrix} r_p \\ r_w \end{pmatrix}_k = \begin{pmatrix} g_L(x_k,\lambda_k) \\ f(x_k) \end{pmatrix} + \begin{pmatrix} H_L(x_k,\lambda_k) & J(x_k) \\ J(x_k)^T & 0 \end{pmatrix} \begin{pmatrix} p_k \\ w_k \end{pmatrix} \tag{3a}$$

satisfies

$$\phi_k \equiv \left\| \begin{pmatrix} r_p \\ r_w \end{pmatrix}_k \right\| / \left\| \begin{pmatrix} g_L \\ f \end{pmatrix}_k \right\| \leq \eta_k, \tag{3b}$$

where η_k is a parameter. Then set

$$\begin{pmatrix} x_{k+1} \\ \lambda_{k+1} \end{pmatrix} = \begin{pmatrix} x_k \\ \lambda_k \end{pmatrix} + \begin{pmatrix} p_k \\ w_k \end{pmatrix}. \tag{3c}$$

Here η_k may depend on (x_k,λ_k); taking $\eta_k \equiv 0$ gives Newton's method. The notation $H_L(x,\lambda)$ is used to denote the Hessian of the Lagrangian function

$$L(x,\lambda) \equiv f_0(x) + \sum_{i=1}^{m} \lambda_i f_i(x) , \tag{4}$$

and $f : \mathbf{R}^n \to \mathbf{R}^m$ is the vector whose components are $\{f_i(x);\ i = 1,2,\ldots,m\}$.

The importance of Inexact Newton methods is apparent from the following result, which may be obtained by direct application of Theorems 2.3 and 3.3 in Dembo, Eisenstat and Steihaug (1980).

<u>Theorem</u> If $(x_k,\lambda_k) \to (x^*,\lambda^*)$ as $k \to \infty$, then the sequence $\{(x_k,\lambda_k)\}$ generated by the Inexact Newton method (3) will be at least:

(i) linearly convergent if $\phi_k \leq c_L < 1$ as $k \to \infty$;

(ii) superlinearly convergent iff $\phi_k \to 0$ as $k \to \infty$; and

(iii) quadratically convergent iff

$$\phi_k \Big/ \left\| \begin{pmatrix} g_L \\ f \end{pmatrix}_k \right\| \le c_Q \quad \text{as } k \to \infty ,$$

where c_L and c_Q are constants.

Since most existing algorithms for solving (ENLP) may be viewed as Inexact Newton methods, the above theorem shows that, no matter what method is used, it must essentially obtain a sufficiently accurate solution to (3) if it is to converge rapidly. By rapid convergence here I mean either fast linear convergence (c_L small) or superlinear or quadratic convergence.

4.2. Primal and Dual Inexact Newton methods There are two basic strategies for computing the step (p,w) in (3a); I refer to them as Primal and Dual Inexact Newton methods, since they operate on the primal and dual quadratic programming problems that are equivalent to (3a).

Any $p \in \mathbf{R}^n$ may be expressed as

$$p = Zv + Ju,$$

where $v \in \mathbf{R}^{n-m}$, where $u \in \mathbf{R}^m$, and where Z is a matrix of rank n-m in $\mathbf{R}^{n\times(n-m)}$ such that $J^T Z = 0$. A *Primal Inexact Newton* method then has the form:

"solve" for $u \in \mathbf{R}^m$ from $f + J^T J u = r_u \approx 0,$ (5a)

"solve" for $v \in \mathbf{R}^{n-m}$ from $Z^T(g_L + H_L J u) + (Z^T H_L Z)v = r_v \approx 0,$ (5b)

compute p from $p = Zv + Ju,$ (5c)

update x by $x_+ = x + p.$ (5d)

Note that the residuals r_u and r_v provide a mechanism for trading off work per iteration with rate of convergence.

A *Dual Inexact Newton* method has the form:

"solve" for w from $J^T H_L^{-1} g_L - f + (J^T H_L^{-1} J)w = r_d \approx 0,$ (6a)

compute p from $p = -H_L^{-1}(g_L + Jw),$ (6b)

update x by $x_+ = x + p,$ (6c)

update λ by $\lambda_+ = \lambda + w.$ (6d)

In a large scale setting, equations (5a), (5b) and (6a) may have to be "solved" using iterative methods with low storage overheads. When iterative methods are used on these equations I prefer to call the above methods *Primal and Dual Truncated Newton* methods, since the residuals r_u, r_v and r_d result because these iterative methods are *truncated* prior to obtaining an exact solution.

To analyse the convergence rate of Primal and Dual Truncated Newton methods using the above theorem, one has to be able to relate the residuals r_u, r_v and r_d to r_p and r_w. For example, it is easily shown that in Dual Truncated Newton methods as given by (6), $\|r_p\| \equiv 0$ and $\|r_w\| \equiv \|r_d\|$. Hence the above theorem provides a *constructive* tool for analysing the convergence rate of such methods. That is, $\|r_d\|$ is a measurable and controllable quantity and the theorem tells us how small to make $\|r_d\|$ depending on the desired rate of convergence. The situation for Primal Truncated Newton methods is somewhat more complex as is the case when errors are permitted in (6b) in dual methods.

There are many other possible variants of the above idea. For example, if only first derivatives are available, one may wish to approximate H_L by finite differences. One can show that in this case the above theorem still holds provided that the difference parameter goes to zero sufficiently rapidly.

Both the primal and dual methods are well-suited (but certainly not limited) to large scale problems, because they allow the possibility of using iterative methods with low storage overhead. In this context there is therefore a natural way to define a dividing line between small and large problems. A problem is said to be large if there is insufficient storage to be able to solve the appropriate linear systems in core, using a direct method. Naturally, this definition is computer dependent as it should be.

<u>4.3. Successive inexact quadratic programming methods</u> It is easy to show that the system (3a), (3b) with $\eta_k \equiv 0$ is identical to the optimality conditions of the subproblems that arise in successive quadratic programming methods. An extremely important consequence of the above characterization theorem, therefore, is that it indicates precisely how accurate a solution is needed in each quadratic subproblem in order to achieve a (preset) desired convergence rate.

Why not simply solve each quadratic subproblem accurately? Firstly, there is no justification (either theoretical or empirical) for using a quadratic model when far from a solution. If our experience in unconstrained optimization is indicative of anything (Dembo and Steihaug, 1980), an inaccurate solution may do just as well or perhaps better. Secondly, if an iterative method is used to solve the quadratic subproblem, then as before there is a direct trade-off between the cost per iteration and the accuracy to which (3a) is solved. It therefore seems plausible to use *successive inexact quadratic programming* (SIQP) algorithms even for small problems. Certainly, in large scale programming, SIQP algorithms are likely to dominate their exact counterparts. They can have all the advantages (i.e. the same asymptotic behaviour), and may result in far less work per iteration without (hopefully) a significant increase in the number of iterations. The limited experience in Dembo and Steihaug (1980) indicates that on large problems an inexact approach can be as much as eight times faster than methods based on an accurate solution to the quadratic subproblem.

<u>4.4. Inexact active set strategies</u> There are other areas of nonlinear programming that may benefit from the philosophy upon which Inexact Newton methods are based; namely, there is no need to com-

pute anything more accurately than can be justified theoretically. For example, active set strategies are a convenient mechanism for solving inequality constrained problems. An *inexact active set strategy* is one in which one would minimize on any given active set only to an accuracy that is essential to guarantee global convergence. Heuristics for varying the optimality tolerance on an active set have been implemented (see Murtagh and Saunders, 1978, for example). What is lacking, however, is a sound theoretical basis.

5. Conclusions

Certainly, the future of large scale programming depends on our ability to generate theory that is constructive. That is, theory, such as in Dembo, Eisenstat and Steihaug (1980), that gives an insight into the various trade-offs that are involved, and that is expressed in terms of measurable quantities.

Inexact successive quadratic programming methods are based on constructive theory, and are likely to play an important role in software and algorithms for large scale nonlinear programming.

6.4

The Current State of Constrained Optimization Software

Klaus Schittkowski

Institut für Angewandte Mathematik und Statistik,
Universität Würzburg, Am Hubland, D-87 Würzburg, W. Germany.

1. The Nonlinear Programming Problem

The constrained nonlinear programming problem can be written in the form

$$x \in \mathbf{R}^n: \left.\begin{array}{lll} \min & f(x) & \\ & c_j(x) = 0, & j = 1,2,\ldots,m_e, \\ & c_j(x) \geq 0, & j = m_e+1,\ldots,m \end{array}\right\} . \qquad \text{(NLP)}$$

In the following, we proceed from the assumptions that

a) the problem is small, i.e. the number of variables does not exceed 100,
b) the problem is smooth, i.e. the functions f and c_j are continuously differentiable,
c) the problem is dense, i.e. the Hessian of the Lagrangian is dense.

In particular, we do not consider problems with special structures (e.g. linearly constrained or least squares problems), and large, discrete, or nonsmooth problems. The reason for the restrictions is found in the topic of this paper, namely to investigate existing software for the general problem (NLP). The other problem types mentioned above are either specializations of (NLP) or generally applicable software does not exist.

2. The Gap between a Mathematical and an Implemented Algorithm

To understand some of the subsequent statements about the availability and usage of optimization programs, let us consider the process of developing a new solution method for (NLP). We can distinguish between the following three levels:

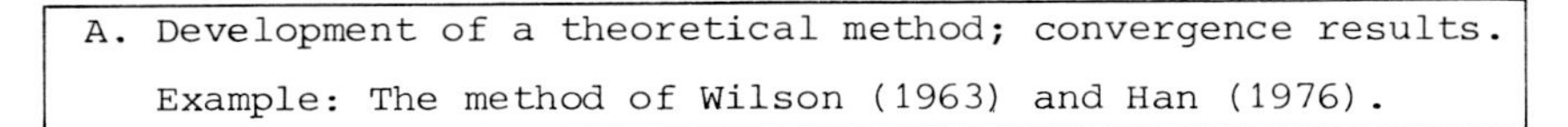

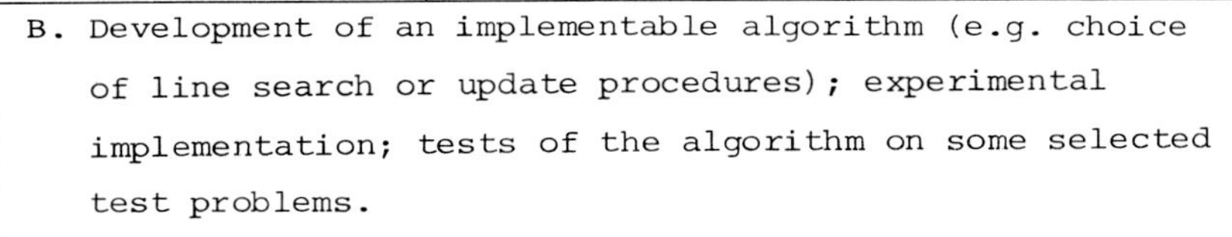

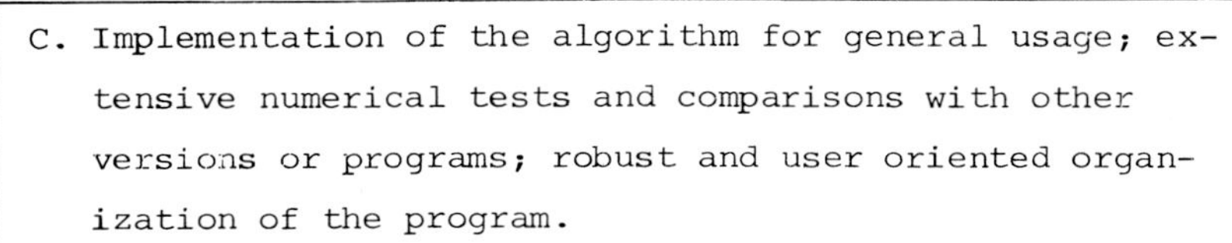

In a natural way, the process starts at A above and ends at C. However, there exists a feedback in the sense that the development of an algorithm or a method can subsequently be influenced by the numerical experience of a lower level. Nevertheless, there is a gap between the levels A and C which can be characterized by two observations:

a) A carefully implemented algorithm can be more reliable than another one for which stronger convergence statements exist, or, conversely an algorithm could fail to solve a problem although the theory guarantees convergence (e.g. because of numerical instabilities due to an increasing penalty parameter).

b) Different implementations of one mathematical algorithm could give computer programs with very different numerical performances (e.g. by using different algorithms to solve quadratic subproblems, see Schittkowski, 1981a).

To overcome the gap, we need more detailed convergence investigations of the mathematical algorithms and, on the other hand, more numerical experience and knowledge about the implementation and performance of the corresponding computer programs.

3. Software for Constrained Nonlinear Programming

The following review of nonlinear programming codes is based on the programs which have been submitted for a comparative study (Schittkowski, 1980b). Most of the codes belong to category C defined in the last section. However, some of them should be considered as experimental programs. This collection can be considered as a representative set of existing optimization programs, although we know that several other codes have been developed and are used for practical applications. But most of the remaining programs are experimental ones, or are outdated, or are not distributed in an unrestricted way.

First, the optimization codes under consideration are classified from the viewpoint of availability. Furthermore, the institution is shown where the program has been developed.

a) Subroutine libraries:

VF01A/VF02AD : Harwell Subroutine Library, AERE, Harwell,

England.

SALQDR/SALQDF/ SALMNF : Optimization Library, National Physical Laboratory, Teddington, England.

b) Programs which can be ordered without reservation:

GRG2 : Department of General Business, University of Texas, Austin, USA.

NLP : Fachgruppe für Automatik, ETH, Zürich, Switzerland.

c) Published programs:

LPNLP : Electronics Research Laboratory, Montana State University, Bozeman, USA (Pierre and Lowe, 1975).

CONMIN : Atomic Energy Board, Pretoria, South Africa (Kuester and Mize, 1973).

SUMT : Research Analysis Corporation, McLean, Virginia, USA (Kuester and Mize, 1973).

d) Institutional programs:

GRGA : Electricité de France, Paris, France.

OPRQP/XROP : Numerical Optimisation Centre, The Hatfield Polytechnic, Hatfield, England.

OPT/BIAS : School of Mechanical Engineering, Purdue University, West Lafayette, Indiana, USA.

FUNMIN/FMIN : Deutsche Forschungs- und Versuchsanstalt für Luftund Raumfahrt, Oberpfaffenhofen, Germany.

GAPFPR/GAPFQL : Department of Chemical Engineering, University of Texas, Austin, Texas, USA.

DFP : Computing Center, S.U.N.Y., New York, USA.

ACDPAC/FCDPAK : Department of Combinatorics and Optimization, University of Waterloo, Waterloo, Ontario, Canada.

The programs of classes a) and b) have been designed for general distribution and usage on different kinds of computer installations. They are well documented, possess a high standard of software quality, and can be obtained by any user for a fee. The programs of class c) are published in the literature, but a user has to do the work of typing them. Most of the submitted optimization codes belong to the institutional codes of class d). They have been developed to solve internal practical optimization problems and are, in general, less well documented and not as user oriented as the codes of the other classes.

Another classification of the above programs is obtained by investigating their mathematical methods. The following list shows the programs and the underlying mathematical methods; furthermore it gives the name of the author, a reference, and the approximate date when the development of the code was finished.

a) Penalty methods:

SUMT : Fiacco and McCormick (1968), 1969.
DFP : Indusi (1972), 1972.
FMIN : Lootsma (1974), 1974 (ALGOL 60 version).
NLP : Rufer (1978), 1978.

b) Multiplier or augmented Lagrangian methods:

CONMIN : Haarhoff and Buys (1970), 1969.
VF01A : Fletcher (1975), 1973.
LPNLP : Pierre and Lowe (1975), 1975.
GAPFPR/GAPFQL : Newell and Himmelblau (1975), 1975.
SALQDR/SALQDF/
SALMNF : Gill and Murray (1974b), 1975.
FUNMIN : Kraft (1977), 1976.
BIAS : Root (1977), 1977.
ACDPAC : Best and Bowler (1978), 1978.

c) Generalized reduced gradient methods:

GRGA : Abadie (1975), 1975.

OPT : Gabriele and Ragsdell (1976), 1976.

GRG2 : Lasdon and Waren (1978), 1978.

d) Quadratic approximation, projected Lagrangian, quasi-Newton, and recursive quadratic programming methods:

OPRQP : Bartholomew-Biggs (Biggs, 1972), 1972.

FCDPAK : Best (1975), 1975.

VF02AD : Powell (1978a), 1978.

XROP : Bartholomew-Biggs (1979), 1979.

Most of the submitted programs apply a multiplier method and have been developed in the years between 1975 and 1978. The time delay between the development of a new theoretical algorithm and its implementation for a practical usage increases from year to year. The reason is that a higher standard of software quality, robustness, and reliability are required today before a broad distribution of a program is allowed.

4. Numerical Performance and Organization of Nonlinear Programming Codes

A detailed investigation of the numerical performance and the organization of nonlinear optimization programs is found in Schittkowski (1980b) and in Hock and Schittkowski (1981). Some conclusions and remarks are summarized below, which are based on the extensive numerical tests of the comparative study of Schittkowski (1980b).

a) Efficiency: The main reason for developing new optimization algorithms has been in the past to improve their efficiency. The efficiency depends on the mathematical method, but could be influenced significantly by the current implementation (see Schittkowski, 1981a). A rough ranking of the mathematical methods with respect to their efficiency is given in the following list:

1) Quadratic approximation, projected Lagrangian, quasi-Newton and recursive quadratic programming methods.
2) Generalized reduced gradient methods.
3) Multiplier or augmented Lagrangian methods.
4) Penalty methods.

The bound between multiplier and generalized reduced gradient methods is not sharp. A carefully implemented multiplier method can be more efficient than a generalized reduced gradient algorithm.

b) Reliability: This criterion is most important for practical applications, and can be considered as a measure of the probability that a given problem will be solved successfully. The reliability of a program depends on both the mathematical method and the numerical implementation. The most reliable codes of the comparative study are GRGA, GRG2, and VF02AD.

c) Ease of use: The organization, structure, and description of a program determine whether a problem can be solved more or less easily by a user. The criterion is independent of the mathematical method. The best ease of use scores of the comparative study (Schittkowski, 1980b) are obtained by the codes in the program libraries (SALQDR/SALQDF/SALMNF, VF01A/VF02AD) and the codes GRG2 and NLP.

d) Program organization: The optimization programs are organized in very different ways. Although the codes have the common property that they are all written in FORTRAN, they use different input strategies (e.g. data are read in or provided in the driving program), they have fixed or variable array dimensions, they differ in the way the user provides the problem functions and their gradients, and they even differ in the formulations of the underlying problem type (NLP). It would be very worthwhile to develop software standards for nonlinear programming algorithms, in particular to facilitate the exchange of programs.

e) Portability: Most of the submitted optimization codes are not portable with respect to the FORTRAN compiler used (standard Telefunken TR440 compiler). However, the few statements that prevented a compilation could be repaired so easily that this effort was negligible compared with the overall working time to check the validity of the driving program, to prepare the provision of the problem data and functions, and to solve the problem in the desired way.

6.5

Software for Constrained Optimization*

Philip E. Gill, Walter Murray,
Michael A. Saunders and Margaret H. Wright

Systems Optimization Laboratory,
Department of Operations Research,
Stanford University, Stanford, California 94305, USA.

1. Introduction

In this paper, we shall consider several aspects of the development of methods and the associated software for constrained optimization problems. Our concern is only with *library-quality general purpose software*. Library-quality software must be robust, reliable and transportable; general purpose software must be able to cope with extreme variations in user sophistication and in the nature of the problem to be solved. Thus, the observations in this paper are not intended to apply to software that is experimental, application dependent, or designed solely to aid research.

The development of mathematical software as a research area has introduced several topics that are largely independent of the problem class for which the software is intended. For example, the issue of the user interface is germane whenever the problem definition involves an (unknown) user-supplied function or set of

*This research was supported by the U.S. Department of Energy Contract DE-AC03-76SF00326, PA No. DE-AT03-76ER72018; National Science Foundation Grants MCS-7926009 and ECS-8012974; the Office of Naval Research Contract N00014-75-C-0267; and the U.S. Army Research Office Contract DAAG29-79-C-0110.

functions; the topics of software organization and modularity are relevant whenever closely related calculations are performed within different methods for the same problem, or methods for separate but related problems. Furthermore, optimization methods are not unique in their flexible form — i.e., the dependence of certain portions of the algorithm on parameters whose optimal values are problem dependent. For further discussion of these and other issues, see, e.g., Cody (1974); Ford and Hague (1974); Smith, Boyle and Cody (1974); Shampine and Gordon (1975); Lyness and Kaganove (1976); and Gill, Murray, Picken and Wright (1979).

Rather than consider the large class of software issues that are not restricted to optimization, we shall concentrate on *the effects of implementation (as software) on constrained optimization algorithms*. We shall illustrate these effects with a few brief examples from constrained optimization. Only "dense" problems will be considered (in which no advantage is taken of sparsity); the issues in software design for sparse problems are even more complicated than for the dense case.

2. Linearly Constrained Optimization

The problem to be considered in this section is

$$\left.\begin{array}{ll} \underset{x\in\mathbf{R}^n}{\text{minimize}} & F(x) \\ \text{subject to} & Ax = b \\ \text{or} & Ax \geq b \end{array}\right\} \qquad (1)$$

,

where F is a smooth function and A is an $m\times n$ matrix.

A very general class of methods for such problems is that of *feasible point active set methods* (see, e.g., Gill and Murray, 1977a). This name reflects two important properties of the methods Firstly, all iterates are feasible after an initial feasible point has been found by a "phase 1" procedure. Secondly, a subset of the constraints — which we shall call the *working set* — is used to define the search direction at each iteration; typically, the working set includes constraints that are exactly satisfied ("active") at the current iterate.

Assume that the working set contains t constraints, and let $\hat{A}$ denote the matrix of coefficients of the constraints in the current working set. The search direction p is constrained to lie within the subspace of vectors that are orthogonal to the rows of $\hat{A}$; in effect, the constraints in the working set are treated as *equalities* for the purpose of defining the search direction p. Let Z denote a matrix whose columns form a basis for the set of vectors orthogonal to the rows of $\hat{A}$, so that $\hat{A}Z = 0$; the search direction p is then of the form

$$p = Zp_Z, \tag{2}$$

where p_Z is an (n-t)-vector. This definition ensures that $\hat{A}p = 0$, and hence moves along p do not alter the values of any of the constraints in the working set.

Depending on the nature of the algorithm, there are various strategies for adding and deleting constraints from the working set. Typically, a constraint is *added* to the working set when its presence restricts the step length to be taken along p. An inequality constraint is *deleted* from the working set when a sufficiently accurate Lagrange multiplier estimate predicts that the constraint is unlikely to be active at the solution.

We now briefly discuss two ways in which considerations of implementation have a substantive effect on such an algorithm.

2.1. Representation of a basis for the null space For dense problems, Z may be taken as a section of the orthogonal matrix Q in an orthogonal factorization of $\hat{A}$ (see, e.g., Gill, Murray and Wright, 1981). Because of the special nature of changes in the working set, this factorization (and hence Z) can be *updated* rather than recomputed after each such change. For example, when a constraint is deleted from the working set, the updated basis for the null space (which we shall denote by $\bar{Z}$) consists of all the columns of the old Z, plus a single new column that is a linear combination of the remaining columns of Q. Obviously, the theoretical properties of $\bar{Z}$ are not affected by the position in which the new column appears. However, it is advantageous in the implementation for the new column of $\bar{Z}$ to be added *in a particular place* with respect to the columns of the old Z, namely as *the last column* of $\bar{Z}$, i.e.

$$\bar{Z} = (Z \quad z). \tag{3}$$

The ordering of the columns of $\bar{Z}$ is relevant because of the need to compute the Cholesky factorization of the projected Hessian (or its approximation). Let G denote the Hessian matrix at the current point. When $\bar{Z}$ is given by (3), the new projected Hessian is

$$\bar{Z}^T G \bar{Z} = \begin{pmatrix} Z^T GZ & Z^T Gz \\ z^T GZ & z^T Gz \end{pmatrix},$$

and only one further step of the Cholesky factorization needs to be carried out to obtain the factors of $\bar{Z}^T G\bar{Z}$ from those of $Z^T GZ$. The housekeeping associated with the update would be more complicated if the new column of $\bar{Z}$ appeared in any position except the last.

Therefore, when implementing active set methods of this type, programming efficiency can be increased by using a variant of the usual orthogonal factorization which we shall call the *TQ factorization*

$$\hat{A}Q = (0 \quad T),$$

where T is a "reverse" triangular matrix such that $t_{ij} = 0$ for $i + j \le n$. (Clearly, T is simply a lower triangular matrix with its columns in reverse order.)

The TQ factorization has an implementation advantage because existing columns of Z need not be reordered after a change in the working set. Note that the *first* n-t columns of Q define Z, i.e.

$$Q = (Z \quad Y),$$

where the columns of Y form a basis for the range space of $\hat{A}^T$. When a row is deleted from $\hat{A}$, the updated matrix $\bar{Z}$ takes the form (3), with its new column z appearing in the correct position as the first column of Y (suitably transformed).

Similarly, when a row is added to $\hat{A}$, the last column of Z could sometimes become the first column of Y, with no change to Q at all. Although rare, this situation can be easily recognized, and again there is no need to move columns of Z.

2.2. Special treatment of bounds A second effect of implementation results from the observation that many problems of the form (1) contain a mixture of general linear constraints and simple bounds on the variables (often with a preponderance of bounds). In a theoretical description of an active set method, it is immaterial whether or not a row of $\hat{A}$ happens to be a row of the identity. However, when a large number of bound constraints are active, improvements in efficiency (both in storage and computation time) can be acnieved if advantage is taken in the implementation of the distinction between simple bounds and general linear constraints. This distinction is fundamental in codes for large scale linearly constrained problems.

At a typical iteration, suppose that $\hat{A}$ contains r bound constraints and t-r general constraints. The variables corresponding to the active bounds are effectively "fixed" on those bounds during the current iteration, while the remaining n-r variables are free to change. Let the suffixes "FX" and "FR" denote vectors or matrices corresponding to the two sets of variables. If we assume for simplicity that the last r variables are fixed, the matrix of constraints in the working set can be written as

$$\hat{A} = \begin{pmatrix} 0 & I_r \\ \hat{A}_{FR} & \hat{A}_{FX} \end{pmatrix}, \tag{4}$$

where I_r is an $r \times r$ identity matrix, and $\hat{A}_{FR}$ is a $(t-r)\times(n-r)$ matrix. When $\hat{A}$ is of the form (4), the relationship $\hat{A}p = 0$ implies that the presence of r bounds in the working set "removes" r columns of the general constraints in the working set. This means that substantial economies can be achieved in the calculation of the search direction and the Lagrange multipliers when r is large relative to t.

In particular, a suitable matrix Z whose columns span the null space of the matrix $\hat{A}$ of (4) is given by

$$Z = \begin{pmatrix} Z_{FR} \\ 0 \end{pmatrix},$$

where Z_{FR} is an $(n-r)\times(n-t)$ matrix whose columns form a basis for the null space of $\hat{A}_{FR}$. Thus, only the orthogonal factorization of $\hat{A}_{FR}$ needs to be computed. The search direction is of the form $p = Zp_{FR}$, where the definition of p_{FR} depends on the particular algorithm. For example, in a Newton-type method p_{FR} satisfies the equations

$$Z_{FR}^T G_{FR} Z_{FR} p_{FR} = -Z_{FR}^T g_{FR}.$$

Similar savings in computation can be achieved during the calculation of the Lagrange multiplier estimates.

The effect on $\hat{A}$ of changes in the working set depends on whether a general or a bound constraint is involved. If a *bound* constraint enters or leaves the working set, the *column* dimension of $\hat{A}_{FR}$ and $\hat{A}_{FX}$ alters; if a *general* constraint enters or leaves the working set, the *row* dimension of $\hat{A}_{FR}$ and $\hat{A}_{FX}$ alters. In both cases, it is possible to update a factorization of $\hat{A}_{FR}$. Although the updating procedures are more complicated when bounds are treated specially, the overall savings can often be substantial.

3. Nonlinearly Constrained Optimization

3.1. Issues in the implementation of QP-based methods For nonlinearly constrained problems, there is no consensus concerning the most effective general algorithm. In this section, we shall consider the implementation of a projected Lagrangian method in which a QP subproblem is solved at each iteration to determine the direction of search (see Gill, Murray and Wright, 1981). Certain crucial issues remain unresolved in the theory as well as the implementation of these methods (for a more detailed discussion, see, e.g., Murray and Wright, 1980). It is of interest that some of the difficulties became apparent only after the methods had been implemented.

The following topics are particularly relevant in practical implementations: the detection and treatment of incompatible or ill-conditioned linear constraints of the QP subproblem; the determination of the active set; and the definition and computation of consistent Lagrange multiplier estimates. To illustrate the nature of the associated questions of implementation, we note that the detection of incompatibility in the linear constraints of the subproblem can be performed directly for certain forms of the subproblem, whereas a relatively expensive iterative procedure is required for others. Furthermore, the efficiency of an implementation will be affected by whether it is possible to apply the results of computation with an incompatible set of linearized constraints in formulating alternative sets of constraints.

Whatever the results of further study of such issues, the implementation of QP-based methods will be affected by the fact that many (if not most) practical problems contain a mixture of simple bounds, linear constraints and nonlinear constraints, and not simply nonlinear constraints. Therefore, the question arises concerning the ways in which software for QP-based methods can be designed to take advantage of the known linearity of certain constraints. In particular, it would be desirable for such a method to be efficient when all the constraints are linear.

In theory, some QP-based algorithms handle linear constraints "naturally", in the sense that under certain conditions the search direction will satisfy the properties mentioned in Section 2 with respect to the linear constraints. Unfortunately, an implementation of such a method will not *automatically* treat linear constraints efficiently. For example, it would be extremely costly in computer time if an implementation failed to exploit the fact that the gradient of a linear constraint is constant and its Hessian is zero, since unnecessary calls to user-provided subroutines would otherwise be generated. Furthermore, some QP-based methods would not necessarily retain feasibility with respect to the linear constraints (for some problems, this would lead to the danger of encountering singularities in the nonlinear functions).

Separate treatment of linear constraints is beneficial even when a feasible point QP-based method is used. For example, the barrier trajectory algorithm (Wright, 1976; Murray and Wright, 1978) produces iterates that remain strictly feasible with respect to *inequality* constraints, but the associated barrier transformation cannot be applied to *equality* constraints. However, this method may be used for problems with *linear* equality constraints if linear constraints are treated separately.

The considerations mentioned in this section are of special importance when solving large scale problems. QP-based methods will be applied successfully to large scale problems only if constraint linearities can be exploited whenever possible to reduce computation and storage.

3.2. Building new software from existing software In order to reduce the effort associated with implementing a new method, software designers often take advantage of existing software. However, the best way to use existing software is not always obvious. Certainly it is highly desirable to use low-level modules as much as possible in developing new software, particularly when adding to a software library in which the number of new routines should be minimized. Certain computations recur in most optimization algorithms — particularly linear algebraic operations — and substantial savings in programming effort and debugging time usually result when existing routines of this type are used without modification. The advantages of using modular software as building blocks will become even more significant due to the increasing trend of producing special purpose optimization methods.

Despite the benefits of using existing software, there are instances in which it is inappropriate to use an existing optimization software package as a "black box" within a new method. Certain algorithms for nonlinearly constrained optimization require the solution of a sequence of unconstrained or linearly constrained subproblems (or a single such subproblem). Although it is conceptually convenient simply to assume that standard methods may be used to solve these subproblems, there are several pitfalls if existing packages are inserted without modification into an implementation.

To illustrate the difficulties, consider an augmented Lagrangian method in which the next iterate is the result of an unconstrained subproblem (for a description of these methods, see, e.g., Fletcher, 1977). Unfortunately, it cannot be guaranteed *a priori* that the unconstrained subproblem has a bounded solution, and hence even the best available "black-box" unconstrained package would be unable to solve the subproblem. Depending on the failure procedures of the unconstrained routine, the augmented Lagrangian method might fail unnecessarily, or might expend a large amount of wasted effort — an undesirable result in either case. In addition, there may be complex programming difficulties in communicating information about the nonlinearly constrained problem to the unconstrained routine.

Similarly, use of some existing QP packages to solve the subproblem in a QP-based method may mean that it is difficult to take advantage of certain efficiencies. For example, it is highly desirable to be able to specify an initial working set in the QP subproblem (in order to save effort in the phase 1 calculations), but most QP software does not allow this facility.

4. General Observations

4.1. Flexibility In developing software libraries, some compromises may be necessary because of the conflicting demands of users. For example, it is well known that some users may prefer an inferior method simply because the software is easier to use. The documentation of a function comparison ("polytope") algorithm in the NAG numerical software library (Mark 8) includes numerous strong warnings that a particular routine should be used only as a last resort. Nonetheless, in one survey of the usage of NAG routines, over half the calls to the optimization routines were to this method!

In view of the inevitable disparity between the aims of the typical user and the software designer, good optimization software should contain a *range* of options. Precisely how such flexibility should be provided remains an open question. Ideally the user should be able to specify values for an arbitrary range of parameters, with the remainder taking default values. However, all mechanisms for achieving this type of flexibility in Fortran, for example, have substantial drawbacks (see, e.g., Gill, Murray, Picken and Wright, 1979).

4.2. The dangers of automatic procedures Given the desirability of making routines easy to use, it can be argued that certain procedures should be included automatically in optimization software. We illustrate some of the resulting difficulties with the example of automatic scaling.

It is unarguable that automatic scaling procedures are helpful for many problems. However, this does *not* imply that automatic scaling procedures should be built into general purpose optimization software. Scaling ultimately reduces to the need to define what is "small". Since there is no universal definition of "small", the "correct" scaling of a problem must depend on the problem, and cannot be determined automatically.

Any automatic scaling procedure will inevitably be harmful on some problems. For example, certain techniques attempt to calibrate the variables of a problem so that they are all of similar size. However, this procedure would be inappropriate when the difference in size of the variables has deliberately been chosen by the user to reflect (in some sense) their significance.

The difficulties become even more complicated in attempting to scale constraints. For example, consider the following matrix:

$$\begin{pmatrix} 1 & 1 \\ 0 & \varepsilon \end{pmatrix} . \tag{5}$$

An estimate of the rank of (5) depends on whether ε may be treated as "negligible", and obviously there is no universal answer.

With this example in mind, consider the constraint matrix

$$\begin{pmatrix} 1 & 1 \\ -10^{-6} & 10^{-6} \end{pmatrix} . \tag{6}$$

If (6) were the matrix of constraint gradients in a general problem, an automatic scaling procedure of the type mentioned above might attempt to re-scale the matrix so that the largest element in each row and column is unity; hence, (6) would become

$$\begin{pmatrix} 1 & 1 \\ -1 & 1 \end{pmatrix} .$$

Such a re-scaling is based on the assumption that the elements of the second row of (6) are fully significant numbers; however, this will not always be true (in some contexts, the elements of the second row may be at "noise level"). The matrix (6) becomes the form (5) after reduction to triangular form, and hence no single scaling procedure is valid for all cases.

Note that, in the unlikely event that the user could specify accurately the meaning of "negligible" for every variable and every combination of constraints, it might be possible to develop a sensible (albeit complicated) automatic scaling procedure. However, it would still be necessary to have a set of default values for users who did not specify these definitions.

The best solution to this difficulty seems to be to provide library service routines that perform a selection of automatic scaling procedures, and then allow the user the option of using some of the automatic scaling procedures, or none at all. In addition, it may be desirable to allow the specification of user-provided scale factors, with suitable default values (see, e.g., Gay, 1980a).

4.3. The treatment of failure exits and numerical stability We believe that library-quality software should take a *conservative* view of failure and numerical stability. In particular, we believe that a routine should not declare that it has converged until "stringent" criteria have been satisfied, and that a "reasonable" level of additional computation to improve reliability is almost always justified. (Obviously, it is difficult to define these terms precisely.)

This view has certain disadvantages. For example, a routine may indicate uncertainty (and even failure) when the correct solution has been found, or may perform unnecessary computation. The first possibility causes distress for users who do not understand a message such as "the method has probably converged", and the second annoys those with an extreme concern about computation time. A less conservative strategy would please some users, since routines would claim to have converged successfully more often, and less computer time might be required to solve well-behaved problems. However, there would inevitably be an increased danger of error or inaccuracy in the computed solution. In a general purpose environment, where the reliability of the optimization routine may be crucial in processes such as monitoring critically ill patients or guiding a rocket, a cautious approach seems preferable. The flavour of the standard of caution that we prefer was given by G.W. Stewart in his suggestion of the "airplane test": "Although this procedure is not infallible, I would fly in an airplane designed using it".

In order to allow a less conservative approach in some circumstances, software can be designed so that procedures that improve reliability are the default option for general use, but may be overridden by the knowledgeable user who is prepared to accept the consequences. Examples of such procedures are: a local search to move away from a possible saddle point; the determination of the proper treatment of very small Lagrange multipliers; the resolution of degeneracy; and the treatment of rank-deficiency.

Discussion on "The Current State of Software"

6.6 Gutterman

I suggest that we should try to define what is meant by software. To me, a FORTRAN compiler, a sorting routine, and a large LP code are all examples of software, but in my view most of Paper 6.1 is on related subjects. A definition that I like is that software is a computer program which can be used effectively by someone who either cannot or will not study the details of any algorithms, any data structures, any heuristics, and any implementations that are internal to the code. I do not mean to imply that every algorithm should be made into software in this sense. However, it seems to me that one of our goals should be to package the best algorithms in this way — perhaps with some automatic switching between algorithms if preliminary computational results indicate that some switching is desirable.

6.7 Powell

When computer codes are being developed for optimization algorithms, I think it is important to test them in cases when the required accuracy cannot be achieved due to computer rounding errors. Of course one would like the inability to achieve the accuracy to be recognised automatically, and for the computer program to provide an "error return" without wasting many function and gradient evaluations. If this is done successfully, then one may have a code that is suitable for "noisy functions", provided that one does not use parameters that depend on the working accuracy of the computer arithmetic. Moreover, this error return sometimes has the advantage that it can be regarded as a termination condition that is independent of scaling.

6.8 Moré

It is certainly possible to design algorithms which perform satisfactorily when the required accuracy cannot be achieved. However, this does not imply that the algorithm is suitable for noisy functions. For instance, if the algorithm estimates derivatives by differences, then this usually requires a knowledge of the noise level of the function.

6.9 Kraft

When an optimization calculation is only one activity in a computer program, there are many advantages in preferring reverse communication to subroutine calls to obtain function and gradient values. Reverse communication can sometimes avoid the use of large common areas with fixed dimension, and it makes programming more flexible when changing from problem to problem. The need to exit from nested subroutine calls when communicating in a reverse manner is a minor difficulty compared to the obvious advantages. An example of reverse communication is subroutine VF02AD in the Harwell library, which implements the sequential quadratic programming algorithm of Han and Powell (Powell, 1978a).

6.10 Schnabel

I am interested in software that provides a choice of modules, for example a line search, a dogleg, and a Levenberg–Marquardt module for the iteration step, from which the user can select and compare. The software that will accompany the Dennis–Schnabel book does this. Because there are wide variations from one problem to another in the relative performance of different algorithms that are obtained from different global modules, I believe it can be useful to provide software with such options, so that the user can select the best algorithm for a particular class of problems.

6.11 Moré

Modularity is an important issue in the development of optimization software. It is very helpful to write modular software, because then it is easier to understand and to modify the computer programs. There is, however, disagreement on how or how far to modularize. There seems to be a tendency to over-modularize optimization software. For example, I know of optimization codes which use over 30 different subroutines. Among the disadvantages of over-modularization are an increase in the amount of documentation, and some obfuscation of the underlying algorithms.

6.12 Ragsdell

I agree that the modularity concept is important, but care must be exercised to tailor the level to the intended audience of the software. For instance we have produced a package, OPTLIB, which is intended for use by students. This package contains a high degree of modularity. For instance a user can select one of many transformation methods, one of several unconstrained methods, one of several line search techniques, as well as associated termination criteria. The educational possibilities are obvious.

We have used a completely different philosophy with our other software which is intended to solve real problems (namely a GRG code called OPT and a method of multipliers code called BIAS). Here we think it wise to limit the users' options, which facilitates communication with users on the performance of their calculations.

6.13 Griewank

The term "scaling" means various things to different people. Paper 6.1 does not make it entirely clear with respect to which affine transformations on range or domain mathematical programming software should ideally be invariant.

6.14 Moré

I use the term scale invariant for invariance under diagonal changes to the variables. More specifically, given two unconstrained minimization problems whose objective functions and initial variables are related by

$$\tilde{f}(x) = f(Dx), \quad \tilde{x}_0 = D^{-1}x_0,$$

where D is a constant nonsingular diagonal matrix, an algorithm is scale invariant if it generates iterates related by

$$\tilde{x}_k = D^{-1}x_k, \quad k > 0.$$

With this terminology, algorithms which use a trust region can be made scale invariant by choosing appropriate ellipsoidal trust regions. As another example, note that the angle test

$$-\nabla f(x)^T p \geq \theta \|\nabla f(x)\| \; \|p\|$$

is invariant with respect to the transformation

$$\tilde{f}(x) = f(Dx), \quad \tilde{x}_0 = D^{-1} x_0,$$

only if D is a multiple of an orthogonal matrix, and therefore is not scale invariant.

The above type of invariance is important because it holds in inexact arithmetic if the elements of D are powers of the base used by the computer. This is not the case if D is not a diagonal matrix.

6.15 Davidon

I agree with the need to distinguish different types of scaling to avoid confusion. For example, while angle tests are not invariant under affine or even just diagonal scaling of the domain of the objective function, they are invariant both under domain dilations and under affine scalings of the range of the objective function. Perhaps we could use the following terms to distinguish among these concepts:

range scaling - affine scalings of the range of the objective function;

diagonal scaling - scaling of the individual variables by possibly different factors;

affine scaling - affine scaling of the domain of the objective function.

Another matter that has been mentioned only in passing is providing users with a variance-covariance matrix at the end of a minimization calculation. While it may not be feasible or even desirable to do this always, I do think it is important to give some information about the size and shape of level sets for the objective function near a minimizer, in order that false conclusions are not drawn about the significance of results obtained.

6.16 Moré

I agree that it is very important to provide the user with information on the reliability and sensitivity of his results. Unfortunately, this is not a common feature of optimization software, but, for example, the users' guide for MINPACK-1 contains information of this type.

6.17 Wolfe

One often finds that the submitter of a problem has intentionally scaled it, for example to put increased emphasis on certain variables or measurements. For this reason, a scale invariant (or automatic scaling) procedure should allow as an option the preservation of the user's scaling, or an option should permit the scaling to be declared.

6.18 Moré

I strongly agree with this remark, and in fact MINPACK software either automatically scales the problem or, optionally, uses the scaling provided by the user.

6.19 Lootsma

A useful way of fixing the scale of the variables of an optimization calculation is the following: ask the user to choose his units in such a manner that all components of the expected solution are of the same order of magnitude. For example, if a chemical engineer works with certain quantities which are related to Avogadro's number, he may be interested in a coefficient that multiplies the number 10^{23}, instead of in a solution with components of the order of magnitude of 10^{23}. This type of rescaling often yields a problem formulation that is much easier to handle, because during the computations the variables will mostly remain in similar ranges.

6.20 Sargent

It seems to me that the prescription of the last contribution is too simple. Often the user has a "feel" for the physical behaviour of the problem, and this is lost if he is presented with results in a scaled form. Often too, some aspects of the behaviour are unimportant in the context of the problem, which is properly reflected in the "natural" scaling. Artificial scaling can then result in much time being spent on computing effects which only marginally influence the results; on the other hand, ill-conditioning can magnify errors in such effects, so that scaling to prevent loss of accuracy becomes important. There is no easy answer for scaling nonlinear problems.

6.21 Rosen

The point, made in Paper 6.2, that in large optimization calculations in industry one seeks improvements rather than optimality, agrees with much of my earlier experience at Shell oil.

6.22 Baker

It is true that the scheduling system that I described strives only for a suboptimal solution. We must compare what is currently done in practice with what is possible. The problem to be solved is very difficult, for the sequencing of operations is a travelling salesman problem, and that is only a part of the complete calculation. I think that taking an approach which is designed to find the global optimum would be impossible to apply in practice.

6.23 Gutterman

Our management is also mainly interested in results that are feasible and that are better than they had before. Further, they would like to have an indication of closeness to optimality.

6.24 Lasdon

The version of sequential linear programming that is given in Paper 6.2 can be expressed as an exact ℓ_1 penalty function calculation. Each linear program in the sequence corresponds to minimizing the objective function that is obtained by replacing each nonlinear function involved in the penalty function by its first order Taylor series linearization, with a trust region defined by upper and lower bounds on the components of the search direction. This viewpoint suggests that one can use the ℓ_1 penalty function in a trust region strategy to force convergence of SLP.

6.25 Wolfe

Please tell us more about the portfolio problem of Paper 6.3 and how it is solved.

6.26 Dembo

I solved the portfolio problem: minimize $\frac{1}{2}x^T V^T Vx$ subject to $Ax = b$ in the following way. Any $x \in \mathbf{R}^n$ satisfying $Ax = b$ can be expressed as

$$x = Zy + A^T(AA^T)^{-1}b, \tag{1}$$

where $y \in \mathbf{R}^{n-m}$, $b \in \mathbf{R}^m$ and $AZ = 0$. I chose to use

$$Z = \begin{pmatrix} -B^{-1}N \\ I \end{pmatrix}, \quad \text{where } A = (B \quad N)$$

with B nonsingular. Since for this problem the number of rows of A (viz m) is 2, the terms $u = A^T(AA^T)^{-1}b$ and $B^{-1}N$ can be computed and stored explicitly. Substituting (1) into the objective function gives the equivalent unconstrained QP problem:

$$\underset{y \in \mathbf{R}^{n-m}}{\text{minimize}} \quad \tfrac{1}{2}y^T(Z^T V^T VZ)y + (Z^T V^T Vu)^T y.$$

I solved this problem using a conjugate gradient method. The advantage of a CG method is twofold. Firstly, it has a low storage overhead, and secondly the reduced Hessian matrix $Z^T V^T VZ$ is never required explicitly: only products of the form $Z^T V^T VZd$ are needed, where d is a direction vector that is generated by the algorithm. Since $B^{-1}N$ and V are stored as dense matrices the product $Z^T V^T VZd$ can be formed easily. One such product is required per CG iteration. Thus, as long as the CG algorithm converges in approximately $\frac{1}{2}n$ or fewer iterations, it will require less work to solve the problem than to form the

Hessian matrix V^TV once! In all the examples solved this was the case. Note also that the vector (1) remains feasible as y is altered.

6.27 Tanabe

Are there any special features for the semi-definite case in the conjugate gradient method of Paper 6.3?

6.28 Dembo

We have used the conjugate gradient method on some semi-definite problems without any modification and with success. Naturally, in the interests of stability, any implementation of a CG method should check for directions of (numerically) zero curvature.

6.29 Davidon

In the example of Paper 6.3 on stock market prices there are $1000 \times 5000 = 5 \times 10^6$ numbers in the input data. How are these stored?

6.30 Dembo

We did not solve problems of that size. However, such problems are currently being contemplated, and I admit that storing the data matrix V presents a major difficulty.

6.31 Beale

Paper 6.3 gives a good example of a dense large problem, but I hope it is agreed that, if there are many constraints, then they can in practice be expressed compactly.

6.32 Powell

I was pleased to hear the favourable comments on VF02AD in Paper 6.4, but I wish to point out that there are cases where this subroutine does not work well. In particular the "Maratos effect" can cause severe inefficiencies. Therefore a new version of VF02AD is being developed. It has been applied to problems with highly nonlinear constraints, and some further inefficiencies have occurred, which can be avoided by the user if he has control over the changes that are made to the variables on each iteration. Therefore I cannot agree that there is not much room for improvement in algorithms for constrained optimization.

6.33 Schittkowski

"Efficiency" has a statistical meaning and should be regarded as a mean value. I do not expect that another version of VF02AD can lead to a drastic improvement of the efficiency score just by solving a few selected problems (with Maratos effect) more efficiently. A small improvement will not convince a user to replace an existing code by another one.

6.34 Bus

I am somewhat amazed that Paper 6.4 gives a *general* ranking of the efficiency of codes. I would expect such a ranking to depend on certain properties of the problems, such as the costs to evaluate the problem functions and their gradients, or the dimensions of the problems. Did you try to find any dependence of the efficiency on such problem characteristics?

6.35 Schittkowski

The ranking scheme of the mathematical methods is presented to give a rough impression of the numerical efficiency. It is valid with respect to calculation time and number of function or gradient evaluations and for nearly all classes of test problems under consideration.

6.36 Sargent

I think we must distinguish two quite different objectives for comparative testing. The user with a problem to solve wants to know how a given piece of software behaves, while the algorithm developer is looking for the underlying behaviour of the algorithm. To illustrate, in Hiebert's (1980) comparison of software for solving nonlinear equations, "Broyden's method" is not very robust. This is all the user needs to know for he will avoid using this particular piece of software, but the algorithm developer wants to know how and why the method fails. Careful reading of Hiebert's report shows that nearly all the failures occur in the routine which inverts the initial finite difference approximation of the Jacobian, i.e. before the Broyden formula is used! Nevertheless, her report has given Broyden's method a bad name, because too many people read only the summarized conclusions, or look at the tables. Similarly the lack of robustness and efficiency (in CPU time) of the Harwell library routine VF02AD (mentioned in Paper 6.4) is almost entirely attributable to the QP subroutine used, and not to the details of the main algorithm.

6.37 Jackson

I certainly agree that it is important to understand the difference between code and algorithm testing, and would point out that we are here only concerned with code testing.

6.38 Polak

Algorithms should be presented in implementable form, with all the significant truncations specified and taken into account in the analysis. That this can be done is clear from the general algorithm theory presented in Polak (1971), and from a number of specific examples of the use of this theory in the literature. The effect of such an approach is to reduce the differences between the algorithm behaviour as predicted by the theory and that delivered by a specific code that implements the algorithm.

6.39 Dembo

I feel that there are a number of different aims that one has when doing computational testing. One of these is to test software that is destined for a library serving a general user. For this class of software a large battery test involving many codes and many test problems and generating many statistics may be appropriate. However, I believe that this group is more concerned with evaluating new algorithmic ideas to benefit future research. To accomplish this a *methodology for testing algorithms* (or components of algorithms) is needed. I do not think that the methodology used in recent computational studies serves this purpose.

6.40 Gutterman

Codes have to be tested because they are what the user needs to solve his problems. After all, that is where the bread and butter is!

6.41 Jackson

I wish to oppose the last remark. It seems to me that assisting users is *one* of the concerns of this group, but I am not convinced that it is the main one. Furthermore I am not sure that I think it *ought* to be the main one. I really believe that some mathematical optimization types ought to "live in a teacup", as Dick Tapia said the other day. If one agrees with me on this then one must agree that another important concern of this assembly is research strictly for the sake of research, even if there does not seem to be any possibility of real application for it. This I believe is the fundamental nature of research.

Also I want to comment on the statement in Contribution 6.39 that we need to do more *algorithm* testing as opposed to *code* testing. I agree in principle, but it seems that we are undecided on where algorithm testing ends and code testing begins. Eventually we may all agree that the *techniques* for testing are the same in the two cases, but until that happens, or until I am proven wrong, I think we should concentrate on code testing first and then move on to algorithm testing. In short, we have to crawl before we can walk.

6.42 Wolfe

I believe that the remark about flying, given in Paper 6.5, must have had its origin in a comment made by Alston Householder at the beginning of a talk at a meeting of the A.C.M. around 1960: "Whenever I take a 'plane to a meeting like this, I am forced to reflect that the aircraft was designed in the light of our present understanding of floating-point arithmetic."

6.43 Schnabel

Moré, Lootsma, Wright and others have all mentioned scaling, and I would like to comment on it too. In my experience, the user almost always knows *a priori* the ranges of his parameters, so a simple diagonal scaling based on ranges supplied by the user may alleviate scaling difficulties. However, if a code has as an option a simple calling sequence where only minimal information is supplied, then it will almost always be used in this way. Therefore the available scaling information may only be supplied to the code by an inexperienced user, if the calculation is so inefficient without scaling that the user seeks the advice of a qualified consultant, who tells him to provide the scaling information to the computer program. For this reason, a back-up *automatic* scaling procedure in a code is also useful, and I wonder if there is any consensus on how to include it.

6.44 Wright

Effective scaling techniques have been developed for certain problem classes and applications. For example, the simple technique of scaling all variables to be of the same "size" considerably improves the performance of optimization methods when applied to many practical problems. However, the state of the art with respect to *general* scaling techniques is not well developed. Even in the case of linear programming, there is no universal agreement as to how to achieve (automatically) "good" scaling.

6.45 Ragsdell

I agree that variable and constraint scaling is really very important, particularly when implementing a method of multipliers code. We have devised and installed an automatic scaling algorithm in BIAS (see Contribution 6.12). We performed quite a few tests to ensure that the scaling would not upset the performance of the method in the worst case, and we have found very often that the improvement in the conditioning of the problem reduces significantly the time of the calculation.

6.46 Wright

There is no doubt that automatic scaling procedures can be helpful in many circumstances. However, they are *not* suitable for every situation, and it sometimes happens that automatic scaling is actually harmful. For example, automatic scaling may destroy a scaling that the user has carefully chosen in posing the problem. This is why I believe that automatic scaling should not be imbedded within an optimization method, but rather should be available outside the method as an option. The information produced by automatic scaling procedures can also be useful in enhancing the user's understanding of his problem.

6.47 Gutterman

In the LP world, where I know much more about the influence of scaling, automatic scaling is also very important. So much so that we default to using automatic scaling and we only try a solution without it when we are encountering serious numerical difficulties which may be due to the automatic scaling. When comparing LP codes, I have often seen results that are strongly influenced by differences between the automatic scalings of the codes.

6.48 Wolfe

There can be no uniform policy on whether to automatically scale a problem or not, even if it is linear. If it is a generated subproblem in a sequential linear programming procedure, then possibly very small entries should be treated as zero and suppressed. However, if the problem has been hand-created, then each entry is serious, even if very large or small entries would lead to strange-looking scaling. For simplex method calculations the data should always be properly scaled first. Good production software does this automatically (transparently to the user), with benefit to performance.

6.49 Lootsma

We have been discussing the rescaling of variables, but there is also the problem of scaling the constraints. An important research question is the following: how do the methods that reduce a constrained problem to a sequence of simpler problems (e.g. penalty function methods and augmented Lagrangian methods) balance the constraints? Exterior and interior penalty functions can have very different effects. For instance the problem

$$\min\{f(x,y) : x \geq 0, 100y \geq 0\}$$

leads to the quadratic exterior function

$$f(x,y) + r^{-1}\{\min^2(0,x) + \min^2(0,100y)\},$$

and to the logarithmic barrier function

$$f(x,y) - r \ln x - r \ln (100y).$$

Therefore, the coefficient of 100 affects the contours of the exterior function, but it has practically no influence on the contours of the interior function.

6.50 Davidon

Contribution 6.46 on the need to compromise on automatic scaling reminds me of an anecdote told about Niels Bohr, who thought that the complementarity between position and momentum in quantum physics could be generalized to many other quantities. When he was once asked "what is complementary to truth", he thought for a while and then replied "clarity".

But perhaps computer technology can help us achieve both truth and clarity in our dialogue with users. For example, it may be possible to use interactive environments to get more useful information from a user about appropriate scaling. Instead of providing just ample written documentation, appropriate questions can be displayed and help provided on a terminal. While this seems to me likely to improve understanding between the user and the writer of software, I emphasize that I have not prepared any such software myself.

6.51 Gay

Regarding interactive computing, I would like to pass on an idea from TROLL, which is a facility for interactive econometric modelling that is maintained at MIT. In TROLL one can interrupt an executing procedure, change certain controls, and then resume the computation. I think it useful to arrange optimization software so that it can be interrupted and restarted similarly. The idea is to have a logical function, called STOPX say, that returns TRUE if and only if the interrupt key has been pressed since the last call of STOPX. Before starting a new iteration or function evaluation (as appropriate), the optimization code calls the system-dependent function STOPX to see whether the user wants to interrupt the computation. In the transportable versions of my codes, I simply replace STOPX by a function that always returns FALSE. Therefore, if these codes are to be interrupted in a restartable manner, it is necessary to restore STOPX to a suitable system-dependent function.

6.52 Conn

I wish to point out that it is possible to use a program interactively without any forethought by the author of the original software. For example I am able to run any FORTRAN program (provided I recompile it at least once), in a way that allows many features to be controlled interactively, including asking for output, changing the values of variables, and redirecting the flow. I believe that this is not an uncommon feature of the debugging routines that are available on time-sharing systems.

6.53 Gay

Many users that I see computing interactively like to let an optimization routine run for a small number of iterations, then they look at the progress of the calculation in order to decide whether to let it run further. I therefore think it useful for optimization software to be easily restartable, without loss of accumulated information, and possibly with convergence tolerances and output levels changed, even if dynamic interruptions are not allowed.

6.54 Gutterman

If a display of the progress of a calculation allows the user to guide the algorithm, then the display must be oriented to the user's knowledge level. The display should give quantities that the user can relate to his understanding of the problem, for he is likely to be confused by more sophisticated information.

6.55 Baker

With regard to the problem of having users of varying levels of expertise interact with a common set of software, perhaps our experience with interactive scheduling systems may be of interest. We found that at the beginning users wanted didactic and conversational interactions. More experienced users, however, did not wish to be slowed down. Moreover the way in which the algorithms reacted to computational difficulties, solution anomalies, and excessive CPU times etc. depended (or should have depended) on several parameters, such as problem type, solution environment, and the experience of the user. Our current thinking is to follow a symbol-state table approach which, in effect, governs the reaction of the software to given events based on the parameters mentioned above. This same approach might be useful in making nonlinear optimization software more adaptable to different users and environments.

PART 7

Future Software and Testing

7.1

Performance Evaluation of Nonlinear Optimization Methods via Multi-Criteria Decision Analysis and via Linear Model Analysis

F.A. Lootsma

Department of Mathematics and Informatics,
Delft University of Technology, 2600 AJ Delft, Netherlands.

Abstract

This report shows how Saaty's priority theory for pairwise comparison can be used to evaluate the performance of nonlinear optimization methods. First, priorities are assigned to the predominant performance criteria: generality, robustness, efficiency, capacity, simplicity of use, and program organisation. Thereafter, the comparative studies of nonlinear optimization codes are used to assign weights to the methods under each of the performance criteria separately. Here, Saaty's eigenvector method is modified, and logarithmic regression is employed in order to deal with incomplete and multiple pairwise comparisons. The procedure outlines what performance evaluation can do and what it cannot do: it demonstrates how multi-criteria decision analysis can be used for an integrated assessment of optimization methods, taking into account both the performance criteria of decision makers and the results of comparative studies. Finally, it is shown that efficiency weights can be obtained from these studies by the analysis of variance. A ln-additive model is introduced; the least squares estimates of certain "contrasts" yield efficiency ratios which can be further employed as pairwise comparisons. This provides a tool for a more rigorous and formal analysis of the results produced by the test batteries of nonlinear optimization.

1. Introduction

Since the pioneering work of Colville (1968), several authors have studied the performance of nonlinear optimization codes on a variety of test problems and computers. Staha and Himmelblau (1973), Eason and Fenton (1974), and Sandgren (1977) were concerned with general nonlinear optimization codes and with test problems which typically occur in research and development laboratories (i.e. highly nonlinear problem functions of a relatively small number of variables). Dembo (1978), Sarma, Martens, Reklaitis and Rijckaert (1978), and Rijckaert and Martens (1978) addressed the question of whether geometric programming codes are more effective on geometric programming problems than general nonlinear optimization codes. Recently, we have seen a startling growth of the so-called test batteries, giving more and more test problems and a variety of starting points. Schittkowski (1980b) developed the most extensive battery for general nonlinear optimization codes. Fattler, Sin, Root, Ragsdell and Reklaitis (1980) compared again the behaviour of geometric programming codes and general codes, but now on a much larger scale than previous authors. Thus, a substantial amount of material is available for computer managers and scientists who want to decide which code (or which underlying method) to use for their particular purposes. Nevertheless, the picture is gloomier than one might expect at first sight. There have been widespread disagreements on the performance criteria and on the test problems to be used (real-life or artificial, randomly generated problems?); the ranking and rating procedures to summarize results and to support decision making processes are underdeveloped; and the reported results are highly condensed or are incomplete. What makes the situation even worse is the basic (and controversial) philosophy underlying the comparative studies; it is tacitly assumed that the conclusions are valid for a wide range of computers, compilers and practical problems — therefore the effect of the computational environment is mostly ignored.

Hence, a decision maker will frequently run into difficulty when he tries to draw conclusions from these studies. Moreover, we have the impression that the authors do not always anticipate the questions that will be asked by the decision makers. Sometimes there is only a verbal interpretation of the results instead of a formal analysis of the recorded execution times or failure rates; important cases, such as the behaviour of codes with numerical or analytical derivatives, are not distinguished; other criteria, such as simplicity of use and program organization, are neglected; finally, attempts are made to estimate the relative performance of codes to an accuracy that is out of proportion to the information that is sought by the decision maker — in many cases he wishes to know only the relative orders of magnitude of performance.

It is the purpose of the present paper to outline the role of performance evaluation by considering the selection of a nonlinear optimization *method* in order to solve a particular class of nonlinear optimization problems. We shall use the results of some recent comparative studies. Of course the studies are concerned with optimization *codes*, that is with computational implementations of certain methods, but we have the firm impression that the published results reveal some sort of clustering which is due to the underlying methods.

In order to make the selection in a structured way, we shall employ a method for multi-criteria decision analysis, namely the priority theory of Saaty (1977, 1980), which has been developed to weigh the significant factors in a decision problem by pairwise comparison. Each ratio that expresses the relative significance of a pair of factors is displayed in a matrix. The weights of the significant factors (the so-called priorities) may be chosen by using normalized row sums, or inverted column sums, or Perron—Frobenius eigenvectors, or logarithmic regression (these four techniques are considered in Section 5). We have a two-level decision problem, and at the first level the significant performance criteria (the factors) are identified and weights (priorities) are assigned to them. At the second level the relative performances of the methods that are being compared are established under each of the performance criteria. Finally, by adding the weighted priorities of the performance criteria, one obtains a score for each method. The highest score determines the method to be selected. At both levels priority theory can be used: first, to assign suitable weights to the relevant performance criteria, and thereafter to determine the relative performances of the methods under each of the criteria separately.

In the selection process at hand, we shall be addressing the following basic question: *what is the best method for solving geometric programming problems in research, development and engineering — is it the geometric programming method or some general nonlinear optimization procedure*? To answer this question, we shall employ particularly the results of the comparative studies of Rijckaert and Martens (1978), Schittkowski (1980b), and Fattler *et al*. (1980), because of the size of their test batteries (120, 370, and 420 runs per code respectively).

2. Methods for Nonlinear Optimization

In its general form a nonlinear optimization problem can be written as

$$\left.\begin{array}{lll} \text{minimize} & f(x) & \\ \text{subject to} & g_i(x) \geq 0, & i = 1,2,\ldots,m \\ & h_j(x) = 0, & j = 1,2,\ldots,p \end{array}\right\} . \tag{1}$$

The objective function f and the constraint functions $\{g_i;\ i = 1, 2,\ldots,m\}$ and $\{h_j;\ j = 1,2,\ldots,p\}$ are real valued functions of the n-vector x. A geometric programming problem has the typical form

$$\left.\begin{aligned}
&\text{minimize} \quad \sum_{k\in K_0} c_k \prod_{j=1}^{n} x_j^{a_{jk}} \\
&\text{subject to} \sum_{k\in K_i} c_k \prod_{j=1}^{n} x_j^{a_{jk}} \le 1, \quad i = 1,2,\ldots,m \\
&\qquad x_j \ge 0, \quad j = 1,2,\ldots,n
\end{aligned}\right\} \qquad (2)$$

The coefficients $\{c_k;\ k\in K\}$ and $\{a_{jk};\ j = 1,2,\ldots,n,\ k\in K\}$, where $K \equiv K_0 \cup K_1 \cup \ldots \cup K_m$ are real but not necessarily positive. Obviously problem (2) is a special case of problem (1). We note that the first and the second derivatives of the functions appearing in problem (2) are easy to calculate, once the coefficients c_k and a_{jk} have been supplied.

The GP (Geometric Programming) method has several variants which can be classified on the basis of (a) the distinction between primal and dual techniques to solve the original GP problem or its dual, and (b) the distinction between techniques to solve the posynomial GP problem where all c_j are positive, and techniques to solve the signomial problem where some c_j are negative. For further details we refer the reader to the comparative studies by Dembo (1978), Sarma *et al.* (1978), Rijckaert and Martens (1978), Fattler *et al.* (1980), and to the original books of Duffin, Peterson and Zener (1967) and Avriel, Rijckaert and Wilde (1973).

Today there is a confusing variety of methods to solve the general problem (1). The more successful ones employ the gradients and sometimes the Hessian matrices of the problem functions f, $\{g_i;\ i = 1,2,\ldots,m\}$ and $\{h_j;\ j = 1,2,\ldots,p\}$. Because an extensive bibliography may be found in Waren and Lasdon (1979) and in Schittkowski (1980b), we mention here only some well-known methods and papers.

In the RG (Reduced Gradient) method the values of certain (basic) variables are obtained numerically from the remaining (non-basic) variables by solving constraint equations. The method has an excellent reputation in the field of nonlinear optimization, which is due particularly to the computer implementations developed by Abadie (1975) and by Lasdon, Waren, Ratner and Jain (1975a, 1975b).

The QA (Quadratic Approximation) method is studied by Robinson (1972, 1974) and Han (1976, 1977b, 1979); there is an excellent implementation by Powell (1978a). Equally successful is the "Recursive Quadratic Programming" method, which is studied and implemented by Biggs (1972, 1975). These algorithms did not appear in any broad comparative study until Schittkowski (1980b) discovered their startling performance.

Since the intensive work by Fiacco and McCormick (1968) the PF (Penalty Function) method has attracted considerable theoretical and computational attention. Penalty function codes appear in almost all comparative studies. Although the original interior and exterior variants are somewhat obsolete now, the ideas are still fruitful in other solution strategies, such as the quadratic approximation methods just mentioned and the methods based on augmented Lagrangians and moving truncations. Moreover, interior PF methods can guarantee that the computed variables remain feasible; this useful feature is not provided by other methods.

The AL (Augmented Lagrangian) method, which is also called the method of multipliers, tries to avoid the disadvantage of the PF method that a certain controlling parameter has to tend to infinity in order to achieve convergence. The underlying ideas, due to Hestenes (1969) and Powell (1969), have been developed further by Rockafellar (1973) and Fletcher (1975).

The method based on MT (Moving Truncations) of the objective function is closely related to Huard's method of centres (Huard, 1967), and to Morrison's minimization via least squares (Morrison

1968). Although promising in the early comparative studies of Staha and Himmelblau (1973) and Sandgren (1977), this method is still somewhat unexplored (see also Groszmann and Kaplan, 1979).

3. Comparative Studies

Some characteristic features of nine leading comparative studies are recorded in Table 1. The studies were concerned

Table 1

Number of codes and number of test problems used in comparative studies of nonlinear optimization software

Study and year	Number of codes							Total codes	Total problems	Runs per code	Max n
	RG	QA	PF	AL	GP	MT	XX				
Colville 1968	4	1	4			1	24	34	8	8	16
Staha and Himmelblau 1973	1		1			1	1	4	24	31	100
Eason and Fenton 1974			17					17	10	10	7
Sandgren 1977	4		27	1		1	2	35	30	30	48
Dembo 1978	2			1	5	1	1	10	8	13	16
Sarma *et al.* 1978			1		4			5	16	21	24
Rijckaert and Martens 1978				1	16			17	24	120	13
Schittkowski 1980b	4	3	6	12			1	26	185	370	20
Fattler *et al.* 1980	1		1	1	6		1	10	42	420	30

mainly with the methods that are mentioned in the previous section, but the column headed XX includes codes that are not in the given classes. The number of test problems in the studies before 1978 is generally small, and there is mostly one starting point per test problem; the idea of using several randomly generated starting points is relatively new. The test problems are also rather small (roughly 2–30 variables). As stated in Section 1, we shall mainly draw our conclusions from the recent studies by Rijckaert and Martens (1978), Schittkowski (1980b) and Fattler *et al.* (1980). Nevertheless, it is interesting to note certain features of the older studies. Staha and Himmelblau (1973), for instance, made a clear distinction between the behaviour of codes in the presence and absence of user-supplied derivatives, and also between the performance of codes on unconstrained, linearly constrained, and nonlinearly constrained problems. Eason and Fenton (1974) and Sandgren (1977) employed numerical derivatives only. Dembo (1978) was the first author who compared both GP codes and general codes. In all the studies the behaviour of the RG codes is excellent, even on GP problems. Rijckaert and Martens (1978) concluded that, even for GP problems, RG codes are equally efficient but less robust than typical GP codes. Schittkowski (1980b) used a large number (185) of randomly generated GP problems (but he left the GP codes out of consideration), and he made a distinction between the performance of general codes on convex, ill-conditioned, degenerate, and indefinite problems. Fattler *et al.* (1980) came to the conclusion (on the basis of 42 test problems and 10 random starting points per problem) that RG codes are somewhat more efficient and more robust on GP test problems than the leading GP codes; they were the first authors to carry out a statistical analysis of their results.

4. Performance Criteria

The following list of performance criteria for nonlinear optimization codes, although not exhaustive, seems to contain the main decision factors in the field of nonlinear optimization (see also the COAL Newsletter, Mathematical Programming Society, January 1980).

F_1: Domain of Applications: the types of problem for which the code has been designed.

F_2: Robustness: the power to solve problems in the domain of applications with the required accuracy.

F_3: Efficiency: the effort (usually measured by the number of equivalent function evaluations or by CPU time) necessary to solve problems in the domain of applications with the required accuracy.

F_4: Capacity: the maximum size of the problems that can generally be solved by the code.

F_5: Simplicity of Use: the presence of high quality documentation and user-oriented features, and the conceptual simplicity of the underlying method.

F_6: Program Organization: the language, length, and structure of the code.

5. Priority Theory

Let us consider a decision problem where s significant factors (performance and decision criteria) have been identified; their respective significances may be different in the situation at hand. A well-known procedure to put the significances on a numerical scale is *ranking and rating*. The factors are ranked in ascending order of significance, and a further refinement is to assign to each factor a numerical value (the weight, or priority) between a given positive, lower bound (for the most insignificant factor) and a given upper bound (for the most important factor). Frequently, however, it is not easy to decide on the numerical values when there are many factors. A decision maker does not readily accept a reduction to a one-dimensional scale, and, when the numerical values are assigned, he may form the opinion that there are several inconsistencies. Priority theory may be used instead. It starts with the idea that it is easier to consider each pair of factors separately. For each pair one decides whether the two factors are equally significant, or whether one of them is (much) more significant than the other in the given situation. Quantification of the *relative significance* for each pair of factors produces a matrix from which suitable priorities can be extracted.

Let us begin with the assumption that the significant factors $\{F_i;\ i = 1,2,\ldots,s\}$ have positive priorities $\{w_i;\ i = 1,2,\ldots,s\}$ which are acceptable to the decision maker; for simplicity, we take these priorities to be normalized to that they sum to unity. Now the elements p_{ij} of the matrix P of priority ratios can be written $p_{ij} = w_i/w_j$. The rank of P is clearly one since each row is a multiple of $(1/w_1\ 1/w_2\ \ldots\ 1/w_s)$. Thus P has only one non-zero eigenvalue. Because Pw = sw, where w is the vector with components $\{w_i;\ i = 1,2,\ldots,s\}$, the nonzero eigenvalue of P is equal to s and w is the corresponding eigenvector.

Let us now return to the decision maker who estimates the relative significance of each pair of factors. Let r_{ij} denote the numerical value which he assigns to the relative significance of the factors F_i and F_j; thus r_{ij} is an *estimate* of the priority ratio $p_{ij} = w_i/w_j$. Equal importance is expressed by setting $r_{ij} = 1$. If F_i is believed to be more important than F_j, then r_{ij} is usually set to 3, and, if F_i is felt to be much more important than F_j, then $r_{ij} = 5$. The intermediate values 2 and 4 are assigned to r_{ij} in cases of doubt between two adjacent values. Occasionally, if F_i is felt to be predominant with respect to F_j, r_{ij} may be set to a value higher than 5, but in this case a helpful comparison of F_i and F_j is probably beyond the scope of the automatic treatment of the decision problem. Of course we have $r_{ij} < 1$ if F_i is less important than F_j, and normally the judgment satisfies the reciprocal condition

$$r_{ij}\, r_{ji} = 1. \tag{3}$$

Finally one sets the diagonal elements $\{r_{ii};\ i = 1,2,\ldots,s\}$ to one. Thus a square positive matrix R is formed with elements r_{ij}. It is called a "reciprocal matrix" if all its elements satisfy equation (3). In what follows we let

$$r_{ij} = p_{ij}\,(1 + \varepsilon_{ij}),$$

where ε_{ij} stands for the relative error in the estimate r_{ij} of p_{ij}. Suitable approximations to the priorities $\{w_i;\ i = 1,2,\ldots,s\}$ can now be obtained in several ways: the main ones are as follows.

(a) Normalized row sums. The equation

$$\sum_{j=1}^{s} p_{ij} = \sum_{j=1}^{s} \frac{w_i}{w_j} = w_i \sum_{j=1}^{s} \frac{1}{w_j} \tag{4}$$

shows that the row sums provide a multiple of the vector w of priorities. Therefore the normalized row sums of R can be used to approximate w, giving

$$\frac{\sum_{j=1}^{s} r_{ij}}{\sum_{i=1}^{s}\sum_{j=1}^{s} r_{ij}} = \frac{\sum_{j=1}^{s} p_{ij}(1+\varepsilon_{ij})}{\sum_{i=1}^{s}\sum_{j=1}^{s} p_{ij}(1+\varepsilon_{ij})} \approx \frac{\sum_{j=1}^{s} p_{ij}}{\sum_{i=1}^{s}\sum_{j=1}^{s} p_{ij}} = w_i .$$

(b) Inverted column sums. Because of the identity

$$\sum_{i=1}^{s} p_{ij} = \sum_{i=1}^{s} \frac{w_i}{w_j} = \frac{1}{w_j}\,, \tag{5}$$

we can approximate the priorities by

$$\frac{1}{\sum_{i=1}^{s} r_{ij}} = \frac{1}{\sum_{i=1}^{s} p_{ij}(1+\varepsilon_{ij})} \approx \frac{1}{\sum_{i=1}^{s} p_{ij}} = w_j,$$

but it is usually necessary to multiply the approximations by a constant, if one requires the estimated priorities to sum to one.

(c) <u>The Perron–Frobenius eigenvector</u>. Observing that R is a square and positive matrix, we can invoke the Perron–Frobenius theorem (see Bellman, 1960, for instance), which states that the eigenvalue of R of largest modulus, λ_{max} say, is real and positive, and that the sign of the corresponding eigenvector, z say, may be chosen so that all the components of z are positive; we normalize z so that its components sum to unity — thus z is unique. This vector can be taken as an approximation to w, since $z \to w$ if all the errors $\{\varepsilon_{ij}\}$ tend to zero. Saaty (1977) obtained the following interesting result for positive reciprocal matrices R. The eigenvalue λ_{max} is never less than the dimension s, and $\lambda_{max} = s$, if and only if we can write $r_{ij} = z_i/z_j$ for some z, which means that the estimates r_{ij} are consistent.

(d) <u>Logarithmic regression</u>. Recently, de Graan (1980) suggested approximating w by the normalized vector α which minimizes

$$\sum_{i<j} (\ln r_{ij} - \ln \alpha_i + \ln \alpha_j)^2 = \sum_{i<j} (y_{ij} - x_i + x_j)^2,$$

where $y_{ij} = \ln r_{ij}$ and $x_i = \ln \alpha_i - \beta$ for some constant β. By solving the associated normal equations, after removing the freedom in x by requiring $\Sigma x_j = 0$, we find the explicit solution

$$x_i = \frac{1}{s} \sum_{j=1}^{s} y_{ij} = \frac{1}{s} \sum_{j=1}^{s} \ln r_{ij} .$$

Hence

$$w_i \approx \alpha_i = \exp(\beta) \left[\prod_{j=1}^{s} r_{ij} \right]^{1/s}, \quad i = 1,2,\ldots,s, \qquad (6)$$

where exp(β) is fixed by the normalization condition $\Sigma\alpha_i = 1$.

The most interesting feature of de Graan's suggestion is that logarithmic regression can also be applied to incomplete pairwise comparisons. Then we minimize

$$\sum_{i<j} \delta_{ij} (y_{ij} - x_i + x_j)^2,$$

where $\delta_{ij} = 0$ if no estimate of p_{ij} has been given by the decision maker, and where $\delta_{ij} = 1$ if an estimate r_{ij} is available. Now, observing that $y_{ij} = -y_{ji}$ and $\delta_{ij} = \delta_{ji}$, we can write the normal equations as

$$\sum_{\substack{j=1\\ j\neq i}}^{s} \delta_{ij} (y_{ij} - x_i + x_j) = 0,$$

or, equivalently, as

$$x_i \sum_{\substack{j=1\\ j\neq i}}^{s} \delta_{ij} - \sum_{\substack{j=1\\ j\neq i}}^{s} \delta_{ij} x_j = \sum_{\substack{j=1\\ j\neq i}}^{s} \delta_{ij} y_{ij}. \tag{7}$$

As before these equations are dependent. In order to approximate the priorities, we take the exponentials of a solution to (7) and we normalize these quantities so that they sum to unity.

A significant, further generalization is to consider *multiple* comparisons by a number of decision makers in a committee which faces a certain decision problem. Now we take r_{ijk} to denote the estimate of p_{ij} expressed by the k-th decision maker, and, allowing for the possibility that some committee members abstain from giving their opinion on the relative significance of the factors F_i and F_j, we reformulate the problem as follows. We let δ_{ij} denote the number of estimates of p_{ij} obtained in the committee, and we let r_{ijk} $(k = 1,2,\ldots,\delta_{ij})$ stand for the k-th estimate of p_{ij}. We approximate the vector w of weights by the normalized vector α which minimizes

$$\sum_{i<j} \sum_{k=1}^{\delta_{ij}} (\ln r_{ijk} - \ln \alpha_i + \ln \alpha_j)^2$$

$$= \sum_{i<j} \sum_{k=1}^{\delta_{ij}} (y_{ijk} - x_i + x_j)^2,$$

where $y_{ijk} = \ln r_{ijk}$ and $x_i = \ln \alpha_i - \beta$ $(i = 1,2,\ldots,s)$. The normal equations can be written as

$$\sum_{\substack{j=1 \\ j\neq i}}^{s} \sum_{k=1}^{\delta_{ij}} (y_{ijk} - x_i + x_j) = 0,$$

or, equivalently, as

$$x_i \sum_{\substack{j=1 \\ j\neq i}}^{s} \delta_{ij} - \sum_{\substack{j=1 \\ j\neq i}}^{s} \delta_{ij}\, x_j = \sum_{\substack{j=1 \\ j\neq i}}^{s} \sum_{k=1}^{\delta_{ij}} y_{ijk}. \tag{8}$$

Formulating the normal equations is easy. We let R be the tableau whose (i,j)-th cell contains the estimates r_{ijk}. In general we have an unequal number of estimates per cell, and some cells may even be empty. The coefficient of x_i in (8) is the number of estimates in the i-th row of R excluding the main diagonal cell, the coefficient of x_j is minus the number of estimates in cell (i,j), and the right hand side element is the sum of the logarithms of all estimates in the i-th row. Obviously formula (8) is a generalization of (7). In fact it covers both the cases of incomplete comparisons (some δ_{ij} equal to zero) and the cases of multiple comparisons (some or all $\delta_{ij} > 1$).

In order to approximate the weights of the decision factors, we take the exponentials of a solution to (8), and we normalize these quantities in the usual manner.

In the examples that follow we shall mainly employ logarithmic regression, and particularly the generalization just described, since we have to deal with various comparative studies. In comparing methods under the performance criterion of efficiency, for instance, we shall take r_{ijk} to be the efficiency ratio of methods i and j estimated in the k-th comparative study.

It is worth noting that priorities are somewhat dependent on the scale that is employed to estimate the priority ratios r_{ij}. We have put the range of the ratios on a numerical scale running from 0.2 to 5, but, if a wider scale were used instead, then the calculated priorities would be more separated. The reason for our choice is that a ratio of 5 for the extreme case in a decision problem is believed to be reasonable; one does not normally compare widely disparate factors.

6. Consistent Judgment

As noted in the previous section, there is an indication whether the decision maker has made some errors in his judgments. The absolute largest eigenvalue λ_{max} of the positive reciprocal matrix R is greater than the order s if and only if the matrix elements r_{ij} cannot be expressed as ratios z_i/z_j. Moreover, defining η_{ij} by

$$r_{ij} = \frac{z_i}{z_j} (1 + \eta_{ij}),$$

we obtain

$$\frac{2(\lambda_{max}-s)}{s-1} \approx \frac{2}{s(s-1)} \sum_{i<j} \eta_{ij}^2 .$$

Therefore the left hand side can be interpreted as the variance in the errors of judgment (see also Lootsma 1979, 1980).

Usually small errors of judgment (perturbations of the priority ratios) are unavoidable, and are not necessarily serious. The so-called "cyclic judgment", however, in which factor A is preferred to factor B, factor B to factor C, and also factor C to factor A, is unacceptable. An error of this kind can easily be detected, even before one starts the priority theory analysis. To clarify matters we consider the estimates of Table 2, and the corresponding directed graph of Figure 1. In the figure each

Table 2

Example of cyclic judgment

Performance criteria	Estimates of priority ratios				
	F_1	F_2	F_3	F_4	F_5
F_1	1	3	1/3	1	2
F_2	1/3	1	1/2	1/2	1/3
F_3	3	2	1	1/3	4
F_4	1	2	3	1	1/2
F_5	1/2	3	1/4	2	1

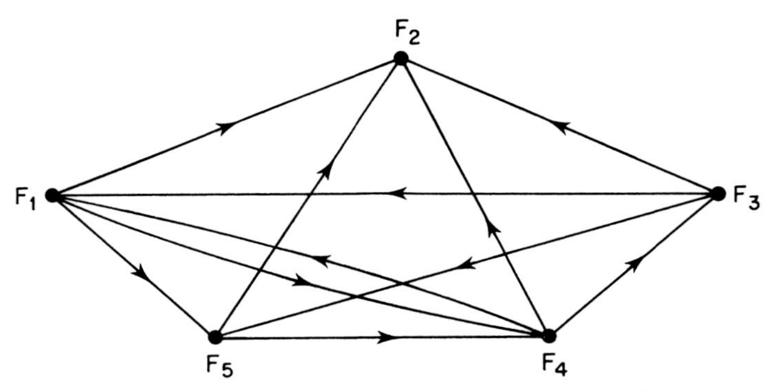

Fig. 1 Directed graph of estimated priority ratios that are not less than one.

node represents a decision factor of Table 2, and a directed arc (i,j) is drawn if $r_{ij} \geq 1$, $i \neq j$. Thus the arcs represent the off-diagonal elements that are not less than one in the matrix R of estimated priority ratios. There are various three-cycles in the graph, such as (1, 5, 4, 1) and (3, 5, 4, 3), and a four-cycle, namely (1, 5, 4, 3, 1), which are all examples of cyclic judgments. Because the two-cycle (1, 4, 1) is due to equal significance, it is in principle acceptable. We see that the decision factors of Table 2 cannot be ordered in ascending or descending order of significance.

7. Priorities of Performance Criteria

As stated in Section 1, we shall be concerned with the selection of the best method for solving GP problems. Because the comparative studies of the last few years will constitute our major source of information, we drop the MT methods from consideration.

Let us now carry out the first step of the evaluation: the assignment of weights to the performance criteria identified in Section 4. The reader will agree that it is not easy to do this, but it will become clear that the assignment is important for a structured analysis of the selection process.

The estimated weights displayed in Table 3 are the personal opinions of the author, and they should be regarded as illustrative examples. Leaving the reader to check for cyclic judgments, we determine the priorities by normalizing the geometric means of the elements in each row; these priorities are given in the last column of Table 3. There may be good reasons for adjusting the priorities — perhaps they should be set to the nearest multiples of 0.05, for example. For the time being, however, we proceed with the numerical values that are given.

It is easy to accept that robustness should have the highest priority. Unsolved problems do more harm than inefficient computations, partly because the subsequent attempts to overcome failures may be more expensive than inefficient computer utilization. Simplicity of use and program organization, however, cannot be neglected in nonlinear optimization. Because experience

Table 3

Priorities of performance criteria

Performance criteria	Estimates of priority ratios						Geometric row means	Priorities
	F_1	F_2	F_3	F_4	F_5	F_6		
F_1: Domain of Applications	1	1/3	2	2	2	2	1.322	0.193
F_2: Robustness	3	1	3	3	2	4	2.449	0.358
F_3: Efficiency	1/2	1/3	1	2	2	2	1.049	0.153
F_4: Capacity	1/2	1/3	1/2	1	1/2	1/2	0.525	0.077
F_5: Simplicity of Use	1/2	1/2	1/2	2	1	1	0.794	0.116
F_6: Program Organization	1/2	1/4	1/2	2	1	1	0.707	0.103

has shown that there is no black box to solve all problems in this field, it may be necessary for the user to have a detailed knowledge of the codes and the underlying methods. It is also desirable sometimes to be able to generalize codes easily, because it often happens that users extend their calculations in ways that do not fit easily into the typical geometric programming formulation. Moreover, there are many small problems (with 2 to 20 variables) in research, development and engineering, for which capacity should have a lower priority.

One should note that Table 3 would change immediately if one considered a different selection problem. In selecting the best method for solving *general* nonlinear optimization calculations, for instance, "domain of applications" would be a dominant requirement (not just a moderately desirable feature); then the GP method would have to be dropped, and we would restrict ourselves to general codes only. Similarly, if we faced large nonlinear optimization calculations, then the priority of the capacity criterion would increase drastically.

8. Priorities of Methods

At the second level of the decision procedure, where priorities are assigned to optimization methods under various performance criteria, it is sometimes easier to quantify relative performance. In the comparative studies we find measurable phenomena, for example: the number of unsolved problems, the elapsed CPU time, etc. Nevertheless, estimating the relative performance remains a nontrivial matter, to which we shall return in Section 10. Let us now consider the performance of the *general* RG, QA, PF and AL methods, and also the *typical* GP methods, under the criteria of Section 4.

F_1: Domain of Applications In our judgment, the domain of applications of the general methods is larger, or even much larger, than the domain of GP algorithms. We quantify this judgment by assigning the value 4 to the corresponding ratios. The assignments and the resulting priorities are displayed in Table 4.

Table 4

Priorities of methods under performance criterion F_1 (domain of applications)

Method	Estimates of priority ratios					Geometric row means	Priorities
	RG	QA	PF	AL	GP		
RG	1	1	1	1	4	1.3195	0.235
QA	1	1	1	1	4	1.3195	0.235
PF	1	1	1	1	4	1.3195	0.235
AL	1	1	1	1	4	1.3195	0.235
GP	1/4	1/4	1/4	1/4	1	0.3299	0.059

F_2: Robustness In estimating the relative robustness of a pair of methods, we consider the ratio of their respective failure rates (the percentage of unsolved problems). For the general methods, we have mainly employed Table 5a of Schittkowski (1980b), which shows that the RG and QA methods are equally robust, and have failure rates which are roughly two and three times lower than the failure rates of the PF and AL methods respectively. On the basis of Table 4 of Fattler *et al.* (1980), we conclude that the GP codes GGP and GPKTC (which were identified as best by Rijckaert and Martens, 1978) and the RG code called OPT are equally robust; we also conclude that the GP codes are somewhat less robust than the AL code called BIAS. No figures are taken from the previous study by Dembo (1978) because it employed an unacceptably small number of test problems, and we have also omitted the comparison of GP codes and general codes in the studies by Sarma *et al.* (1978) and Rijckaert and Martens (1978) because the general codes in these test batteries are obsolete. It is useful, however, to include the results of Table 3 of Sandgren and Ragsdell (1979), where several current RG, PF and AL codes are compared. We did not find a suitable direct comparison between the leading GP codes and the leading QA or PF codes. The relative robustness estimates that are obtained from each study are given in Table 5. We judge that method i is more

Table 5

Priorities of methods under performance criterion F_2 (robustness)

Method	Estimates of priority ratios					Priorities
	RG	QA	PF	AL	GP	
		1 (S)	2 (S)	3 (S)	–	
RG	1	–	–	1/2 (F)	1 (F)	.224
		–	2 (SR)	1 (SR)	–	
	1 (S)		2 (S)	3 (S)	–	
QA	–	1	–	–	–	.321
	–		–	–	–	
	1/2 (S)	1/2 (S)		2 (S)	–	
PF	–	–	1	–	–	.143
	1/2 (SR)	–		1/2 (SR)	–	
	1/3 (S)	1/3 (S)	1/2 (S)		–	
AL	2 (F)	–	–	1	2 (F)	.173
	1 (SR)	–	2 (SR)			
	–	–	–	–		
GP	1 (F)	–	–	1/2 (F)	1	.139
	–	–	–	–		

robust or much more robust than method j if the failure rates have ratios between 1 and 3 or 3 and 5 respectively, which gives numerical values to the comparisons. In this table the figures obtained from the different studies are distinguished by S, F, and SR (which stand for Schittkowski, Fattler *et al.*, and Sandgren and Ragsdell respectively), and we have the general case where there are an unequal number of estimates per cell. The normal equations (8) give the system

$$\left.\begin{array}{rcl} 7x_1 - x_2 - 2x_3 - 3x_4 - x_5 & = & 1.7918 \\ -x_1 + 3x_2 - x_3 - x_4 & = & 1.7918 \\ -2x_1 - x_2 + 5x_3 - 2x_4 & = & -2.0794 \\ -3x_1 - x_2 - 2x_3 + 7x_4 - x_5 & = & -0.8109 \\ -x_1 - x_4 + 2x_5 & = & -0.6931 \end{array}\right\} .$$

The equations are dependent (they sum to the zero equation), and one of their solutions is

$$\left.\begin{array}{l} x_1 = 0.4768 \\ x_2 = 0.8381 \\ x_3 = 0.0291 \\ x_4 = 0.2165 \\ x_5 = 0.0000 \end{array}\right\} .$$

By normalizing the exponentials $\{\exp(x_i);\ i = 1,2,\ldots,5\}$, as in Section 5, we obtain the priorities that are given in the last column of Table 5.

F_3: Efficiency Our main source of information is again Schittkowski (1980b). His Table 3a shows that QA codes are roughly two times faster than RG codes, three times faster than AL codes, and five times faster than PF codes. Fattler *et al.* (1980) (see their Tables 5B, 6B and 11B) show that the RG code is roughly two times faster than the leading GP codes, which in turn are some-

Table 6

Priorities of methods under performance criterion F_3 (efficiency)

Method	Estimates of priority ratios					Priorities
	RG	QA	PF	AL	GP	
		1/2 (S)	3 (S)	2 (S)	–	
RG	1	–	–	4 (F)	2 (F)	.275
		–	5 (Sa)	–	–	
		–	5 (SH)	–	–	
	2 (S)		5 (S)	3 (S)	–	
	–		–	–	–	
QA	–	1	–	–	–	.384
	–		–	–	–	
	1/3 (S)	1/5 (S)		1/2 (S)	–	
PF	–	–	1	–	–	.065
	1/5 (Sa)	–		–	–	
	1/5 (SH)	–		–	–	
	1/2 (S)	1/3 (S)	2 (S)		–	
	1/4 (F)	–	–		1/2 (F)	
AL	–	–	–	1	–	.106
	–	–	–		–	
	–	–	–	–		
GP	1/2 (F)	–	–	2 (F)	1	.171
	–	–	–	–		
	–	–	–	–		

what faster than the AL code. We also include the results of Staha and Himmelblau (1973) and Sandgren (1977), who show that RG codes are roughly five times faster than PF codes. Again a direct comparison of GP codes versus QA and PF codes is not available. We have used ratios of execution times to assign numerical values to the pairwise comparisons, which gives the data in Table

6. Here the notation (S) and (F) is the same as in Table 5, but (Sa), and (SH) stand for Sandgren, and Staha and Himmelblau respectively. The computations to find the priorities of the methods under Criterion F_3 (efficiency) proceed in the same manner as under Criterion F_2 (robustness). Formula (8) gives the normal equations

$$\left.\begin{array}{rl} 7x_1 - x_2 - 3x_3 - 2x_4 - x_5 &= 6.3969 \\ -x_1 + 3x_2 - x_3 - x_4 &= 3.4012 \\ -3x_1 - x_2 + 5x_3 - x_4 &= -6.6201 \\ -2x_1 - x_2 - x_3 + 5x_4 - x_5 &= -3.1781 \\ -x_1 - x_4 + 2x_5 &= 0.0000 \end{array}\right\} ,$$

and one of the solutions is

$$\left.\begin{array}{rl} x_1 &= 0.4770 \\ x_2 &= 0.8100 \\ x_3 &= -0.9713 \\ x_4 &= -0.4770 \\ x_5 &= 0.0000 \end{array}\right\} .$$

Thus, as before, we obtain the priorities that are shown in Table 6.

We note that, in producing the estimates of Table 6, we did not make a distinction between test batteries using analytical or numerical derivatives of the functions in the test problems. Because the large comparative studies by Schittkowski and by Fattler *et al*. were all based on GP test problems, the first derivatives could be generated easily for the codes which do not have facilities for numerical differentiation. Nevertheless, it remains an obvious weakness that most comparative studies do not properly distinguish the behaviour of codes in the presence or absence of analytical derivatives.

F_4: Capacity The comparative studies do not evaluate properly the codes for increasing dimensionality of the test problems. In fact, the test problems are all rather small (see the maximum dimensions in Table 1). Because there is no information about the storage requirements either, we decided to assign equal priorities (0.20) to each of the methods under consideration.

F_5: Simplicity of Use Many user-oriented features (input/output facilities, numerical differentiation, high quality documentation, etc.) bear practically no relationship to the underlying method. Therefore, since it is our intention to compare methods, we have chosen instead to take the conceptual simplicity in order to rate simplicity of use. We feel that PF methods are somewhat easier to understand than RG and AL methods, which in turn are somewhat simpler than QA methods (at least for unsophisticated users). GP methods are definitely more complicated than the general methods. This leads to the estimates and results exhibited in Table 7.

Table 7

Priorities of methods under performance criterion F_5 (simplicity of use)

Method	Estimates of priority ratios					Geometric row means	Priorities
	RG	QA	PF	AL	GP		
RG	1	2	1/2	1	3	1.2457	0.218
QA	1/2	1	1/3	1/2	2	0.6988	0.122
PF	2	3	1	2	4	2.1689	0.379
AL	1	2	1/2	1	2	1.1487	0.201
GP	1/3	1/2	1/4	1/2	1	0.4611	0.081

F_6: Program Organization From Schittkowski's study (1980b), we have the impression that there is a relationship between the length of a code (the number of lines of coding), and the underlying method. The leading RG codes (some 4000 lines) are much longer than the PF, AL and GP codes (roughly 1000 lines) and the QA codes (some 750 lines). This leads to the estimates and results of Table 8. Note that a ratio of 1 to 6 for program

Table 8

Priorities of methods under performance criterion F_6 (program organization)

Method	Estimates of priority ratios					Geometric row means	Priorities
	RG	QA	PF	AL	GP		
RG	1	1/6	1/4	1/4	1/4	0.3042	0.051
QA	6	1	2	2	2	2.1689	0.366
PF	4	1/2	1	1	1	1.1487	0.194
AL	4	1/2	1	1	1	1.1487	0.194
GP	4	1/2	1	1	1	1.1487	0.194

length almost excludes the RG methods from further consideration, since we mainly work on a scale running from 1 to 5.

9. Evaluation

The final results are displayed in Table 9, where most entries are pairs of numbers, and where the priorities of the performance criteria, taken from Table 3, are given in the last column. The upper number of a pair is the priority of a method according to the performance criterion of the row, and the lower number is the upper number scaled by the priority of the performance criterion. The final score of each method is obtained by summing the lower

Table 9

Final scores of nonlinear optimization methods

Performance criteria	Priorities of methods					Priorities of criteria
	RG	QA	PF	AL	GP	
F_1: Domain of Applications	.235 .045	.235 .045	.235 .045	.235 .045	.059 .011	.193
F_2: Robustness	.224 .080	.321 .115	.143 .051	.173 .062	.139 .050	.358
F_3: Efficiency	.275 .042	.384 .059	.065 .010	.106 .016	.171 .026	.153
F_4: Capacity	.200 .016	.200 .016	.200 .016	.200 .016	.200 .016	.077
F_5: Simplicity of use	.218 .025	.122 .014	.379 .044	.201 .023	.081 .009	.116
F_6: Program Organization	.051 .005	.366 .038	.194 .020	.194 .020	.194 .020	.103
Final score	.213	.287	.186	.182	.132	1.000

numbers in its column. However, addition is a questionable operation here which needs further consideration. We see that, according to our calculation, the QA methods are leading and the RG methods are second when GP problems have to be solved, but we do not expect future users to immediately obey these numerical results by selecting a QA code. Although a direct comparison between QA and GP methods has not been carried out, Table 9 provides a structured analysis of the situation, which may help the selection of an algorithm. The table demonstrates clearly both the weights that are given to the performance criteria and the strengths and weaknesses of the methods, thus drawing attention to the features that are of most importance in the decision process. This is typically the objective of priority theory.

The example also shows clearly what performance evaluation

can and cannot do. The comparative studies have been used to assign priorities to the methods under various performance criteria (Section 8), but it is up to the decision maker to identify and to give weights to the relevant performance criteria (Section 7). Priority theory provides the framework for an integrated assessment of methods.

As in many decision or selection problems, there is a general vagueness in the significance of the performance criteria. Therefore the calculated priorities should be used only as a guideline for deliberations. Obviously one is concerned with orders of magnitude rather than accurate values of priorities. This implies that, for performance evaluation of methods, rough estimates of relative performance must be established under various criteria.

10. The Determination of Weights for Efficiency and Robustness

The procedure described in Sections 7 and 8 is based mainly on the judgments of a decision maker. It is not easy to apply sensitivity analysis in order to check the significance of the results, because generally we do not know how well human judgment fits reality. Two exceptions are the calculated priorities of the codes under the performance criteria of robustness and efficiency. In these cases, if a statistical analysis of the priorities is required, one can obtain a sufficiently large number of observations, because the data of the calculations are provided by test batteries.

In Section 8 we used the results of some test batteries in a somewhat superficial manner, in order to assign priorities to the underlying methods. In the present section we show that a more precise calculation of priorities of codes is a nontrivial matter. For ease of exposition, we consider a comparative study where several unconstrained minimization codes were tested in a penalty function procedure for solving constrained minimization problems (Lootsma 1972, 1975, 1978). The coded methods were Powell's conjugate direction method (P64), the variable metric method with the Broyden–Fletcher–Shanno* updating formula (BFS), and Newton's method as modified by Fiacco and McCormick (NFM). The BFS algorithm was applied with difference approximations to the first derivatives, and also with user-supplied first derivatives. Similarly, the NFM algorithm was employed in two versions: with user-supplied first derivatives and difference approximations to the second derivatives, and also with user-supplied first and second derivatives. Table 10 shows the execution times t_{ik} of code i on test problem k. The numbers in brackets after each code indicate the orders of derivatives that are supplied by the user. For each test problem, n is the number of variables and m is the number of constraints.

Assuming that there are no interactions between test problems and optimization codes (or that possible interactions can be ignored), one may try to explain the execution times by an additive model. The usual procedure is to write

$$t_{ik} = \mu + \alpha_i + \beta_k + \varepsilon_{ik}, \tag{9}$$

where the ε_{ik} stand for identically distributed, independent perturbations with zero mean. If there were no failures (i.e. if we had an observation t_{ik} for each i and k), we might estimate the parameters μ, $\{\alpha_i\}$ and $\{\beta_k\}$ in the least squares sense, that is we might calculate the values of m, $\{a_i\}$ and $\{b_k\}$ to minimize the sum of squares

$$\sum_i \sum_k (t_{ik} - m - a_i - b_k)^2, \tag{10}$$

*This formula is usually known as the Broyden–Fletcher–Goldfarb–Shanno formula. My apologies are due to D. Goldfarb. I have used the acronym BFS in my programs and papers since 1970, i.e. before all contributions to the establishment of the BFGS formula were known.

Table 10

Execution times of five unconstrained minimization codes on eleven test problems

	Execution times of codes (F = fail)				
	P64(O)	BFS(O)	BFS(1)	NFM(1)	NFM(1,2)
Problem 1 n=2, m=0	5	3	3	2	2
Problem 2 n=2, m=0	4	2	2	1	1
Problem 3 n=4, m=0	14	8	6	5	5
Problem 4 n=4, m=0	7	11	10	8	7
Problem 5 n=3, m=3	12	10	9	5	5
Problem 6 n=3, m=3	23	19	16	11	8
Problem 7 n=4, m=3	F	34	24	19	14
Problem 8 n=6, m=4	44	32	26	13	12
Problem 9 n=9, m=13	F	287	152	270	199
Problem 10 n=5, m=10	265	120	71	39	39
Problem 11 n=15, m=5	F	F	793	313	234
Failure rates	27%	9%	0%	0%	0%

having introduced some additional, linear constraints to force uniqueness. Unfortunately it is a common experience (Sieben, private communication) that this procedure frequently produces unacceptable results, such as negative least squares estimates of some of the terms μ, $\{\alpha_i\}$ and $\{\beta_k\}$.

Satisfactory results, however, are often obtained by a multiplicative model of the form

$$t_{ik} = \mu \, \alpha_i \, \beta_k \, (1 + \varepsilon_{ik}), \tag{11}$$

which can also be written as a *ln-additive* model since

$$\ln t_{ik} = \ln \mu + \ln \alpha_i + \ln \beta_k + \ln (1 + \varepsilon_{ik}). \tag{12}$$

In this case we minimize the sum of squares

$$\sum_i \sum_k (\ln t_{ik} - \ln m - \ln a_i - \ln b_k)^2. \tag{13}$$

Although the normal equations derived from expressions (10) and (13) do not have unique solutions (Scheffé, 1959), certain linear combinations of the parameters (such as the *contrasts* $\alpha_i - \alpha_j$ in the additive model or the *contrasts* $\ln \alpha_i - \ln \alpha_j$ in the ln-additive model) do have unique, unbiased least squares estimates (Gauss–Markov theorem). The contrasts $\ln \alpha_i - \ln \alpha_j$, for instance, are estimated by

$$\ln a_i - \ln a_j = \frac{1}{K} \sum_{k=1}^{K} \ln t_{ik} - \frac{1}{K} \sum_{k=1}^{K} \ln t_{jk},$$

where K is the number of test problems. We let g_i denote the geometric mean of the i-th row

$$g_i = \left[\prod_{k=1}^{K} t_{ik} \right]^{1/K},$$

in order to obtain immediately that the ratios α_i/α_j (representing the relative effects of the codes on the execution time) are estimated by the ratios g_i/g_j. Since we normally express a better performance by a higher rate, we take the inverse to estimate the efficiency ratio. Therefore the efficiency ratio of code i to code j is given the value

$$\frac{g_j}{g_i} = \left[\prod_{k=1}^{K} \frac{t_{jk}}{t_{ik}} \right]^{1/K}. \tag{14}$$

Thus, if there are no failures in the test battery, it is easy to derive priorities under the criterion of efficiency. Further, the vector with components $\{g_i^{-1};\ i = 1,2,\ldots,I\}$ (where I is the number of codes) can be taken as a multiple of the vector of priorities, and then suitable normalization will produce the desired priorities. Obviously there is no need for a scaling procedure that first replaces execution times by "minimum time ratios" (see Lootsma, 1980).

If the table of results includes some failures (see Table 10), we proceed as in the case of incomplete judgments (Section 5). Therefore we minimize

$$\sum_i \sum_k \delta_{ik} \left(\ln t_{ik} - \ln m - \ln a_i - \ln b_k\right)^2, \tag{15}$$

where $\delta_{ik} = 0$ if code i fails on test problem k, and where $\delta_{ik} = 1$ otherwise. Thus, estimates of the contrasts $\ln \alpha_i - \ln \alpha_j$ are obtained from the normal equations associated with (15). This calculation has been solved recently by de Kovel (paper in preparation and de Kovel, 1981) for the data of Table 10 using standard software (the IMSL routine AGLMOD for General Linear Model Analysis). The results are given in Table 11, where as before the calculated ratio a_j/a_i is the estimate of the efficiency ratio of code i to code j. The table also shows the priorities that are obtained by normalizing the geometric row means. Of course an important question is whether the contrasts $\ln \alpha_i - \ln \alpha_j$ are significantly different from 0, i.e. whether the efficiency ratios are significantly different from 1. This question was also analyzed by de Kovel (using routine AGLMOD again). His results are displayed in Table 12, where "+" and "-" signs denote efficiency ratios that are significantly greater than and less than one respectively.

Lootsma (1980) proposed an alternative method to calculate efficiency ratios in the presence of failures, by considering the cross-section of successfully solved test problems for each pair of codes. This approach gives priorities that are practical-

Table 11

Estimated efficiency ratios and priorities of five unconstrained minimization codes

Code	Estimated efficiency ratios					Geometric row means	Priorities
	P64 (0)	BFS (0)	BFS (1)	NFM (1)	NFM (1,2)		
P64(0)	1.00	0.70	0.58	0.39	0.34	0.56	0.10
BFS(0)	1.42	1.00	0.82	0.55	0.49	0.79	0.15
BFS(1)	1.73	1.22	1.00	0.67	0.59	0.96	0.18
NFM(1)	2.57	1.81	1.48	1.00	0.88	1.43	0.27
NFM(1,2)	2.93	2.06	1.69	1.14	1.00	1.63	0.30

Table 12

Significance of efficiency ratios at 95% level

Code	Significance of efficiency ratios				
	P64 (0)	BFS (0)	BFS (1)	NFM (1)	NFM (1,2)
P64(0)	.	0	–	–	–
BFS(0)	0	.	0	–	–
BFS(1)	+	0	.	–	–
NFM(1)	+	+	+	.	0
NFM(1,2)	+	+	+	0	.

ly the same as those in Table 11.

The calculation of weights for robustness presents some unexpected difficulties, which can also be illustrated by the data of Table 10. Using the percentage of successfully solved test problems (73%, 91%, 100%, 100%, 100%), and normalizing these quantities, we would obtain the respective weights 0.16, 0.20, 0.22, 0.22, and 0.22. It is probable that optimization specialists and sophisticated users would reject these weights on the grounds that their preferences are not reflected adequately: the P64 method with a success rate of 73% (hardly acceptable in practice) is not distinguished sufficiently from the methods with a 100% success rate. Because only a high success rate (at least 90%, for instance) is acceptable, we drop the P64 method, but we retain the BFS method with difference approximations to first derivatives, and proceed as follows.

In Section 5 we introduced a scale, running from 1 to 5, which will be used now to estimate the robustness ratio of any two methods i and j. As before the ratio is set to 1 if both methods are felt to be equally robust, and to 3 or 5 if method i is felt to be more robust or much more robust than method j. The ratio of 5 is practically the maximum; a higher ratio means that method i dominates method j and that method j can be dropped from consideration. Let us now decide to consider only methods with a success rate of not less than 90%. Then, in order to assign the ratio 5 to two methods at opposite ends of our range, if two methods i and j have success rates of α% and β% respectively, we let the robustness ratio have the value

$$(\alpha - 87.50)/(\beta - 87.50).$$

Thus the second and third methods of Table 10 have an estimated robustness ratio of 0.28, and the complete matrix of estimated robustness ratios is

$$\begin{pmatrix} 1.00 & 0.28 & 0.28 & 0.28 \\ 3.57 & 1.00 & 1.00 & 1.00 \\ 3.57 & 1.00 & 1.00 & 1.00 \\ 3.57 & 1.00 & 1.00 & 1.00 \end{pmatrix} .$$

The reader may verify that priority theory would now assign the robustness weights (0.09, 0.30, 0.30, 0.30) to the methods (2, 3, 4, 5) respectively, which is usually a more acceptable rating than the set of weights given in the previous paragraph.

11. Concluding Remarks

In the comparative studies of nonlinear optimization codes, conclusions are always drawn in a verbal, somewhat informal manner. Given the massive amount of results that exists already, and anticipating future studies, a more rigorous and formal analysis is indispensable. Therefore we have used multi-criteria decision analysis to present a general framework where the decision process is broken down into a number of transparent steps (separate pairwise comparisons for each study), in order that human judgment, expressed as numerical values on a one-dimensional scale, can be used to rank and rate the codes in order of preference. A further refinement is to employ linear model analysis and significance tests to establish efficiency ratios for each pair of codes; this procedure generates suitable efficiency weights. Until recently, such a statistical analysis of the results from test batteries was inconceivable, simply because the number of test problems was too small. Since the appearance of the comparative studies by Schittkowski (1980b) and Fattler *et al.* (1980), however, it is obvious that a large scale experiment, where several codes are applied to many test problems with a variety of starting points, is feasible.

Acknowledgment

It is a great pleasure to thank my colleague J. Sieben for his excellent advice, and my student W. de Kovel for his careful assistance in carrying out the statistical analysis of several computational experiments.

7.2

Testing MP Software: A History and Projection

Richard H.F. Jackson

Center for Applied Mathematics,
National Bureau of Standards,
Washington, D.C. 20234, USA.

According to the literature survey reported in Jackson and Mulvey (1978), the earliest published work on the computational testing of mathematical programming software was in 1953 by Hoffman, Mannos, Sokolowsky and Wiegmann (1953). Although the results of many computational testing efforts were published from 1953 to 1976, Jackson and Mulvey (1978) found that by 1976 the development of a sound methodology for testing MP software was not much advanced from the state of the art in 1953. All too frequently, in that interim period, papers were published with overlapping and conflicting claims of performance.

Partly in response to a perceived need to improve testing methodologies (and partly because the time was right), the Mathematical Programming Society formed the Committee on Algorithms (COAL) in 1973. The three major charges of COAL are (1) to ensure the existence of a suitable basis for comparing algorithms, (2) to act as a focal point for information about software and test problems, and (3) to encourage those who distribute programs to meet certain standards of portability, ease of use, testing, and documentation. Since its formation, the members of COAL have been active in many areas, including: the investigation of techniques for removing timing, compiler, and computer variability from computational results; the establishment of reliable performance evaluation criteria for comparing mathematical programming software; and the development of a sound methodology for testing and comparing MP software. Guidelines for reporting results of computational experiments were developed and published in the journals Mathematical Programming, JORSA, and TOMS. COAL members have organized sessions presenting results of software testing research at conferences sponsored by MPS, IFAC, EURO, TIMS, and ORSA, and, in 1978, there was a NATO Advanced Study Research Institute on the Design and Implementation of Optimization Software (see Greenberg, 1978). Continuing that tradition, a conference on Testing and Validating MP Software was held in Boulder, Colorado in January 1981, whose proceedings will be published (Mulvey, 1981). Special issues of the SIGMAP Newsletter have been assembled by COAL members on MP software (SIGMAP Bulletin, No. 29, April, 1980) and on mixed integer programming software (SIGMAP Bulletin, to be published). Also COAL has conducted three surveys: on relevant literature (Hoffman and Jackson, 1981; Jackson and Mulvey, 1978), on users' preferences for performance criteria (Crowder and Saunders, 1980), and on available codes (COAL Code Availability Survey, to be published).

What is the state of the art in computational testing, and where are we heading? Computational testing has changed significantly in the past 10 years. Automatic generators of test problems with known solutions and controllable structures have emerged, and hand-picked problems are becoming more accessible to a larger audience. The performance measures that were most prior to 1970 (i.e. CPU time, function evaluations, and iteration counts) are being augmented with new measures related to accuracy, ease of use, core storage requirements, and input/output requirements. It is likely that this shift towards more concern for user time and effort will become even more prevalent in the 1980's, due to the continual growth in the storage capacity and computational speed of computers. With this change comes the need for quantitative measures for the "ease-of-use" and "set-up-time" of codes and algorithms. Also the challenge of finding a consistent measure for computational effort continues to exist as well.

Future papers on mathematical software testing will probably rely more heavily on the statistical techniques of "experiment design". By using the procedures described in the statistical literature, one can be more assured of the reproducibility and statistical validity of the results obtained. A related statistical field, that of multi-criteria analysis and clustering, can also make analysis easier, and can give results that are more understandable to potential users of test results (see Lootsma's paper 7.1 in this volume). Certainly, anyone publishing results on computational testing should adhere to the guidelines presented by Crowder, Dembo and Mulvey (1979).

Another important development is software which makes testing easier. There are currently packages which:

(1) test the portability of software (see Feiber, Taylor and Osterweil, 1980; Fosdick, 1979; Ryder, 1973),
(2) "tidy up" software so that it is more readable (see Dyer, 1972; Crowder and Saunders, 1980),
(3) perform routine counts of function and derivative evaluations (see Buckley, 1981),
(4) supply users with large sets of test problems in a standard format (see Buckley, 1981),
(5) count the number of times each portion of code is executed (see Lyon and Stillman, 1974; Buckley, 1981), and
(6) scan and summarize the output of general MP software packages (see Buckley, 1981).

In addition, researchers at the University of Colorado are develing a system called TOOLPACK (Osterweil, 1981), which will incorporate the features described above into one package, thereby simplifying their use. These packages are expected to make large scale test efforts available to a much larger group of researchers.

I will conclude this paper with a few observations about the testing of mathematical programming software. Firstly, testing is extremely costly and time-consuming. Secondly, one must be very careful in designing the experiment if one hopes to obtain reproducible results (especially because of the variations that are caused by compilers and computing environments). Thirdly, since the methodologies for testing are still at an embryonic stage, one can expect many new approaches to emerge in the future.

On the positive side, there are many researchers (organized through the Committee on Algorithms of the Mathematical Programming Society) who are willing to assist those wishing to perform such testing. Test problems and test problem generators have been collected, a survey of available software is currently being made, guidelines for reporting test results exist, and the Committee is currently investigating the organization of a testing laboratory which would assist in funding large test efforts. Furthermore, this Committee has received a very positive and supportive response from the applications community. There are many people who wish to know which of the available software packages are capable of solving their problems quickly and easily. The practical importance and intellectual challenge of MP software testing should assure continuing progress in this difficult research area.

7.3

Parallel Computers and their Impact on Numerical Algorithms and Related Software

J.S. Kowalik

Systems and Computing, Washington State University, Pullman, Washington 99164, USA.

1. Parallel Computers

The recent appearance of pipelined and multiprocessor computers will have a significant impact on numerical algorithms and related software. So far no two high-speed parallel computers have been similar from the architectural viewpoint, but some general trends in parallel computer architecture can be identified. We briefly describe the following four broad categories of parallel computers: pipelined processors, systolic arrays, SIMD, and MIMD machines.

1.1. Pipelined processors In pipelined processors pairs of operands are streamed from memory through one or several arithmetic units. The arithmetic unit is segmented, and several pairs of operands are processed in the pipe in various stages of the arithmetic execution. To perform efficiently the pipelined computer must operate on sufficiently long vectors whose elements are stored contiguously in the memory (or the elements are a fixed increment apart). One of the principal concerns in using pipelined machines, such as the CRAY-1 or CYBER 205, is the average length of vectors to be processed. For long vectors the start-up time to initialize and fill the pipe becomes less significant than for short vectors. It has been found, for example, that the pipelined CDC STAR computer ran two to four times faster than the CDC 7600 computer on programs which contained vectors with hundreds of elements. However, straightforward implementations of non-vectorized programs on the STAR resulted in performance which was substantially slower than that of the CDC 7600. Some pipelined computers have vectorizing compilers, but "automatic vectorization is no substitute for a total rethinking of algorithms, mathematics, and programming techniques" (M. Rowe, NASA).

1.2. Systolic arrays These are two-dimensional arrays of interconnected processors akin to the cellular automata studied by John von Neumann thirty years ago. The reappearance of this architecture can be attributed to the advent of the LSI and VLSI technology, and to the constantly decreasing cost of computer chips. Computational algorithms for systolic array processors have been designed in several areas of application. They include pattern recognition, graph problems, sorting, and numerical linear algebra. The systolic arrays are most suitable for problems which have very regular structure defined in terms of vector operations. Each processor in the interconnected network performs some simple calculations and pumps (hence the word systolic in the name of the processor) data out to its neighbours. Geometry of the communication paths and the operations performed by the processors determine the global function of a particular systolic device. It is envisioned that the systolic array processors will be built as special purpose machines attached to general purpose host computers.

1.3. Single-instruction multiple-data (SIMD) parallel machines

In this architecture there are several to several hundred identical processors with one or more memory modules, and (usually) a permutation network (a switch) that permits information to be exchanged between the processors. Each processor executes the same instruction at any given time, but it can operate on different pairs of operands, i.e. all the processors work in "lock-step". However, certain local autonomy can be achieved by the technique known as mask indexing. Several SIMD computers have been built, and among them ILLIAC-IV, BURROUGHS BSP and ICL DAP are well known. The Mathematics Institute of Queen Mary College, London has been active in investigating various applications of the DAP (Distributed Array Processor) in numerical analysis. In the last eighteen months (1980–1981) several algorithms have been developed for solving nonlinear algebraic equations, finding eigenvalues, singular value decompositions, and other calculations. Initial evaluation of the algorithms for SIMD computers has resulted in surprising conclusions. Perhaps the most important conclusion is that an efficient algorithm for a standard serial computer may lead to an inferior algorithm for a SIMD computer. This fundamental conclusion applies also to the pipelined and the MIMD machines. Clearly the SIMD computer can be applied efficiently only to problems where the major part of the computational process consists of identical operations applied simultaneously to different data. Hence the SIMD computers are not fully flexible general purpose machines, and many algorithms of numerical analysis are ill-matched to their capabilities and requirements. An important task for numerical analysts will be to examine the current sequential algorithms, in order to identify the classes of problems and related algorithms that are suitable for the SIMD computers.

1.4. Multiple-instruction multiple-data (MIMD) parallel machines

This is an architectural design where multiple cooperating programs (also called processes), which are not necessarily identical, are run concurrently by multiple functional units belonging to one or more processors. It is felt that these computers will find wide application in problems of substantial complexity and size which are not suitable for vector computation. The MIMD computers open up new vistas in designing parallel algorithms, since many computational problems can be broken up into similar or dissimilar component parts which can be computed concurrently. The numerical algorithms based on problem partitioning (decomposition, tearing, etc) are very attractive for the MIMD machines. On the other hand, the job of designing efficient algorithms for the MIMD computers requires a thorough and complete re-evaluation of the sequential methods for a given class of problems. An important issue in this design is that of partitioning a problem into many processes (cooperating subprograms) that can be executed concurrently on a MIMD machine. This partitioning is not unique (in many cases), and can have a significant impact on the efficiency of the resulting algorithm.

The MIMD computers allow one to invent and design completely new and truly parallel algorithms. By this we mean algorithms that are not derived from existing sequential algorithms, and that would be inappropriate (i.e. inefficient) for sequential computation.

At present, there are only a few MIMD computers in operation. At the Carnegie-Mellon University there is an experimental multiprocessor called Cm*. A commercial MIMD machine called HEP (Heterogeneous Element Processor) has been built by Denelcor Inc. in Denver, U.S.A. The HEP processor has been used to solve various numerical and combinatorial problems by a team of researchers at the Washington State University in Pullman, Washington. The results of this work (Lord, Kowalik and Kumar, 1980; Lord and Kumar, 1980; Deo, Pang and Lord, 1980) suggest that it is possible to design efficient MIMD algorithms for common problems of numerical analysis and graph theory.

2. Optimization

So far not much work has been done to develop optimization algorithms which exploit capabilities of parallel machines. One of the early papers on a non-gradient parallel version of the Powell—Zangwill method is due to Chazan and Miranker (1970). Straeter (1973) of NASA has proposed a parallel variable metric method, and simulated its numerical performance on a standard serial computer. In another report Straeter and Markos (1975) have described a parallel version of the Jacobson—Oksman method based on the so called homogeneous model function.

3. Conclusion

Many current numerical algorithms cannot run efficiently (or at all) on the new parallel high-performance computers. Algorithmic methodologies for parallel machines are seriously lagging behind architectural advances. It is reasonable to expect that parallel computers will be commonly available to a significant segment of computer users in the second half of the present decade (1985—1990). These machines might be underutilized due to the lack of compatible algorithms and software. It takes five to ten years to develop a conceptual algorithm into a software implementation for a commonly available subroutine library. In view of this substantial time delay, there is an urgent need to re-evaluate the existing sequential algorithms, to establish their suitability for new architectures, and to develop numerical algorithms for parallel computers. This suggests that numerical analysts should become familiar with the state-of-the-art in computer architecture, and should provide the manufacturers of computer hardware with some guidelines relative to their requirements.

The current and future advances in computer architecture will allow one to use different computers for different computational problems, in order to obtain efficient and cost-effective solutions. Our task will be to analyse the familiar problems and algorithms and match them with the emerging variants of parallel machines.

Finally, it should be pointed out that parallel computing offers new challenges for numerical algorithm designers, applied mathematicians and computer scientists. Parallel computing is more than a different way of executing existing sequential programs.

7.4

A Proposal for the Establishment of an Optimization Software Testing Center

K.M. Ragsdell

School of Mechanical Engineering, Purdue University, West Lafayette, Indiana 47906, USA.

1. The Past

At this time, it is now relatively well known that several comparative experiments of optimization algorithms and software have been completed. The comparative studies of Sandgren (1977), Fattler, Sin, Root, Ragsdell and Reklaitis (1980), and Schittkowski (1980b) stand out for at least two reasons: first because they give an almost exhaustive review of previous comparative work up to this time, and secondly because of the very useful results which they supply. Recently, Ragsdell (1981) reviewed the first two experiments, Sandgren (1981) considered the statistical implication of his previous work, and Schittkowski (1981a) gave results for the latest successive quadratic programming algorithms, and proposed a model (Schittkowski, 1981b) for the conduct of future comparative experiments. Although there is great similarity in the work of Sandgren and Schittkowski, there are fundamental differences. Sandgren compares codes on the basis of robustness and efficiency at prescribed accuracy levels (Himmelblau, 1972), whereas Schittkowski uses Saaty's priority theory as outlined by Lootsma (1980), and Paper 7.1 in these proceedings), with termination point accuracy as one of many measures of code merit. I believe that termination point accuracy is an experiment parameter which must be controlled (for instance as suggested by Eason and Fenton, 1974) as a prerequisite to meaningful comparison. In addition, Schittkowski does not give results on real problems from practice as did Sandgren (1977), Eason and Fenton (1974) and Colville (1968) before him. The Schittkowski test problem set is large and varied, but the performance of an algorithm on "real" problems is of pivotal interest.

2. The Present

There is currently a keen interest in test results, in the design of numerical experiments, and in the conduct of comparative studies in general. This is healthy and hopefully will continue for some time to come. For example, see the recent monograph by Van der Hoek (1980), which gives extensive, remarkably complete, and beautifully precise comparative results for a number of "reduction methods". Hoek defines the family of reduction methods as those which involve subproblems which are "simpler" in some sense (say less nonlinear, fewer constraints, etc.) than the original nonlinear program. Most (if not all) of the latest algorithms would fit in this family. But all is not goodness and light. For instance, Eason (1977) has pointed out some fundamental difficulties in code comparisons, and recently gave more complete information on these difficulties (Eason, 1981). He asks some important questions concerning the effect of machine word length, compiler optimization level, and other machine dependent parameters, which suggest that specific results (and trends) in one computing environment will not necessarily hold in another. These difficulties along with the great expense associated with testing would lead many to abandon testing as a research activity. But just the opposite action is needed! These and other questions on the relative value of new methods are worthy of answers. I believe that in the near term (10—20 years) the answers will flow from well designed experiments and not from pure analysis. Accordingly, *the present need is for additional tests*. At the very least we need to:

a) build better test problem sets.
b) study the effect of machine paramaters (such as word length, compiler, standard function library, etc.) on code performance.
c) study the effect of multiple starting points.
d) investigate the effect of various gradient approximation schemes.
e) devise and use additional and more useful code ranking schemes.
f) conduct the experiments such that appropriate statistical measures of confidence can be attached to the results.
g) develop meaningful measures of problem difficulty.

Again, *much additional testing is needed*. The situation is complex, but certainly not hopeless as some would lead us to believe.

3. The Future

I propose the establishment of a coordinating testing body, a center if you will, which should encourage and support the testing efforts of individual investigators. I see this *Optimization Software Testing Center* as a focal point which will not only encourage testing efforts to go forth, but will provide a degree of communication and continuity which is now genuinely needed. Hopefully, the center would be established with the blessing and sponsorship (and possibly minor material support) of the Mathematical Programming Society, through its Committee on Algorithms. Major long term support (250,000$ — 350,000$ per year for 5—10 years) would come from a variety of sources including NSF, ONR, NASA, NBS, private industry and elsewhere. The center must be located in proximity to plentiful, low-cost computing power. In addition, physical accommodation should be provided for one or two senior visiting investigators each year in order to ensure proper communication with other testing efforts. This would be in addition to the normal resident graduate student/ research associate testing activity (approximately three full time positions per year). The center would develop, maintain and distribute appropriate test problem sets. In addition, experiments would be conducted continuously and performed in such a manner that the results of new codes could be added conveniently to existing results and rankings. Periodic reports would be issued as new results became available, say every six months. Both algorithm and software tests would be conducted.

The Testing Center would provide a valuable service to code and algorithm developers and to the user community alike. The center would provide a "half-way house" interface between emerging codes and potential users. In this way algorithm development would be enhanced by the availability of immediate comparative results, and users would be spared the ordeal of using untested experimental versions of new algorithms.

Discussion on "Future Software and Testing"

7.5 Rosen

Please comment on the parameters that have to be adjusted by the user for the numerical tests of Paper 7.1.

7.6 Lootsma

In recent studies, the starting point is varied, and sometimes the penalty parameters also in order to study the sensitivity of the code to these variations. The reduced gradient codes appear to be less sensitive to parameters which control the computation than other codes, since they have been designed to be black boxes.

7.7 Powell

The techniques that are used in Paper 7.1 to draw conclusions about algorithms have so many disadvantages that I am doubtful that they can give useful results. For instance they are highly dependent on the test problems that are selected, on the conclusions of the separate tests that are amalgamated, and on the chosen weights and criteria. Does the speaker agree that his methodology has to be developed much more before it can give useful conclusions?

7.8 Lootsma

Of course the results are dependent on the test problems that have been selected, and they will always be so. It was my objective, however, to do a formal analysis of the leading comparative studies that have been published in the last ten years. In fact, our current knowledge of the performance of nonlinear optimization algorithms is based on these studies only. The criteria and the weights depend largely on the feelings of decision makers, and they may vary from one individual to another. I feel strongly, however, that multi-criteria analysis is an excellent tool to map out the performances, and that a statistical analysis of the observed failures and execution times will nicely support the analysis. Of course we have crude results, since we are only interested in orders of magnitude. I would like to add that authors of comparative studies should anticipate some sort of multi-criteria analysis. Therefore I would recommend that they try to rank and rate the codes in their studies according to various criteria.

7.9 Mifflin

We should classify test problems as convex or nonconvex, so that if a method passes the set of convex function test problems it is worthwhile expending effort in trying different approaches to make the method robust with respect to general functions.

7.10 Lootsma

I agree that the study of classes of test problems is important. Therefore we have to refine our criteria of efficiency and robustness. We might consider, for instance, the robustness for convex problems and for certain classes of nonconvex problems. This can all be done within the same framework of priority theory. Similarly, one might consider the efficiency in the absence or the presence of analytical derivatives, and efficiency measured in execution time or in function evaluations.

7.11 Rijckaert

I would like to point out that codes based on GRG, augmented Lagrangians, and penalty functions are all using a somewhat uniform and standardized approach. This is in contrast to codes in geometric programming, where one has the possibility of using a primal or a dual code. The major differences between these two approaches are likely to give different characteristics, so it is doubtful if they should be compared by automatic techniques.

7.12 Lootsma

Previous comparative studies of Dembo (1978) and Rijckaert and Martens (1978) identified the GGP and the GPKTC codes as the leading geometric programming codes. The final comparison by Fattler, Sin, Root, Ragsdell and Reklaitis (1980) was based only on these codes. Therefore I hope that justice has been done to several approaches in geometric programming. Of course, I have concentrated in my talk on what the comparative studies tell us now, and on the conclusions that can be drawn formally.

7.13 Dixon

I agree with Contribution 7.11 that the conclusions on geometric programming that are given in Paper 7.1 are far too sweeping. One of my students, Diana Kieley made tests on posynomial functions to compare REQP codes used in their general form with their application to the reduced dual geometric programming formulation (Kieley, 1976). We found that the specialised techniques are always superior when the GP has "degree of difficulty" equal to zero, and they are usually better when the degree of difficulty is less than the dimension of the original problem. When the degree of difficulty is greater than the dimension of the original problem the general method is always superior.

7.14 Lootsma

I have not found this in the comparative studies of GP codes. Therefore, either Diana Kieley's observation is incorrect, or you have pointed out a weakness of my studies.

7.15 Baker

Is anyone in this community doing work on the characterisation of nonlinear optimization problems? It seems that the given categories, namely sparse, convex, nonlinearly constrained etc, are too loose to have real meaning in the analysis and comparison of algorithms.

7.16 Bus

I have tried to test programs for solving nonlinear systems

of equations in such a way as to relate program behaviour to various properties of the problems. In my thesis (Bus, 1980) I actually give a set of relevant properties for these problems, and I perform testing relative to these properties as far as possible. I wish to emphasise that, in my opinion, such testing is very important from the point of view of people developing algorithms. Testing of codes on large representative sets of test problems (whatever they may be) is sensible only for those codes that are presented to the user-community as complete, robust, reliable, and well-documented software. It is not suitable for programs that are still at the development stage.

7.17 Schnabel

SUMT has solved and continues to solve many practical numerical calculations, for instance in civil engineering. This is an example of a technique that is suited to special problem classes, and it also shows that, when users find an approach that solves their problems, it is very hard to convince them to prefer newer, improved methods.

7.18 Meyer

Software that may have been state-of-the-art at its initial release may soon become obsolete relative to newly developed methods. If this indeed occurs, then there is the possibility that obsolete software will give the field a "bad name" because it fails in the hands of inexperienced practitioners. A drastic approach that will prevent this is to build a self-destruct mechanism in the software that will destroy the software after 1 or 2 years of use. This feature has been included in some systems software, but not, to my knowledge, in any optimization software. A less dramatic approach might be to have the software print out a warning of the following sort: "This software is now t [age in years] years old — it may be obsolete as a result of new developments in the field". (This would be done when the internal clock in the software calculated that the age was at least 1 or 2.)

7.19 Schittkowski

How are failures of algorithms to solve test problems included in the statistics of Paper 7.1?

7.20 Lootsma

The efficiency ratios are based on the problems that have been solved by the codes. The failures are taken into account in the robustness ratios, where the ratios of the failure rates are considered.

7.21 Davidon

It is highly desirable that, if an algorithm fails, the user should be told about the failure. Is this feature included in the comparisons of robustness in Paper 7.1?

7.22 Lootsma

Several studies record two types of failures: (1) full accuracy not obtained, and (2) no reasonable approximation to a local minimum or an excessive amount of computation. In my evaluation

I have only considered failures of the second type. In the first case it is not easy to decide whether an inaccurate solution should be taken as a failure or not, and I wish that we could give some "credit points" for the graceful termination of a run.

7.23 Polak

Algorithms with significant rates of failure obviously can only be used in situations where failure does not have catastrophic effects and where back-up algorithms exist. It is sensible to use such an algorithm if it is much faster than the robust back-up algorithm, and if massive repeated computing is carried out. In such a situation, a failing algorithm can be evaluated by distributing the cost of failure in terms of computer time, human labour and delays incurred, over the successful runs it has performed.

7.24 Rijckaert

I agree that in measuring efficiency one should in some way account for the time that is spent on failures. In general, however, it is difficult to allocate the time of a failure in a unique way. One difficulty is the dependence of the conclusions on the criterion for deciding when to classify a run as a failure.

7.25 Powell

In my opinion the attention that is given to failures in algorithm comparisons is far too slight. In the case of current algorithms I think it is important to discover why any failures occur. If the reason is that the algorithm is being applied to a test problem of a kind that is totally unsuitable, for example an algorithm for smooth optimization may be applied to a function that has large discontinuities, then the results should be excluded from the comparison. Otherwise research should be done on the cause of the failure, in order to modify the algorithm, or at least to ensure that the failure is automatically brought to the attention of the user.

7.26 Ragsdell

Code failures are an important problem, particularly in post-analysis of the performance data. These failures create holes in the data which can cause difficulty in statistical analysis. I think Professor Polak's idea (Contribution 7.23) is a good one. In addition I agree with Professor Powell (Contribution 7.25) that failure should be carefully examined with an eye towards algorithm improvement.

7.27 Gutterman

When a code fails on a large problem, it may be valuable to construct a small problem which fails in the same way. Such a problem would be a useful test case, because it would be an economical partial substitute for the large problem.

7.28 Lemaréchal

Because the lack of test problems is mentioned in Paper 7.2, it seems appropriate for me to do a little self-advertising. We are currently trying to develop software in France. Considering that the existing test problems are more suitable for testing algorithms and theorems rather than for testing software, we try to balance the effort that is given to algorithms and to the selection of test problems. Therefore our group includes people from outside (e.g. physicists and economists), who bring their own problems to enrich our library.

7.29 Sorensen

On the need for more and better test problems, please clarify what is meant by "better". It seems to me that we would like to have problems which are economical to solve and which are representative of problems that are encountered in practice.

7.30 Jackson

While we live in times of a dearth of test problems, I think better must just mean more — more of the hand-picked problems that I mentioned. Perhaps I did not make it clear in my talk that these hand-picked problems typically are of two types: those that arise naturally in practice and those that are specially constructed to test a particular aspect of an algorithm or code. I think it is particularly important to have better *collections* of test problems. Here I have in mind test problem sets that are large, complete in some sense, acceptable to the algorithm-development community, and widely available. Later, when we live in a world full of test problems, we will probably have to choose among them in a fashion similar to the way we choose among codes now. In that regard I recommend the work of O'Neill (1981) and myself (Jackson, 1979).

7.31 Sargent

The definition of reliability in Paper 7.2 is very dependent on the set of test problems, and in particular on the starting points. Often algorithms have success rates that are close to 100% with standard starting points. Therefore we have applied the following technique to compare algorithms. We use 100 randomly generated starting points within a certain range, which is chosen by picking a "standard" algorithm and enlarging the range until this algorithm achieves about 50% success on the 100 points. Then these same points are used for the other algorithms.

7.32 Lootsma

Please will the speaker give his impression of the algorithms of Straeter for unconstrained optimization that are mentioned in Paper 7.3.

7.33 Kowalik

Straeter from NASA proposed two parallel optimization algorithms and tested them on a serial machine (Straeter, 1973; Straeter and Markos, 1975). He claimed that both algorithms performed very well. We have not replicated his work, and I am not able to comment on the accuracy of Straeter's conclusions.

7.34 Rosen

Why would one prefer a SIMD (single-instruction multiple-data) machine to a MIMD (multiple-instruction multiple-data) machine for parallel computation?

7.35 Dixon

In my opinion SIMD machines should sometimes be preferred to MIMD machines when one has the problem of dividing tasks between say 4000 completely general subtasks. This problem is so complex that most MIMD systems in general use are limited to about 50 parallel processors, but the problem of devising parallel tasks for the "step-locked" SIMD machines, for instance the 4096 processors on the ICL-DAP, can be relatively easy. In my research paper I show that the conceptual implementation of a parallel Newton method on that machine is quite straightforward, and can give major savings in users' time.

7.36 Kowalik

SIMD computers are more suitable for solving certain classes of applied problems, for example signal processing and dense matrix manipulations.

7.37 Boggs

In large scale parallel processing there is a need for appropriate languages which allow the natural expression of parallel algorithms. Such languages may also provide useful ways of thinking about parallelism. FORTRAN is clearly not suitable and its extensions may be somewhat limited. The language ADA does incorporate parallelism, and there are some others.

7.38 Kowalik

Two of the currently available parallel computers (ICL DAP and

HEP) have extended FORTRAN compilers. These extensions are justified by the popularity of FORTRAN and the very substantial investment in existing FORTRAN software. Also a PASCAL compiler will be available for the HEP machine.

7.39 Mifflin

Are we now in a position in parallel processing to test decomposition schemes for solving large scale optimization problems, where the independent inner optimization subproblems are solved simultaneously? This approach would give some good test problems for nonsmooth optimization, because recent advances in nonsmooth optimization are useful for solving the outer problem. See also Discussion Contribution 5.31.

7.40 Kowalik

The HEP processor is suitable for decomposition methods, but our experience with this machine is limited to solving ordinary differential equations, linear algebraic equations, and certain graph theoretic problems.

7.41 Ragsdell

It is an exciting prospect that it may now be possible to solve very large problems on parallel machines by decomposition.

7.42 Beale

When working on decomposition in the mid 1960's (Beale, Hughes and Small, 1965), I became very conscious of the difficulties that are caused by dense and nearly linearly dependent relations between the variables in the master problem. These difficulties may not be insuperable, but do not imagine that it will necessarily be easy to solve them.

7.43 Lemaréchal

I wish to emphasise that the progress that has been made in nondifferentiable optimization during the last few years brings new ways for treating the master program, and it can avoid the degeneracy problem. Further, the decomposition of nonlinear problems has become much more tractable.

7.44 Moré

In terms of speed and reliability, how does the algorithm for the solution of systems of linear equations, mentioned in Contribution 7.40, compare with Gaussian elimination with partial pivoting on an IBM 3033?

7.45 Kowalik

The LU and QR matrix decompositions can be implemented efficiently on parallel MIMD processors such as HEP. In the case of LU, if the ratio of the number of processors to the number of linear equations is less than 0.1, then the efficiency of our algorithm is better than 0.9, which means that the speed-up over serial computation is better than a factor of 0.9r, where r is the number of processors. The numerical stability of our algorithm is exactly the same as that of the corresponding serial algorithm.

7.46 Powell

Will the testing centre, that is proposed in Paper 7.4, collaborate with COAL?

7.47 Ragsdell

My proposal is so new that I do not yet have the formal approval or support of any organisation. I will be making a presentation to the other members of COAL in August, and I hope to have their support and the support of the Mathematical Programming Society. Requests for funding will begin this fall.

7.48 Jackson

I also wish to reply to Contribution 7.46 on whether there will be collaboration between COAL and the proposed Central Testing Center. As chairman of the Committee on Algorithms, I, unfortunately, can neither endorse nor deny the possibility of collaboration, since it has not yet been brought before the full committee. It is bound to be a controversial issue, and I am sure that some COAL members will want to provide input to the proposal for a Central Testing Center before it receives support from COAL and is submitted to the Mathematical Programming Society for their consideration. I can say, however, that I agree in principle with the *concept* of a Central Testing Center, although I feel that in pressing for its creation we must proceed with caution, and consider carefully its organisation, charter, funding and goals.

7.49 Wolfe

I am happy to learn that the Mathematical Programming Society's Committee on Algorithms will study the proposal for an Optimization Software Testing Center. The Society is dedicated to the promotion of all aspects of mathematical programming, and is ready to give (at least moral) support to any well-founded project in this area. It invites anyone who could use the kind of help it can provide to a discussion of the possibilities.

7.50 Bus

What is precisely the task that is intended for the testing centre? In my opinion most of the people here, who are developing algorithms, have a need for tools to make their own selective experimental testing easy. Therefore the provision of carefully analyzed test problems, and maybe even some programs for doing testing and analysis of the results, might be a nice task for the testing centre. However, I do not think it is sensible to provide a centre to which everybody can send the code which he/she is developing for testing. The centre would spend a lot of time and money just to conclude that small changes have to be made to the program, and that the program may have to be resubmitted for a new test run.

7.51 Ragsdell

I feel that the centre should carry out all the activities that have just been mentioned, but with less emphasis on judging new algorithms. Certainly the centre should do all that is reasonable to support the comparative work of others. In addition I hope that the centre will provide a productive environment for the collection and analysis of comparative results at all levels.

7.52 Boggs

From the point of view of a funding agency, certain questions must be raised on Ragsdell's proposal, and I will play devil's advocate and raise them. A significant fraction of the total support in nonlinear programming would have to go into such an effort. Thus, it is important to assess the trade-off between this work and traditional research. More particularly, what is the value of such testing and what useful information do we really get? I think that such questions must be clearly thought out in making any formal proposals.

7.53 Ragsdell

I think that the last speaker is too pessimistic about the consequences of financing the proposed centre. We are talking about funding at the level of 250—350 thousand dollars per year, which should be an acceptable perturbation to the funding situation. Secondly, part of these funds may come from other sources, in particular various industrial groups. Support from government and industry is vital. We have enjoyed much support in the past and hope to once again. Finally, it is important to recognise that this work desperately needs to be done, and is not a luxury! Recall my analogy to theoretical and experimental physics — this is experimental mathematics. To my mind experimental does not mean unimportant.

7.54 Lootsma

More testing is absolutely necessary. In the last ten years we have seen only 10 or 20 comparative studies, but hundreds of papers on the theoretical development of algorithms. All we know about the relative performance of codes is due to this limited amount of testing!

7.55 Final Comments by Introductory Speaker (Lootsma)

Several viewpoints on software development and software testing have been expressed during Session 6 (The Current State of Optimization Software) and the present session. In many cases, no final answer could be given, only initial steps (in parallel computing, for instance) or continuing progress (in the amount of testing, the collection or generation of test problems, and the analysis of test results, for instance) could be reported. Briefly summarized, the main issues that have been raised during the lively discussions are the following ones:

1. Objectives of Testing What do we want to test: fool-proof commercial optimization codes with all their user-oriented features and safety provisions to serve the general user, or the underlying algorithmic ideas to support future research of algorithm designers? Should we concern ourselves with complete optimization codes for solving (classes of) nonlinear optimization problems, or with the "building bricks" such as linear searches, LP subroutines, and QP subroutines?

2. Selection of Test Problems The question of how to select artificial or real-life test problems is a point of major concern, and various ideas have been brought forward: the selection on the basis of particular *properties* (dimension, sparsity, ill-condition-

ing, ...), and the selection of certain *categories* of test problems (convex, linearly constrained, posynomial problems, sums of squares, ...).

<u>3. Performance Criteria</u> There is a growing consensus on the performance criteria to be considered (particularly efficiency, robustness, simplicity of use, and program organization), but quantification of the performance remains cumbersome. Efficiency can be expressed in execution times (highly machine-dependent) or equivalent function evaluations (ignoring the overheads of the algorithm). Both efficiency and robustness are very sensitive to parameters which control the course of the computations (e.g. tuning of nested iterative processes). Simplicity of use (documentation, input/output facilities, numerical differentiation, graceful termination) and program organization (modularity, portability) are not easy to assess.

<u>4. The Computational Environment</u> The test results are highly dependent on the computers and compilers that have been used. The development of optimization algorithms is seriously challenged by the appearance of parallel computers. Interactive optimization software will be required, both for inexperienced and sophisticated users so that they can follow the course of the computations and the degree of improvement.

<u>5. Analysis of Test Results</u> A neglected area has hitherto been the formal analysis of the test results (statistical analysis, multi-criteria analysis, significance tests), and the ranking and rating of codes under various performance criteria.

In the seventies, we have seen an increasing concern with optimization software and testing methodologies. The Committee on Algorithms (COAL) of the Mathematical Programming Society played a pioneering role in raising the critical issues. This will hopefully lead to the establishment of sound testing practices in the eighties.

List of Research Seminars

M.C. Bartholomew-Biggs, "Continuing developments of nonlinear programming methods using recursive quadratic programming".

E.M.L. Beale, "Conjugate gradient approximation programming" and "Chains of linked ordered sets".

D.P. Bertsekas, "Projected Newton methods for linearly constrained problems".

P.T. Boggs, "Merit functions for implementing nonlinear programming algorithms".

C.A. Botsaris, "Differential nonlinear programming".

A. Buckley, "Programs for testing unconstrained minimization algorithms".

J.C.P. Bus, "Descent secant update algorithms for nonlinear equations".

T.F. Coleman, "Estimation of large sparse Jacobians and Hessians and graph coloring problems".

T.F. Coleman and A.R. Conn, "A successive quadratic programming method that is quasi-Newton and uses only projected Hessians".

R.S. Dembo, "Truncated-Newton methods in large-scale NLP".

J.E. Dennis, "Secant methods for noisy functions".

L.C.W. Dixon, "The place of parallel processing in nonlinear optimization".

A. Drud, "Design of a large scale dynamic GRG algorithm".

J.G. Fiorot, "On the convergence of methods using parametric variable metric formulae".

D.M. Gay, "An approach to computing perturbation bounds on the solutions to some optimization problems".

P.E. Gill, "Topics in large-scale unconstrained optimization".

D. Goldfarb, "Numerically stable dual and primal-dual methods for quadratic programming".

A. Griewank and **Ph.L. Toint**, "Quasi-Newton methods for large unconstrained optimization problems".

S.P. Han, "Use of dual penalty functions in computational algorithms".

K. Jittorntrum, "Strict complementarity: a redundant condition in nonlinear programming".

J.S. Kowalik, "Impact of parallel computers on numerical computation".

D. Kraft, "Solving optimal control problems by direct collocation with B-splines and nonlinear programming".

J.L. Kreuser, "The application of optimization software in a particular environment and the implication for software development" and "Computational experience in solving the least-norm linear programming problem via iterative SOR techniques".

L. Lasdon, "Solving large sparse nonlinear programs by successive quadratic programming, successive linear programming, and GRG algorithms".

C. Lemaréchal, "A challenging test problem for constrained optimization".

F.A. Lootsma, "Multi-criteria decision analysis".

O.L. Mangasarian, "Differentiable dual exact penalty functions".

L. McLinden, "Successive approximation and linear stability involving convergent sequences of optimization problems".

R.R. Meyer, "Large-scale optimization, piecewise-linear approximation, and network flows".

R. Mifflin, "A superlinearly convergent algorithm for one-dimensional constrained minimization problems with convex functions".

J.J. Moré, "On the estimation of noise".

M.L. Overton, "Minimizing a sum of Euclidean norms: an algorithm with quadratic convergence to both differentiable and non-differentiable solutions".

E. Polak, "On the extension of differentiable optimization algorithms to the nondifferentiable case".

S.M. Robinson, "An elementary treatment of stability, tangency, and first-order optimality".

J.B. Rosen, "Global minimization of a linearly constrained concave function by partition of feasible domain".

R.W.H. Sargent, "Recursive quadratic programming algorithms with global superlinear convergence".

R.B. Schnabel, "Nonstandard models for unconstrained optimization and nonlinear equations".

T. Steihaug, "Globally convergent methods in large-scale unconstrained optimization".

K. Tanabe, "Feasibility-improving-gradient-acute-projection methods and compatible Lagrangian multipliers".

R.A. Tapia, "Convergence rates of quasi-Newton methods for constrained optimization".

P. Wolfe, "Experience with the ellipsoid algorithm on nonlinear problems".

M.H. Wright, "Algorithms for problems with a mixture of constraint types".

References

(The numbers in brackets at the end of each reference are the numbers of the papers and discussion contributions that include the reference.)

J.O. Aasen (1971), "On the reduction of a symmetric matrix to tridiagonal form", *BIT*, Vol. 11, pp. 233–242; (1.31), (3.2).

J. Abadie (1975), "Méthode du gradient réduit généralisé: le code GRGA", Note HI/1756/00, Electricité de France, Paris; (6.4), (7.1).

J. Abadie and J. Carpentier (1969), "Generalization of the Wolfe reduced gradient method to the case of nonlinear constraints", in *Optimization*, ed. R. Fletcher, Academic Press (London); (4.5), (5.1).

J.P. Abbott and R.P. Brent (1975), "Fast local convergence with single and multistep methods for nonlinear equations", *J. Austral. Math. Soc. Ser. B*, Vol. 19, pp. 173–199; (1.5).

S. Agmon (1954), "The relaxation method for linear inequalities", *Canad. J. Math.*, Vol. 6, pp. 382–392; (3.2).

E. Allgower and K. Georg (1980), "Simplicial and continuation methods for approximating fixed points and solutions to systems of equations", *SIAM Rev.*, Vol. 22, pp. 28–85; (3.3), (6.1).

F. Aluffi, S. Incerti and F. Zirilli (1980a), "Systems of equations and A-stable integrations of second order o.d.e.'s", in *Numerical Optimization and Dynamic Systems*, eds. L.C.W. Dixon and G. Szegö, North Holland Publishing Co. (Amsterdam); (1.5).

F. Aluffi, S. Incerti and F. Zirilli (1980b), "Systems of simultaneous equations and second order differential equations", in *Optimizzazione Non-lineare e Applicazioni*, eds. S. Incerti and G. Treccani, Pitagora Editrice; (1.5).

D.H. Anderson and M.R. Osborne (1977), "Discrete, nonlinear approximation problems in polyhedral norms: a Levenberg-like algorithm", *Numer. Math.*, Vol. 28, pp. 157—170; (2.4).

T.W. Anderson and H. Rubin (1956), "Statistical inference in factor analysis", in *Proceedings of the Third Berkeley Symposium on Mathematical Statistics and Probability (Vol. 5)*, University of California Press (Berkeley); (2.12).

K.J. Arrow, L.Hurwicz and H. Uzawa (1958), *Studies in Linear and Nonlinear Programming*, Stanford University Press (Stanford, California); (1.37).

K.J. Åström: *see* Bellman and Åström (1970).

M. Avriel, M.J. Rijckaert and D.J. Wilde (eds.) (1973), *Optimization and Design*, Prentice Hall (Englewood Cliffs, N.J.); (7.1).

S.A. Awoniyi (1980), "A piecewise-linear homotopy algorithm for computing zeros of certain point-to-set mappings", Ph.D. dissertation, School of Operations Research, Cornell University; (3.3).

S.A. Awoniyi and M.J. Todd (1981), "An efficient simplicial algorithm for computing a zero of a convex union of smooth functions", in preparation; (3.3).

J.W. Bandler: *see* Charalambous and Bandler (1976).

P. Barrera and J.E. Dennis (1979), "When to stop making quasi-Newton updates", presented at the Tenth International Symposium on Mathematical Programming (Montreal); (6.1).

I. Barrodale and F.D.K. Roberts (1973), "An improved algorithm for discrete ℓ_1 linear approximation", *SIAM J. Numer. Anal.*, Vol. 10, pp. 839—848; (2.4), (2.33).

R.H. Bartels (1980), "A penalty linear programming method using reduced gradient basis-exchange techniques", *Linear Algebra Appl.*, Vol. 29, pp. 17—32; (4.4).

R.H. Bartels and A.R. Conn (1981), "An approach to nonlinear ℓ_1 data fitting", Report CS-81-17, Computer Science Department, University of Waterloo (to be published in *Proceedings of 3rd Workshop on Numerical Analysis, Cocoyoc, Mexico*, Springer—Verlag); (2.1), (2.4), (4.4).

R.H. Bartels, A.R. Conn and C. Charalambous (1978), "On Cline's direct method for solving overdetermined linear systems in the

ℓ_∞ sense", *SIAM J. Numer. Anal.*, Vol. 15, pp. 255–270; (4.4).

R.H. Bartels, A.R. Conn and J.W. Sinclair (1978), "Minimization techniques for piecewise differentiable functions: the ℓ_1 solution to an overdetermined linear system", *SIAM J. Numer. Anal.*, Vol. 15, pp. 224–241; (2.4), (4.4).

M.C. Bartholomew-Biggs (1979), "An improved implementation of the recursive quadratic programming method for constrained minimization", Technical Report No. 105, The Numerical Optimization Centre, The Hatfield Polytechnic; (6.4).

M.C. Bartholomew-Biggs (1981a), "Line search procedures for nonlinear programming algorithms using quadratic programming subproblems", Technical Report No. 116, The Numerical Optimization Centre, The Hatfield Polytechnic; (4.2).

M.C. Bartholomew-Biggs (1981b), "The use of penalty parameters in quadratic programming subproblems for constrained minimization", Technical Report No. 115, The Numerical Optimization Centre, The Hatfield Polytechnic; (4.2).

A.S.J. Batchelor and E.M.L. Beale (1980), "A revised method of conjugate gradient approximation programming", in *Survey of Mathematical Programming*, ed. A. Prékopa, Akademie Kiadó (Budapest); (5.1).

D.M. Bates and D.G. Watts (1980), "Relative curvature measures of nonlinearity", *J. Roy. Statist. Soc. Ser. B*, Vol. 42, pp. 1–25; (2.10).

E.M.L. Beale (1967), "Numerical methods", in *Nonlinear Programming*, ed. J. Abadie, North-Holland Publishing Co. (Amsterdam); (5.48).

E.M.L. Beale (1972), "A derivation of conjugate gradients", in *Numerical Methods for Nonlinear Optimization*, ed. F.A. Lootsma, Academic Press (London); (1.2).

E.M.L. Beale (1974), "A conjugate gradient method of approximation programming", in *Optimization Methods for Resource Allocation*, eds. R.W. Cottle and J. Krarup, English Universities Press (London); (4.5).

E.M.L. Beale (1978), "Nonlinear programming using a general mathematical programming system", in *Design and Implementation of Optimization Software*, ed. H.J. Greenberg, Sijthoff and Noordhoff (Alphen aan de Rijn, Netherlands); (5.1).

E.M.L. Beale (1980), "Branch and bound methods for numerical optimization of non-convex functions", in *COMPSTAT 80 Proceedings in Computational Statistics*, eds. M.M. Barritt and D. Wishart, Physica Verlag (Vienna); (5.1).

E.M.L. Beale and J.J.H. Forrest (1976), "Global optimization using special ordered sets", *Math. Programming*, Vol. 10, pp. 52–69; (1.1), (5.1), (5.43), (5.46).

E.M.L. Beale and J.J.H. Forrest (1978), "Global optimization as an extension of integer programming", in *Towards Global Optimization 2*, eds. L.C.W. Dixon and G.P. Szegö, North-Holland Publishing Co. (Amsterdam); (5.1).

E.M.L. Beale, P.A.B. Hughes and R.E. Small (1965), "Experiences in using a decomposition program", *Comput. J.*, Vol. 8, pp. 13–18; (5.10), (7.42).

E.M.L. Beale and J.A. Tomlin (1970), "Special facilities in a general mathematical programming system for non-convex problems using ordered sets of variables", in *Proceedings of the Fifth International Conference on Operational Research*, ed. J. Lawrence, Tavistock Publications (London); (5.1).

E.M.L. Beale: *see also* Batchelor and Beale (1980).

P. Beck, L.S. Lasdon and M. Engquist (1981), "A reduced gradient algorithm for nonlinear network problems", working paper, Department of General Business, University of Texas; (4.5).

R. Bellman (1960), *Introduction to Matrix Analysis*, McGraw-Hill Book Co. (New York); (7.1).

R. Bellman and K.J. Åström (1970), "On structural identifiability", *Math. Biosci.*, Vol. 7, pp. 329–339; (2.7).

D.P. Bertsekas (1975), "Nondifferentiable optimization via approximation", *Math. Programming Stud.*, Vol. 3 (Nondifferentiable Optimization), pp. 1–25; (2.3).

D.P. Bertsekas (1976), "Multiplier methods: a survey", *Automatica – J. IFAC*, Vol. 12, pp. 133–145; (4.3).

D.P. Bertsekas (1979), "Algorithms for optimal routing of flow in networks", in *International Symposium on Systems Optimization and Analysis*, eds. A. Bensoussan and J.L. Lions, Springer-Verlag (New York); (3.18).

D.P. Bertsekas (1980a), "Enlarging the region of convergence of Newton's method for constrained optimization", LIDS Report R-985, Massachusetts Institute of Technology, (to be published in *J. Optim. Theory Appl.*); (4.3).

D.P. Bertsekas (1980b), "Variable metric methods for constrained optimization based on differentiable exact penalty functions", in *Proceedings of the Eighteenth Allerton Conference on Communication, Control and Computing*, published at the University of Illinois; (4.3).

D.P. Bertsekas (1982), *Constrained Optimization and Lagrange Multiplier Methods*, Academic Press (New York); (4.3).

D.P. Bertsekas and E. Gafni (1981), "Projected Newton methods and optimization of multicommodity flows", LIDS Report P-1140, Massachusetts Institute of Technology; (3.18).

D.P. Bertsekas, E. Gafni and K.S. Vastola (1979), "Validation of algorithms for optimal routing of flow in networks", *Proceedings of the 1978 IEEE Conference on Decision and Control*, IEEE Publications (New York); (3.18).

M.J. Best (1975), "FCDPAK: a FORTRAN IV subroutine to solve differentiable mathematical programmes — users' guide — Level 3.1", Research Report CORR 75—24, Department of Combinatorics and Optimization, University of Waterloo; (6.4).

M.J. Best and A.T. Bowler (1978), "ACDPAC: a FORTRAN IV subroutine to solve differentiable mathematical programmes — users' guide — Level 2.0", Research Report CORR 75—26, Department of Combinatorics and Optimization, University of Waterloo; (6.4).

M.J. Best, J. Bräuninger, K. Ritter and S.M. Robinson (1981), "A globally and quadratically convergent algorithm for general nonlinear programming problems", *Computing*, Vol. 26, pp. 141—153; (4.17).

M.J. Best and K. Ritter (1976), "A class of accelerated conjugate direction methods for linearly constrained minimization problems", *Math. Comp.*, Vol. 30, pp. 478—504; (5.12).

M.C. Biggs (1972), "Constrained minimization using recursive equality quadratic programming", in *Numerical Methods for Nonlinear Optimization*, ed. F.A. Lootsma, Academic Press (London); (2.4), (4.2), (6.4), (7.1).

M.C. Biggs (1975), "Constrained minimization using recursive quadratic programming: some alternative subproblem formulations", in *Towards Global Optimization*, eds. L.C.W. Dixon and G.P. Szegö, North-Holland Publishing Co. (Amsterdam); (4.2), (7.1).

A. Bihain: *see* Lemaréchal, Strodiot and Bihain (1981).

R.E. Bixby and W.H. Cunningham (1980), "Converting linear programs to network problems", *Math. Oper. Res.*, Vol. 5, pp. 321–357; (3.2).

P. Bjorstad and J. Nocedal (1979), "Analysis of a new algorithm for one-dimensional minimization", *Computing*, Vol. 22, pp. 93–100; (1.3), (1.17), (1.20).

P.T. Boggs (1971), "The solution of nonlinear systems of equations by A-stable integration techniques", *SIAM J. Numer. Anal.*, Vol. 8, pp. 767–785; (1.5).

P.T. Boggs and J.W. Tolle (1980), "Augmented Lagrangians which are quadratic in the multiplier", *J. Optim. Theory Appl.*, Vol. 31, pp. 17–26; (4.3).

P.T. Boggs and J.W. Tolle (1981), "Merit functions for nonlinear programming problems", Operations Research and Systems Analysis Report No. 81–2, University of North Carolina at Chapel Hill (to be published in the proceedings of the Third Workshop on Numerical Analysis held in Cocoyoc, Mexico); (4.3).

P.T. Boggs, J.W. Tolle and P. Wang (1979), "On quasi-Newton methods for constrained optimization", Operations Research and Systems Analysis Report No. 79–8, University of North Carolina at Chapel Hill; (4.41).

J. Bollen (1980), "Round-off error analysis of descent methods for solving linear equations", Ph.D. thesis, Technische Hogeschool Eindhoven, The Netherlands; (3.59).

H. Bourgeon and J. Nocedal (1981), "An algorithm for optimization based on a rational approximating function", IIMAS Report, Universidad Nacional Autónoma de México, Mexico; (1.18).

A.T. Bowler: *see* Best and Bowler (1978).

J.M. Boyle: *see* Smith, Boyle and Cody (1974).

A. Brandt (1977), "Multi-level adaptive solutions to boundary-value problems", *Math. Comp.*, Vol. 31, pp. 333–390; (5.3).

F.H. Branin (1972), "Widely convergent method for finding multiple solutions of simultaneous nonlinear equations", *IBM J. Res. Develop.*, Vol. 16, pp. 504–522; (1.5).

J. Bräuninger: *see* Best, Bräuninger, Ritter and Robinson (1981).

R.P. Brent: *see* Abbott and Brent (1975).

K.W. Brodlie (1977), "An assessment of two approaches to variable metric methods", *Math. Programming*, Vol. 12, pp. 344–355; (1.1).

G.G. Brown and G.W. Graves (1979), "Computational tools for solving large mixed integer optimization problems", presented at the ORSA/TIMS May, 1979 Meeting (New Orleans); (6.2).

C.G. Broyden (1965), "A class of methods for solving nonlinear simultaneous equations", *Math. Comp.*, Vol. 19, pp. 577–593; (6.1).

C.G. Broyden (1970), "The convergence of a class of double rank minimization algorithms, 2. The new algorithm", *J. Inst. Math. Appl.*, Vol. 6, pp. 222–231; (1.1).

C.G. Broyden, J.E. Dennis and J.J. Moré (1973), "On the local and superlinear convergence of quasi-Newton methods", *J. Inst. Math. Appl.*, Vol. 12, pp. 223–245; (1.1).

A.G. Buckley (1975), "An alternate implementation of Goldfarb's minimization algorithm", *Math. Programming*, Vol. 8, pp. 207–231; (3.6).

A.G. Buckley (1978a), "A combined conjugate gradient quasi-Newton minimization algorithm", *Math. Programming*, Vol. 15, pp. 200–210; (1.1), (1.2).

A.G. Buckley (1978b), "Extending the relationship between the conjugate gradient and BFGS algorithms", *Math. Programming*, Vol. 15, pp. 343–348; (1.2).

A.G. Buckley (1981), "Testing minimization codes", *COAL Newsletter*, No. 5, pp. 49–52; (7.2).

J.R. Bunch and L. Kaufman (1977), "Some stable methods for calculating inertia and solving symmetric linear systems", *Math. Comp.*, Vol. 31, pp. 163–179; (3.2).

J.R. Bunch and B.N. Parlett (1971), "Direct methods for solving symmetric indefinite systems of linear equations", *SIAM J. Numer. Anal.*, Vol. 8, pp. 639–655; (1.31), (1.33), (3.2).

J.R. Bunch: *see also* Dongarra, Moler, Bunch and Stewart (1979).

J.C.P. Bus (1980), *Numerical Solution of Systems of Nonlinear Equations*, Mathematisch Centrum (Amsterdam); (6.1), (7.16).

J.D. Buys: *see* Haarhoff and Buys (1970).

R.H. Byrd and D.A. Pyne (1979), "On the convergence of iteratively reweighted least squares algorithms", Technical Report No. 313,

Math. Science Department, Johns Hopkins University (to be published in *SIAM J. Sci. Statist. Comput.*); (2.1), (2.2).

P.H. Calamai and A.R. Conn (1980), "A stable algorithm for solving the multifacility location problem involving Euclidean distances", *SIAM J. Sci. Statist. Comp.*, Vol. 1, pp. 512–526; (4.4).

P.H. Calamai and A.R. Conn (1981), "A second-order method for solving the continuous multifacility location problem", in *Numerical Analysis, Dundee 1981*, ed. G.A. Watson, Springer-Verlag (to be published); (4.4).

J. Carpentier: *see* Abadie and Carpentier (1969).

R.M. Chamberlain, C. Lemaréchal, H.C. Pedersen and M.J.D. Powell (1980), "The watchdog technique for forcing convergence in algorithms for constrained optimization", Report DAMTP 80/NA9, University of Cambridge (to be published in *Math. Programming Stud.*); (4.1), (4.2), (4.3), (4.4).

H. Chao: *see* Manne, Chao and Wilson (1980).

C. Charalambous (1979), "Acceleration of the least p-th algorithm for minimax optimization with engineering applications", *Math. Programming*, Vol. 17, pp. 270–297; (2.3).

C. Charalambous and J.W. Bandler (1976), "Nonlinear minimax optimization as a sequence of least p-th optimization with finite values of p", *Internat. J. Systems Sci.*, Vol. 7, pp. 377–391; (2.4).

C. Charalambous and A.R. Conn (1978), "An efficient method to solve the minimax problem directly", *SIAM J. Numer. Anal.*, Vol. 15, pp. 162–187; (2.3), (2.4), (4.4).

C. Charalambous and O. Moharram (1978), "A new approach to minimax optimization", in *Large Engineering Systems 2*, eds. G.J. Savage and P.H. Roe, Pergamon Press (New York); (2.4).

C. Charalambous: *see also* Bartels, Conn and Charalambous (1978).

A. Charnes and C.E. Lemke (1954), "Minimization of nonlinear separable convex functionals", *Naval Res. Logist. Quart.*, Vol. 1, pp. 302–312; (5.4).

D. Chazan and W.L. Mirankar (1970), "A nongradient and parallel algorithm for unconstrained minimization", *SIAM J. Control*, Vol. 8, pp. 207–217; (7.3).

F.H. Clarke (1975), "Generalized gradients and applications",

Trans. Amer. Math. Soc., Vol. 205, pp. 247—262; (2.3).

F.H. Clarke (1976), "A new approach to Lagrange multipliers", *Math. Oper. Res.*, Vol. 1, pp. 165—174; (2.3).

W.J. Cody (1974), "The construction of numerical subroutine libraries", *SIAM Rev.*, Vol. 16, pp. 36—46; (6.5).

W.J. Cody: *see also* Smith, Boyle and Cody (1974).

T.F. Coleman and A.R. Conn (1980a), "Nonlinear programming via an exact penalty function method: asymptotic analysis", Report CS-80-30, Computer Science Department, University of Waterloo (to be published in *Math. Programming*); (4.1), (4.2).

T.F. Coleman and A.R. Conn (1980b), "Nonlinear programming via an exact penalty function: global analysis", Report CS-80-31, Computer Science Department, University of Waterloo (to be published in *Math. Programming*); (2.4), (4.2), (4.4).

T.F. Coleman and A.R. Conn (1980c), "Second-order conditions for an exact penalty function", *Math. Programming*, Vol. 19, pp. 178—185; (4.4).

T.F. Coleman and A.R. Conn (1981), "A successive quadratic programming method, that is quasi-Newton but updates only projected Hessians", in preparation; (4.4).

T.F. Coleman and J.J. Moré (1981), "Estimation of sparse Jacobian matrices and graph coloring problems", Report ANL-81-39, Argonne National Laboratory, Illinois; (1.1), (6.1).

M. Collins, L. Cooper, R. Helgason, J. Kennington and L. LeBlanc (1978), "Solving the pipe network analysis problem using optimization techniques", *Management Sci.*, Vol. 24, pp. 747—760; (6.3).

A.R. Colville (1968), "A comparative study on nonlinear programming codes", Report 320-2949, IBM Scientific Center, New York; (3.3), (4.24), (7.1), (7.4).

A.R. Conn (1976), "Linear programming via a nondifferentiable penalty function", *SIAM J. Numer. Anal.*, Vol. 13, pp. 145—154; (4.4).

A.R. Conn (1979), "An efficient second-order method to solve the (constrained) minimax problem", Report CORR-79-5, Department of Combinatorics and Optimization, University of Waterloo; (2.4).

A.R. Conn and J.W. Sinclair (1975), "Quadratic programming via a non-differentiable penalty function", Technical Report 75/15,

Department of Combinatorics and Optimization, University of Waterloo; (3.1), (3.24), (4.4).

A.R. Conn: *see also* Bartels and Conn (1981); Bartels, Conn and Charalambous (1978); Bartels, Conn and Sinclair (1978); Calamai and Conn (1980, 1981); Charalambous and Conn (1978); Coleman and Conn (1980a, 1980b, 1980c, 1981).

L. Cooper: *see* Collins, Cooper, Helgason, Kennington and LeBlanc (1978).

R. Courant (1943), "Variational methods for the solution of problems of equilibrium and vibrations", *Bull. Amer. Math. Soc.*, Vol. 49, pp. 1–23; (4.1).

H.S.M. Coxeter (1964), *Projective Geometry*, Blaisdell Publishing Co. (New York); (1.3).

H.P. Crowder, R.S. Dembo and J.M. Mulvey (1979), "On reporting computational experiments with mathematical software", *ACM Trans. Math. Software*, Vol. 5, pp. 193–203; (7.2).

H.P. Crowder and P.B. Saunders (1980), "Results of a survey on MP performance indicators", *COAL Newsletter*, January issue, pp. 2–6; (7.2).

H.P. Crowder: *see also* Held, Wolfe and Crowder (1974).

W.H. Cunningham: *see* Bixby and Cunningham (1980).

A.R. Curtis, M.J.D. Powell and J.K. Reid (1974), "On the estimation of sparse Jacobian matrices", *J. Inst. Math. Appl.*, Vol. 13, pp. 117–119; (1.1), (6.1).

Y.M. Danilin: *see* Pschenichny and Danilin (1975).

G.B. Dantzig: *see* Perold and Dantzig (1978).

D.F. Davidenko (1953), "On a new method of numerical solution of systems of nonlinear equations", *Dokl. Akad. Nauk. SSSR*, Vol. 88, pp. 601–602; (1.5).

W.C. Davidon (1959), "Variable metric method for minimization", Report ANL 5990 (rev.), Argonne National Laboratory; (1.1), (1.3), (4.4).

W.C. Davidon (1975), "Optimally conditioned optimization algorithms without line searches", *Math. Programming*, Vol. 9, pp. 1–30; (1.1), (5.3).

W.C. Davidon (1980), "Conic approximations and collinear scalings for optimizers", *SIAM J. Numer. Anal.*, Vol. 17, pp. 268–281; (1.1), (1.3).

M. Davies (1967), "Linear approximation using the criterion of least total deviations", *J. Roy. Statist. Soc. Ser. B*, Vol. 29, pp. 101–109; (2.33).

A. Dax (1980), "Partial pivoting strategies for symmetric Gaussian elimination", Report DAMTP 1980/NA8, University of Cambridge; (3.2).

A. Dax and S. Kaniel (1977), "Pivoting techniques for symmetric Gaussian elimination", *Numer. Math.*, Vol. 28, pp. 221–241; (1.31), (3.2).

R.S. Dembo (1976), "A set of geometric programming test problems and their solutions", *Math. Programming*, Vol. 10, pp. 192–213; (6.1).

R.S. Dembo (1978), "Current state of the art of algorithms and computer software for geometric programming", *J. Optim. Theory Appl.*, Vol. 26, pp. 149–183; (7.1), (7.12).

R.S. Dembo (1979), "Second order algorithms for the posynomial geometric programming dual, part I: analysis", *Math. Programming*, Vol. 17, pp. 156–175; (2.2), (6.3).

R.S. Dembo, S.C. Eisenstat and T. Steihaug (1980), "Inexact Newton methods", Report 47, School of Organisation and Management, Yale University (to be published in *SIAM J. Numer. Anal.*); (1.4), (3.1), (4.30), (6.1), (6.3).

R.S. Dembo and T. Steihaug (1980), "Truncated Newton algorithms for large scale unconstrained optimization", Report 48, School of Organisation and Management, Yale University; (1.4), (6.1), (6.3).

R.S. Dembo: *see also* Crowder, Dembo and Mulvey (1979).

V.F. Demjanov and V.N. Malozemov (1974), *Introduction to Minimax*, John Wiley & Sons (New York); (2.3).

A.P. Dempster, N.M. Laird and D.B. Rubin (1977), "Maximum likelihood from incomplete data via the EM algorithm", *J. Roy. Statist. Soc. Ser. B*, Vol. 39, pp. 1–38; (2.1).

J.E. Dennis (1977), "Nonlinear least squares and equations", in *The State of the Art in Numerical Analysis*, ed. D.A.H. Jacobs, Academic Press (London); (2.1), (2.2).

J.E. Dennis, D.M. Gay and R.E. Welsch (1981), "An adaptive nonlinear least-squares algorithm", *ACM Trans. Math. Software*, Vol. 7, pp. 348–368; (2.1), (2.9), (2.10), (6.1).

J.E. Dennis and H.H.W. Mei (1979), "Two new unconstrained optimization algorithms which use function and gradient values", *J. Optim. Theory Appl.*, Vol. 28, pp. 453–482; (1.1), (1.4).

J.E. Dennis and R.B. Schnabel (1979), "Least change secant updates for quasi-Newton methods", *SIAM Rev.*, Vol. 21, pp. 443–459; (1.1), (2.1).

J.E. Dennis and R.B. Schnabel (1981), "A new derivation of symmetric positive definite secant updates", in *Nonlinear Programming 4*, eds. O.L. Mangasarian, R.R. Meyer and S.M. Robinson, Academic Press (New York); (1.1).

J.E. Dennis and R.B. Schnabel (1982), *Unconstrained Optimization and Nonlinear Equations*, Prentice-Hall (Englewood Cliffs, N.J.); (1.1).

J.E. Dennis and H.F. Walker (1980), "Convergence theorems for least-change secant update methods", Report No. TR 476-(141-171-163)-2, Rice University, Houston (to be published in *SIAM J. Numer. Anal.*); (2.1), (4.26).

J.E. Dennis: *see also* Barrera and Dennis (1979); Broyden, Dennis and Moré (1973).

N. Deo, C.Y. Pang and R.E. Lord (1980), "Two parallel algorithms for shortest path problems", presented at the 1980 International Conference on Parallel Processing; (7.3).

P. Deuflhard (1974), "A modified Newton method for the solution of ill-conditioned systems of nonlinear equations with application to multiple shooting", *Numer. Math.*, Vol. 22, pp. 289–315; (6.1).

G. Di Pillo and L. Grippo (1979), "A new class of augmented Lagrangians in nonlinear programming", *SIAM J. Control Optim.*, Vol. 17, pp. 618–628; (4.3).

G. Di Pillo, L. Grippo and F. Lampariello (1980), "A method for solving equality constrained optimization problems by unconstrained minimization", in *Optimization Techniques Part 2, Lecture Notes in Control and Information Sciences 23*, eds. K. Iracki, K. Malanowski and S. Walukiewicz, Springer-Verlag (Berlin); (4.3).

L.C.W. Dixon (1980), "On the convergence properties of variable metric recursive quadratic programming methods", Technical Report No. 110, The Numerical Optimization Centre, The Hatfield Polytechnic; (4.3), (4.52).

L.C.W. Dixon and G.P. Szegö (eds.) (1975), *Towards Global Optimization*, North-Holland Publishing Co. (Amsterdam); (1.1).

L.C.W. Dixon and G.P. Szegö (eds.) (1978), *Towards Global Optimization 2*, North-Holland Publishing Co. (Amsterdam); (1.1).

J.J. Dongarra, C.B. Moler, J.R. Bunch and G.W. Stewart (1979), *LINPACK Users' Guide*, SIAM Publications (Philadelphia); (1.4).

I.S. Duff and J.K. Reid (1976), "A comparison of some methods for the solution of sparse overdetermined systems of linear equations", *J. Inst. Math. Appl.*, Vol. 17, pp. 267–280; (3.11).

I.S. Duff and J.K. Reid (1979), "Performance evaluation of codes for sparse matrix problems", in *Performance Evaluation of Numerical Software*, ed. L.D. Fosdick, North-Holland Publishing Co. (Amsterdam); (6.1).

R.J. Duffin, E.L. Peterson and C. Zener (1967), *Geometric Programming — Theory and Application*, John Wiley & Sons (New York); (7.1).

S.R.K. Dutta: *see* El-Attar, Vidyasagar and Dutta (1979).

R. Dutter (1977), "Numerical solution of robust regression problems: computational aspects, a comparison", *J. Statist. Comput. Simulation*, Vol. 5, pp. 207–238; (2.2).

R. Dutter: *see also* Huber and Dutter (1974).

J.S. Dyer (1972), "Interactive goal programming", *Management Sci.*, Vol. 19, pp. 62–70; (7.2).

E.D. Eason (1977), "Validity of Colville's time standardization for comparing optimization codes", presented at ASME Design Engineering Technical Conference (Chicago); (7.4).

E.D. Eason (1981), "Evidence of fundamental difficulties in nonlinear optimization code comparisons", presented at the COAL Meeting on Testing and Validating Algorithms and Software (Boulder, Colorado); (7.4).

E.D. Eason and F.G. Fenton (1974), "A comparison of numerical optimization methods for engineering design", *Trans. ASME Ser. B J. Engrg. Indust.*, Vol. 96, pp. 196–200; (7.1), (7.4).

B.C. Eaves (1976), "A short course in solving equations with PL homotopies", in *Nonlinear Programming, SIAM-AMS Proceedings Vol.*

IX, eds. R.W. Cottle and C.E. Lemke, SIAM Publications (Philadelphia); (3.3).

S.C. Eisenstat: *see* Dembo, Eisenstat and Steihaug (1980).

R.A. El-Attar, M. Vidyasagar and S.R.K. Dutta (1979), "An algorithm for ℓ_1-norm minimization with application to nonlinear ℓ_1-approximation", *SIAM J. Numer. Anal.*, Vol. 16, pp. 70–86; (2.4).

B. Engquist and T. Smedsaas (1980), "Automatic computer code generation for hyperbolic and parabolic differential equations", *SIAM J. Sci. Statist. Comput.*, Vol. 1, pp. 249–259; (5.17).

M. Engquist: *see* Beck, Lasdon and Engquist (1981); Palacios-Gomez, Lasdon and Engquist (1981).

J. Eriksson (1976), "A note on the decomposition of systems of sparse nonlinear equations", *BIT*, Vol. 16, pp. 462–465; (6.1).

J. Eriksson (1980), "A note on solution of large sparse maximum entropy problems with linear equality constraints", *Math. Programming*, Vol. 18, pp. 146–154; (6.3).

G.C. Everstine (1979), "A comparison of three resequencing algorithms for the reduction of matrix profile and wavefront", *Internat. J. Numer. Methods Engrg.*, Vol. 14, pp. 837–853; (6.1).

J.E. Fattler, Y.T. Sin, R.R. Root, K.M. Ragsdell and G.V. Reklaitis (1980), "On the computational utility of posynomial geometric programming solution methods", report of the School of Chemical Engineering, Purdue University (to be published in *Math. Programming Stud.*); (7.1), (7.4), (7.12).

J. Feiber, R.N. Taylor and L.J. Osterweil (1980), "NEWTON — a dynamic testing system for Fortran 77 programs — preliminary report", Technical Note, Department of Computer Science, University of Colorado at Boulder; (7.2).

F.G. Fenton: *see* Eason and Fenton (1974).

A.V. Fiacco and G.P. McCormick (1968), *Nonlinear Programming: Sequential Unconstrained Minimization Techniques*, John Wiley & Sons (New York); (4.1), (6.3), (6.4), (7.1).

M.L. Fisher and F.J. Gould (1974), "A simplicial algorithm for the nonlinear complementarity problem", *Math. Programming*, Vol. 6, pp. 281–300; (3.3).

R. Fletcher (1970a), "A new approach to variable metric algorithms", *Comput. J.*, Vol. 13, pp. 317–322; (1.1).

R. Fletcher (1970b), "A class of methods for nonlinear programming with termination and convergence properties", in *Integer and Nonlinear Programming*, ed. J. Abadie, North-Holland Publishing Co. (Amsterdam); (4.3), (4.41).

R. Fletcher (1971), "A general quadratic programming algorithm", *J. Inst. Math. Appl.*, Vol. 7, pp. 76–91; (3.1).

R. Fletcher (1972), "An algorithm for solving linearly constrained optimization problems", *Math. Programming*, Vol. 2, pp. 133–165; (3.1).

R. Fletcher (1973), "An exact penalty function for nonlinear programming with inequalities", *Math. Programming*, Vol. 5, pp. 129–150; (4.2).

R. Fletcher (1975), "An ideal penalty function for constrained optimization", in *Nonlinear Programming 2*, eds. O.L. Mangasarian, R.R. Meyer and S.M. Robinson, Academic Press (New York); (4.2), (6.4), (7.1).

R. Fletcher (1977), "Methods for solving nonlinearly constrained optimization problems", in *The State of the Art in Numerical Analysis*, ed. D.A.H. Jacobs, Academic Press (London); (6.5).

R. Fletcher (1980), *Practical Methods of Optimization, Vol. 1: Unconstrained Optimization*, John Wiley & Sons (Chichester); (1.1), (1.4), (6.1).

R. Fletcher (1981a), "Numerical experiments with an exact ℓ_1 penalty function method", in *Nonlinear Programming 4*, eds. O.L. Mangasarian, R.R. Meyer and S.M. Robinson, Academic Press (New York); (4.1), (4.7).

R. Fletcher (1981b), *Practical Methods of Optimization, Vol. 2: Constrained Optimization*, John Wiley & Sons (Chichester); (4.1), (4.9).

R. Fletcher (1981c), "A model algorithm for composite nondifferentiable problems" (to be published in *Math. Programming Stud.*); (1.4), (4.1).

R. Fletcher (1981d), "Second order corrections for nondifferentiable optimization", Report NA/50, Department of Mathematics, University of Dundee (to be published in *Numerical Analysis, Dundee 1981*, ed. G.A. Watson, Springer-Verlag); (4.1).

R. Fletcher and M.J.D. Powell (1963), "A rapidly convergent descent method for minimization", *Comput. J.*, Vol. 6, pp. 163–168; (1.1), (4.4).

R. Fletcher and C.M. Reeves (1964), "Function minimization by conjugate gradients", *Comput. J.*, Vol. 7, pp. 149–154; (1.2).

R. Fletcher and G.A. Watson (1980), "First and second order conditions for a class of nondifferentiable optimization problems", *Math. Programming*, Vol. 18, pp. 291–307; (2.3), (4.4).

M.A. Florian (ed.) (1976), *Traffic Equilibrium Methods, Lecture Notes in Economics and Mathematical Systems 118*, Springer-Verlag (Berlin); (6.3).

B. Ford and S.J. Hague (1974), "The organization of numerical algorithms libraries", in *Software for Numerical Mathematics*, ed. D.J. Evans, Academic Press (London); (6.5).

J.J.H. Forrest: *see* Beale and Forrest (1976, 1978).

L.D. Fosdick (ed.) (1979), *Performance and Evaluation of Numerical Software*, North-Holland Publishing Co. (Amsterdam); (7.2).

P. Frank and R.B. Schnabel (1981), "Calculation of the initial Hessian approximation in secant algorithms", in preparation; (1.1).

G.A. Gabriele (1980), "Large scale nonlinear programming using the generalized reduced gradient method", Ph.D. dissertation, Department of Mechanical Engineering, Purdue University; (4.5).

G.A. Gabriele and K.M. Ragsdell (1976), "OPT: a nonlinear programming code in FORTRAN IV — users' manual", Technical Report, Purdue University; (6.4).

G.A. Gabriele and K.M. Ragsdell (1977), "The generalized reduced gradient method: a reliable tool for optimal design", *Trans. ASME Ser. B J. Engrg. Indust.*, Vol. 99, pp. 394–400; (4.5).

E. Gafni: *see* Bertsekas and Gafni (1981); Bertsekas, Gafni and Vastola (1979).

B.S. Garbow: *see* Moré, Garbow and Hillstrom (1980, 1981a, 1981b).

C.B. Garcia and W.I. Zangwill (1980), "Global continuation methods for finding all solutions to polynomial systems equations in N variables", in *Extremal Methods in Systems Analysis*, Springer-Verlag (New York); (1.5).

D.M. Gay (1980a), "Subroutines for unconstrained minimization using a model trust region approach", Technical Report 18, Center for Computational Research in Economics and Management Science, Massachusetts Institute of Technology; (1.1), (6.1), (6.5).

D.M. Gay (1980b), "On solving robust and generalized linear regression problems", in *Optimizzazione Non-lineare e Applicazioni*, eds. S. Incerti and G. Treccani, Pitagora Editrice; (2.1), (2.2).

D.M. Gay (1981), "Computing optimal locally constrained steps", *SIAM J. Sci. Statist. Comput.*, Vol. 2, pp. 186–197; (1.1), (1.4), (6.1).

D.M. Gay: *see also* Dennis, Gay and Welsch (1981).

K. Georg: *see* Allgower and Georg (1980).

A. George (1977), "Solution of linear systems of equations. Direct methods for finite element problems", in *Sparse Matrix Techniques, Copenhagen, 1976, Lecture Notes in Mathematics 572*, ed. V.A. Barker, Springer-Verlag (Berlin); (5.3).

A. George and J.W.H. Liu (1979), "A quotient graph model for symmetric factorization", in *Sparse Matrix Proceedings 1978*, eds. I.S. Duff and G.W. Stewart, SIAM Publications (Philadelphia); (5.3).

P.E. Gill, G.H. Golub, W. Murray and M.A. Saunders (1974), "Methods for modifying matrix factorizations", *Math. Comp.*, Vol. 28, pp. 505–535; (1.1), (3.2).

P.E. Gill and W. Murray (1974a), "Newton type methods for unconstrained and linearly constrained optimization", *Math. Programming*, Vol. 7, pp. 311–350; (1.4), (1.28), (1.29), (1.30), (3.6), (6.1).

P.E. Gill and W. Murray (eds.) (1974b), *Numerical Methods for Constrained Optimization*, Academic Press (London); (4.1), (4.5), (6.4).

P.E. Gill and W. Murray (1977a), "Linearly constrained problems including linear and quadratic programming", in *The State of the Art in Numerical Analysis*, ed. D.A.H. Jacobs, Academic Press (London); (3.1), (5.1), (6.5).

P.E. Gill and W. Murray (1977b), "The computation of Lagrange multiplier estimates for constrained minimization", Report NAC 77, National Physical Laboratory, England; (4.4).

P.E. Gill and W. Murray (1978a), "Numerically stable methods for quadratic programming", *Math. Programming*, Vol. 14, pp. 349-372; (3.1).

P.E. Gill and W. Murray (1978b), "Algorithms for the solution of the nonlinear least squares problem", *SIAM J. Numer. Anal.*, Vol. 15, pp. 977–992; (6.1).

P.E. Gill and W. Murray (1979), "Conjugate gradient methods for large scale nonlinear optimization", Technical Report SOL 79–15, Department of Operations Research, Stanford University; (1.8), (1.13), (5.3), (6.1).

P.E. Gill, W. Murray, S.M. Picken and M.H. Wright (1979), "The design and structure of a Fortran program library for optimization", *ACM Trans. Math. Software*, Vol. 5, pp. 259–283; (6.1), (6.5).

P.E. Gill, W. Murray and M.A. Saunders (1975), "Methods for computing and modifying the LDV factors of a matrix", *Math. Comp.*, Vol. 29, pp. 1051–1077; (3.2).

P.E. Gill, W. Murray and M.H. Wright (1981), *Practical Optimization*, Academic Press (New York); (1.1), (3.1), (4.4), (6.5).

W. Gochet, E. Loute and D. Solow (1974), "Comparative computer results of three algorithms for solving prototype geometric programming problems", *Cahiers Centre Études Recherche Opér.*, Vol. 16, pp. 469–486; (3.3).

D. Goldfarb (1969), "Extension of Davidon's variable metric method to maximization under linear inequality and equality constraints", *SIAM J. Appl. Math.*, Vol. 17, pp. 739–764; (3.1).

D. Goldfarb (1970), "A family of variable metric methods derived by variational means", *Math. Comp.*, Vol. 24, pp. 23–26; (1.1).

D. Goldfarb (1976), "Matrix factorizations in optimization of nonlinear functions subject to linear constraints", *Math. Programming*, Vol. 10, pp. 1–31; (3.2).

D. Goldfarb (1980a), "Curvilinear path steplength algorithms for minimization which use directions of negative curvature", *Math. Programming*, Vol. 18, pp. 31–40; (1.1), (1.4).

D. Goldfarb (1980b), "The use of negative curvature in minimization algorithms", Report TR 80–412, Department of Computer Science, Cornell University; (1.31).

S.M. Goldfeld, R.E. Quandt and H.F. Trotter (1966), "Maximization by quadratic hill-climbing", *Econometrica*, Vol. 34, pp. 541–551; (1.1), (1.4), (1.42), (1.43).

G.H. Golub: *see* Gill, Golub, Murray and Saunders (1974).

M.K. Gordon: *see* Shampine and Gordon (1975).

F.J. Gould: *see* Fisher and Gould (1974).

J.G. de Graan (1980), "Extensions to the multiple criteria analysis method of T.L. Saaty", Report of the National Institute for Water Supply, Voorburg, The Netherlands; (7.1).

G.W. Graves: *see* Brown and Graves (1979).

H.J. Greenberg (ed.) (1978), *Design and Implementation of Optimization Software*, Sijthoff and Noordhoff (Alphen aan de Rijn, Netherlands); (7.2).

A. Griewank (1981), "Generalized descent for global minimization", *J. Optim. Theory Appl.*, Vol. 34, pp. 11–39; (1.5).

A. Griewank and Ph. L. Toint (1981a), "Partitioned variable metric updates for large structured optimization problems", Report 81/7, Department of Mathematics, University of Namur (to be published in *Numer. Math.*); (5.3).

A. Griewank and Ph. L. Toint (1981b), "Local convergence analysis for partitioned quasi-Newton updates in the Broyden class", Report 81/9, Department of Mathematics, University of Namur; (5.3).

R.E. Griffith and R.A. Stewart (1961), "A nonlinear programming technique for the optimization of continuous processing systems", *Management Sci.*, Vol. 7, pp. 379–392; (3.1), (5.1).

L. Grippo: *see* Di Pillo and Grippo (1979); Di Pillo, Grippo and Lampariello (1980).

G. Groszmann and A.A. Kaplan (1979), *Strafmethoden und Modifizierte Lagrangefunktionen in der Nichtlinearen Optimierung*, Teubner (Leipzig); (7.1).

P.C. Haarhoff and J.D. Buys (1970), "A new method for the optimization of a nonlinear function subject to nonlinear constraints", *Comput. J.*, Vol. 13, pp. 178–184; (6.4).

S.J. Hague: *see* Ford and Hague (1974).

J. Hald and K. Madsen (1981), "Combined LP and quasi-Newton methods for minimax optimization", *Math. Programming*, Vol. 20, pp. 49–62; (2.3), (2.4), (2.26).

A.D. Hall: *see* Ryder and Hall (1979).

R.W. Hamming (1971), *Introduction to Applied Numerical Analysis*, McGraw-Hill Book Co. (New York); (6.1).

S.P. Han (1976), "Superlinearly convergent variable metric algorithms for general nonlinear programming problems", *Math. Programming*, Vol. 11, pp. 263-282; (6.4), (7.1).

S.P. Han (1977a), "A globally convergent method for nonlinear programming", *J. Optim. Theory Appl.*, Vol. 22, pp. 297-309; (2.4), (4.1), (4.2), (4.3), (4.4).

S.P. Han (1977b), "Dual variable metric algorithms for constrained optimization", *SIAM J. Control*, Vol. 15, pp. 546-565; (7.1).

S.P. Han (1978), "Superlinear convergence of a minimax method", Report TR-78-336, Department of Computer Science, Cornell University; (2.4).

S.P. Han (1979), "Penalty Lagrangian methods via a quasi-Newton approach", *Math. Oper. Res.*, Vol. 4, pp. 291-302; (7.1).

S.P. Han (1981), "Variable metric methods for minimizing a class of nondifferentiable functions", *Math. Programming*, Vol. 20, pp. 1-13; (2.3), (2.4).

S.P. Han and O.L. Mangasarian (1981), "A dual differentiable exact penalty function", Report 434, Computer Science Department, University of Wisconsin; (3.20), (4.3), (4.49).

E. Hansen (1980), "Global optimization using interval analysis — the multi-dimensional case", *Numer. Math.*, Vol. 34, pp. 247-270; (1.1).

T. Hansen: *see* Scarf and Hansen (1973).

M.T. Heath (1978), "Numerical algorithms for nonlinearly constrained optimization", Report CS-78-656, Computer Science Department, Stanford University; (4.2).

M.D. Hebden (1973), "An algorithm for minimization using exact second derivatives", Report TP 515, A.E.R.E., Harwell; (1.1), (1.4), (6.1).

M. Held, P. Wolfe and H.P. Crowder (1974), "Validation of subgradient optimization", *Math. Programming*, Vol. 6, pp. 62-88; (4.56).

R. Helgason: *see* Collins, Cooper, Helgason, Kennington and LeBlanc (1978).

D.R. Heltne and J.M. Littschwager (1975), "Users' guide for GRG73 and technical appendices to GRG73", Technical Report, College of Engineering, University of Iowa; (4.5).

G. Herman (1980), *Construction from Projections: The Fundamentals of Computerized Tomography*, Academic Press (New York); (6.3).

M.R. Hestenes (1969), "Multiplier and gradient methods", *J. Optim. Theory Appl.*, Vol. 4, pp. 303-320; (4.1), (4.3), (7.1).

M.R. Hestenes and E. Stiefel (1952), "Methods of conjugate gradients for solving linear systems", *J. Res. Nat. Bur. Standards*, Vol. 49, pp. 409-436; (1.3), (3.58).

K.L. Hiebert (1979), "A comparison of nonlinear least squares software", Report SAND 79-0483, Sandia Laboratory, Albuquerque; (6.1).

K.L. Hiebert (1980), "A comparison of software which solves systems of nonlinear equations", Report SAND 80-0181, Sandia Laboratory, Albuquerque; (6.1), (6.36).

K.L. Hiebert (1981), "An evaluation of mathematical software that solves nonlinear least squares problems", *ACM Trans. Math. Software*, Vol. 7, pp. 1-16; (6.1).

K.E. Hillstrom: *see* Moré, Garbow and Hillstrom (1980, 1981a, 1981b).

D.M. Himmelblau (1972), *Applied Nonlinear Programming*, McGraw-Hill Book Co. (New York); (3.4), (7.4).

D.M. Himmelblau: *see also* Newell and Himmelblau (1975); Staha and Himmelblau (1973).

J.B. Hiriart-Urruty (1978), "On optimality conditions in non-differentiable programming", *Math. Programming*, Vol. 14, pp. 73-86; (2.3).

J.B. Hiriart-Urruty (1981), "Optimality conditions for discrete nonlinear norm-approximation problems", in *Optimization and Optimal Control, Lecture Notes in Control and Information Sciences 30*, eds. A. Auslender, W. Oettli and J. Stoer, Springer-Verlag (Berlin); (2.3).

M.W. Hirsch and S. Smale (1979), "On algorithms for solving f(x) = 0", *Comm. Pure Appl. Math.*, Vol. 32, pp. 281-312; (1.5).

W. Hock and K. Schittkowski (1981), *Test Examples for Nonlinear Programming Codes, Lecture Notes in Economics and Mathematical Systems 187*, Springer-Verlag (Berlin); (6.1), (6.4).

A. Hoffman, M. Mannos, D. Sokolowsky and N. Wiegmann (1953), "Computational experience in solving linear programs", *J. Soc. Indust. Appl. Math.*, Vol. 1, pp. 17-33; (7.2).

K.L. Hoffman and R.H.F. Jackson (1981), "Testing MP software: progress and problems", in preparation; (7.2).

P. Huard (1967), "Resolution of mathematical programming with nonlinear constraints by the method of centres", in *Nonlinear Programming*, ed. J. Abadie, North-Holland Publishing Co. (Amsterdam); (4.45), (7.1).

P.J. Huber and R. Dutter (1974), "Numerical solution of robust regression problems", in *Compstat 1974, Proceedings of the Symposium on Computational Statistics*, ed. G. Bruchmann, Physika Verlag (Vienna); (2.2).

P.A.B. Hughes: *see* Beale, Hughes and Small (1965).

L. Hurwicz: *see* Arrow, Hurwicz and Uzawa (1958).

S. Incerti, V. Parisi and F. Zirilli (1979), "A new method for solving nonlinear simultaneous equations", *SIAM J. Numer. Anal.*, Vol. 16, pp. 779–789; (1.5).

S. Incerti, V. Parisi and F. Zirilli (1981), "A FORTRAN subroutine for solving systems of nonlinear simultaneous equations", *Comput. J.*, Vol. 24, pp. 87–91; (1.5).

S. Incerti: *see also* Aluffi, Incerti and Zirilli (1980a, 1980b).

J.P. Indusi (1972), "A computer algorithm for constrained minimization", in *Minimization Algorithms*, ed. G.P. Szegö, Academic Press (New York); (6.4).

Y. Ishizaki and H. Watanabe (1968), "An iterative Chebyshev approximation method for network design", *IEEE Trans. Circuit Theory*, Vol. 15, pp. 326–336; (2.4).

R.H.F. Jackson (1979), "Measures of similarities for comparing test problems", presented at the 10th International Conference on Mathematical Programming (Montreal); (7.30).

R.H.F. Jackson and J.M. Mulvey (1978), "A critical review of comparisons of mathematical programming algorithms and software (1953–1977)", *J. Res. Nat. Bur. Standards*, Vol. 83, pp. 563–584; (7.2).

R.H.F. Jackson: *see also* Hoffman and Jackson (1981).

S.L.S. Jacoby and J.S. Kowalik (1980), *Mathematical Modeling with Computers*, Prentice-Hall Inc. (Englewood Cliffs, N.J.); (6.3).

M.N. Jacovlev (1964), "The solution of systems of nonlinear equations by a method of differentiation with respect to a parameter", *U.S.S.R. Computational Math. and Math. Phys.*, Vol. 4, No. 1, pp. 198–203; (1.5).

A. Jain: see Lasdon, Waren, Jain and Ratner (1978); Lasdon, Waren, Ratner and Jain (1975a, 1975b).

R.I. Jennrich and S.M. Robinson (1969), "A Newton–Raphson algorithm for maximum likelihood factor analysis", *Psychometrika*, Vol. 34, pp. 111–123; (2.12).

K.G. Jöreskog (1967), "Some contributions to maximum likelihood factor analysis", *Psychometrika*, Vol. 32, pp. 443–482; (2.12).

J.J. Kaganove: *see* Lyness and Kaganove (1976).

S. Kaniel: *see* Dax and Kaniel (1977).

C.Y. Kao and R.R. Meyer (1981), "Secant approximation methods for convex optimization", *Math. Programming Stud.*, Vol. 14 (Mathematical Programming at Oberwolfach), pp. 143–162; (5.4).

A.A. Kaplan: *see* Groszmann and Kaplan (1979).

L. Kaufman: *see* Bunch and Kaufman (1977).

H.B. Keller (1978), "Global homotopies and Newton methods", in *Recent Advances in Numerical Analysis*, eds. C. de Boor and G.H. Golub, Academic Press (New York); (1.5).

R.B. Kellogg, T.Y. Li and J.A. Yorke (1977), "A method of continuation for calculating a Brouwer fixed point", in *Computing Fixed Points with Applications*, ed. S. Karamadian, Academic Press (New York); (1.5).

J. Kennington: *see* Collins, Cooper, Helgason, Kennington and LeBlanc (1978).

D.E. Kiely (1976), "On geometric programming", Mathematics Degree Project Report, The Hatfield Polytechnic; (7.13).

R. Klessig and E. Polak (1973), "An adaptive precision gradient method for optimal control", *SIAM J. Control*, Vol. 11, pp. 80–93; (4.70).

M. Kojima (1974), "Computational methods for solving the nonlinear complementarity problem", *Keio Engrg. Rep.*, Vol. 27, pp. 1–41; (3.3).

W. de Kovel (1981), "The Algol 60 procedure MINITRUNC for solving nonlinear optimization problems via moving exterior truncations", M.Sc. thesis, Department of Mathematics and Informatics, Delft University of Technology, The Netherlands; (7.1).

J.S. Kowalik: *see* Jacoby and Kowalik (1980); Lord, Kowalik and Kumar (1980).

D. Kraft (1977), "Nichtlineare Programmierung-grundlagen, Verfahren, Beispiele", Report DLR-FB 77-68. DFVLR, Oberpfaffenhofen; (6.4).

J.L. Kreuser (1981), "Computational experience with successive over-relaxation methods for linear programming problems", Computing Activities Department, Technical Report, World Bank, Washington, D.C.; (3.20).

J.L. Kreuser: *see also* Rosen and Kreuser (1972).

J.L. Kuester and J.H. Mize (1973), *Optimization Techniques with FORTRAN*, McGraw-Hill Book Co. (New York); (6.4).

S.P. Kumar: *see* Lord, Kowalik and Kumar (1980); Lord and Kumar (1980).

N.M. Laird: *see* Dempster, Laird and Rubin (1977).

F. Lampariello: *see* Di Pillo, Grippo and Lampariello (1980).

L.S. Lasdon (1980), "A survey of nonlinear programming algorithms and software", working paper, Department of General Business, University of Texas; (4.5).

L.S. Lasdon and A.D. Waren (1978), "Generalized reduced gradient software for linearly and nonlinearly constrained problems", in *Design and Implementation of Optimization Software*, ed. H. Greenberg, Sijthoff and Noordhoff (Alphen aan de Rijn, Netherlands); (4.5), (6.4).

L.S. Lasdon and A.D. Waren (1980), "Survey of nonlinear programming applications", *Oper. Res.*, Vol. 28, pp. 1029–1073; (4.5).

L.S. Lasdon, A.D. Waren, A. Jain and M.W. Ratner (1978), "Design and testing of a generalized reduced gradient code for nonlinear programming", *ACM Trans. Math. Software*, Vol. 4, pp. 34–50; (4.5).

L.S. Lasdon, A.D. Waren, M.W. Ratner and A. Jain (1975a), "GRG system documentation", Technical Memorandum CIS-75-01, Cleveland State University; (7.1).

L.S. Lasdon, A.D. Waren, M.W. Ratner and A. Jain (1975b), "GRG users' guide", Technical Memorandum CIS-75-02, Cleveland State University; (7.1).

L.S. Lasdon: *see also* Beck, Lasdon and Enquist (1981); Palacios-Gomez, Lasdon and Engquist (1981); Waren and Lasdon (1979).

L. LeBlanc: *see* Collins, Cooper, Helgason, Kennington and LeBlanc (1978).

C. Lemaréchal (1975), "An extension of Davidon methods to non-differentiable problems", *Math. Programming Stud.*, Vol. 3 (Non-differentiable Optimization), pp. 95–109; (2.3).

C. Lemaréchal (1978a), "Nonlinear programming and nonsmooth optimization, a unification", Raport Laboria No. 232, INRIA, France; (2.3).

C. Lemaréchal (1978b), "Nonsmooth optimization and descent methods", Report RR 78-4, IIASA, Laxenburg, Austria; (2.3).

C. Lemaréchal (1980a), "Nondifferentiable optimization" in *Non-linear Optimization: Theory and Algorithms*, eds. L.C.W. Dixon, E. Spedicato and G.P. Szegö, Birkhauser (Boston); (2.3).

C. Lemaréchal (1980b), "Extensions diverse des méthodes de gradient et applications", Ph.D. thesis, University of Paris IX; (2.28).

C. Lemaréchal and R. Mifflin (eds.) (1979), *Nonsmooth Optimiz-ation*, Pergamon Press (Oxford); (2.3).

C. Lemaréchal and R. Mifflin (1981), "A globally and superlinearly convergent algorithm for one-dimensional minimization of convex functions", Report TR-81-3, Department of Pure and Applied Mathematics, Washington State University, Pullman; (2.3).

C. Lemaréchal and E.A. Nurminskii (1980), "Sur la différentiabilité de la fonction d'appui du sous-différentiel approché", *C. R. Acad. Sci. Paris Sér. A*, Vol. 290, pp. 855–858; (2.3).

C. Lemaréchal, J.J. Strodiot and A. Bihain (1981), "On a bundle algorithm for nonsmooth optimization", in *Nonlinear Programming 4*, eds. O.L. Mangasarian, R.R. Meyer and S.M. Robinson, Academic Press (New York); (2.3).

C. Lemaréchal: *see also* Chamberlain, Lemaréchal, Pedersen and Powell (1980).

C.E. Lemke (1962), "A method of solution for quadratic programs",

Management Sci., Vol. 8, pp. 442–453; (3.2).

C.E. Lemke: *see also* Charnes and Lemke (1954).

M.L. Lenard (1979), "A computational study of active set strategies in nonlinear programming with linear constraints", *Math. Programming*, Vol. 16, pp. 81–97; (3.1), (3.28).

A. LeNir (1981), "A variable storage conjugate gradient algorithm", M.Sc. thesis, Mathematics Department, Concordia University, Montreal; (1.2).

A.V. Levy and A. Montalvo (1979), "Algoritmo de tunelización para la optimización global de funciones", Comunicaciones Téchnicas No. 204 (Serie NA), IIMAS-UNAM, Mexico City; (1.1), (6.1).

T.Y. Li: *see* Kellogg, Li and Yorke (1977).

J.M. Littschwager: *see* Heltne and Littschwager (1975).

J.W.H. Liu: *see* George and Liu (1979).

F.A. Lootsma (1972), "Penalty function performance of several unconstrained minimization techniques", *Philips Res. Rep.*, Vol. 27, pp. 358–385; (7.1).

F.A. Lootsma (1974), "Convergence rates of quadratic exterior penalty function methods for solving constrained minimization problems", *Philips Res. Rep.*, Vol. 29, pp. 1–12; (4.45), (6.4).

F.A. Lootsma (1975), "The design of a nonlinear optimization programme for solving technological problems", in *Optimization and Optimal Control, Lecture Notes in Mathematics 477*, eds. R. Bulirsch, W. Oettli and J. Stoer, Springer-Verlag (Berlin); (7.1).

F.A. Lootsma (1978), "The Algol 60 procedure Minifun for solving nonlinear optimization problems", in *Design and Implementation of Optimization Software*, ed. H.J. Greenberg, Sijthoff and Noordhoff (Alphen aan de Rijn, Netherlands); (7.1).

F.A. Lootsma (1979), "Performance evaluation of nonlinear programming codes from the viewpoint of a decision maker", in *Performance Evaluation of Numerical Software*, ed. L.D. Fosdick, North-Holland Publishing Co. (Amsterdam); (7.1).

F.A. Lootsma (1980), "Ranking of nonlinear optimization codes according to efficiency and robustness", in *Konstruktive Methoden der Nichtlinearen Optimierung*, eds. L. Collatz, G. Meinardus and W. Wetterling, Birkhäuser (Basel); (7.1), (7.4).

R.E. Lord, J.S. Kowalik and S.P. Kumar (1980), "Solving linear algebraic equations on a MIMD computer", Report CS-80-058, Washington State University, Pullman; (7.3).

R.E. Lord and S.P. Kumar (1980), "Parallel solution of flight simultaneous equations", presented at the 1980 Summer Computer Simulation Conference (Seattle); (7.3).

R.E. Lord: *see also* Deo, Pang and Lord (1980).

E. Loute: *see* Gochet, Loute and Solow (1974).

M.J. Lowe: *see* Pierre and Lowe (1975).

D.G. Luenberger (1970), "Control problems with kinks", *IEEE Trans. Automat. Control*, Vol. 15, pp. 570—575; (4.4).

J.N. Lyness (1976), "An interface problem in numerical software", in *Proceedings Sixth Manitoba Conference on Numerical Mathematics*, eds. B.L. Hartnell and H.C. Williams, Utilitas Mathematica Publishing Inc. (Winnipeg); (6.1).

J.N. Lyness (1977), "Has numerical differentiation a future?", in *Proceedings Seventh Manitoba Conference on Numerical Mathematics and Computing*, eds. D. McCarthy and H.C. Williams, Utilitas Mathematica Publishing Inc. (Winnipeg); (6.1).

J.N. Lyness and J.J. Kaganove (1976), "Comments on the nature of automatic quadrature routines", *ACM Trans. Math. Software*, Vol. 2, pp. 65—81; (6.5).

G. Lyon and R.B. Stillman (1974), "A FORTRAN analyzer", NBS Technical Note 849, National Bureau of Standards, Washington, D.C.; (7.2).

K. Madsen (1975), "An algorithm for minimax solution of overdetermined systems of nonlinear equations", *J. Inst. Math. Appl.*, Vol. 16, pp. 321-328; (2.4).

K. Madsen: *see also* Hald and Madsen (1981).

A.K. Mallin-Jones (1978), "Nonlinear algebraic equations in process engineering calculations", in *Numerical Software — Needs and Availability*, ed. D.A.H. Jacobs, Academic Press (London); (6.1).

V.N. Malozemov: *see* Demjanov and Malozemov (1974).

L.J. Mancini and G.P. McCormick (1979), "Bounding global minima with interval arithmetic", *Oper. Res.*, Vol. 27, pp. 743—754; (1.1).

O.L. Mangasarian (1977), "Solution of symmetric linear complementarity problems by iterative methods", *J. Optim. Theory Appl.*, Vol. 22, pp. 465–485; (3.20).

O.L. Mangasarian (1980), "Least-norm linear programming as an unconstrained minimization problem", Report 2147, Mathematics Research Center, University of Wisconsin; (3.20).

O.L. Mangasarian (1981a), "Iterative solution of linear programs", *SIAM J. Numer. Anal.*, Vol. 18, pp. 606–614; (3.20).

O.L. Mangasarian (1981b), "Sparsity-preserving SOR algorithms for separable quadratic and linear programming", Report No. 438, Computer Science Department, University of Wisconsin; (3.20), (5.4).

O.L. Mangasarian: *see also* Han and Mangasarian (1981).

A.S. Manne, H. Chao and R. Wilson (1980), "Computation of competitive equilibria by a sequence of linear programs", *Econometrica*, Vol. 48, pp. 1595–1615; (3.39).

M. Mannos: *see* Hoffman, Mannos, Sokolowsky and Wiegmann (1953).

N. Maratos (1978), "Exact penalty function algorithms for finite dimensional and control optimization problems", Ph.D. thesis, University of London; (4.2), (4.3).

A.T. Markos: *see* Straeter and Markos (1975).

R.E. Marsten (1978), "XMP: a structured library of subroutines for experimental mathematical programming", Technical Report 351, Department of Management Information Systems, University of Arizona at Tucson; (3.4).

R.E. Marsten and F. Shepardson (1978), "A double basis simplex method for linear programs with complicating variables", Technical Report 531, Department of Management Information Systems, University of Arizona at Tucson; (5.9).

R.E. Marsten: *see also* Shanno and Marsten (1979).

X.M. Martens: *see* Rijckaert and Martens (1978); Sarma, Martens, Reklaitis and Rijckaert (1978).

E.S. Marwil (1978), "Exploiting sparsity in Newton-like methods", Ph.D. thesis, Report TR-78-335, Department of Computer Science, Cornell University; (1.1).

D.Q. Mayne and E. Polak (1978), "A superlinearly convergent

algorithm for constrained optimization problems", Research Report 78-52, Department of Computing and Control, Imperial College, University of London; (4.3).

D.Q. Mayne: *see also* Polak, Trahan and Mayne (1979).

G.P. McCormick (1976), "Computability of global solutions to factorable nonconvex programs: part 1 — convex underestimating problems", *Math. Programming*, Vol. 10, pp. 147–175; (1.1), (5.1).

G.P. McCormick (1979), "A modification of Armijo's step-size rule for negative curvature", *Math. Programming*, Vol. 13, pp. 111–115; (1.1).

G.P. McCormick: *see also* Fiacco and McCormick (1968); Mancini and McCormick (1979).

R.A. McLean and G.A. Watson (1980), "Numerical methods for non-linear discrete ℓ_1 approximation problems", in *Numerical Methods of Approximation Theory: Excerpts of the Conference at Oberwolfach, 1979*, eds. L. Collatz, G. Meinardus and H. Werner, Birkhauser-Verlag (Basel); (2.4).

H.H.W. Mei: *see* Dennis and Mei (1979).

O.H. Merrill (1972), "Applications and extensions of an algorithm that computes fixed points of certain upper semi-continuous point to set mappings", Ph.D. dissertation, Department of Industrial Engineering, University of Michigan; (3.3).

G.H. Meyer (1968), "On solving nonlinear equations with a one parameter operator imbedding", *SIAM J. Numer. Anal.*, Vol. 5, pp. 739–752; (1.5).

R.R. Meyer: *see* Kao and Meyer (1981).

R. Mifflin (1977a), "Semi-smooth and semi-convex functions in constrained optimization", *SIAM J. Control Optim.*, Vol. 15, pp. 959–972; (2.3).

R. Mifflin (1977b), "An algorithm for constrained optimization with semi-smooth functions", *Math. Oper. Res.*, Vol. 2, pp. 191–207; (2.3).

R. Mifflin (1980), "A modification and extension of Lemaréchal's algorithm for nonsmooth minimization", Report 80-1, Department of Pure and Applied Mathematics, Washington State University, Pullman (to be published in *Math. Programming Stud.*); (2.3), (4.36).

R. Mifflin (1981), "A superlinearly convergent algorithm for one-dimensional constrained minimization problems with convex functions", Report TR-81-4, Department of Pure and Applied Mathematics, Washington State University, Pullman; (4.36).

R. Mifflin: *see also* Lemaréchal and Mifflin (1979, 1981).

W.L. Mirankar: *see* Chazan and Mirankar (1970).

J.H. Mize: *see* Kuester and Mize (1973).

O. Moharram: *see* Charalambous and Moharram (1978).

C.B. Moler: *see* Dongarra, Moler, Bunch and Stewart (1979).

A. Montalvo: *see* Levy and Montalvo (1979).

J.J. Moré (1978), "The Levenberg–Marquardt algorithm: implementation and theory", in *Numerical Analysis, Dundee 1977, Lecture Notes in Mathematics 630*, ed. G.A. Watson, Springer-Verlag (Berlin); (1.1), (1.4), (2.1), (6.1).

J.J. Moré (1979), "Implementation and testing of optimization software", in *Performance Evaluation of Numerical Software*, ed. L.D. Fosdick, North-Holland Publishing Co. (Amsterdam); (6.1).

J.J. Moré (1980), "On the design of optimization software", in *Optimazzazione Non-lineare e Applicazioni*, eds. S. Incerti and G. Treccani, Pitagora Editrice; (6.1).

J.J. Moré, B.S. Garbow and K.E. Hillstrom (1980), "User guide for MINPACK-1", Report ANL-80-74, Argonne National Laboratory, Illinois; (6.1).

J.J. Moré, B.S. Garbow and K.E. Hillstrom (1981a), "Testing unconstrained optimization software", *ACM Trans. Math. Software*, Vol. 7, pp. 17–41; (1.1), (6.1).

J.J. Moré, B.S. Garbow and K.E. Hillstrom (1981b), "Fortran subroutines for testing unconstrained optimization software", *ACM Trans. Math. Software*, Vol. 7, pp. 136–140; (6.1).

J.J. Moré and D.C. Sorensen (1979), "On the use of directions of negative curvature in a modified Newton method", *Math. Programming*, Vol. 16, pp. 1–20; (1.1), (1.4), (5.3).

J.J. Moré and D.C. Sorensen (1981), "Computing a trust region step", Report ANL-81-83, Argonne National Laboratory, Illinois; (1.4), (1.33), (6.1).

J.J. Moré: *see also* Broyden, Dennis and Moré (1973); Coleman and Moré (1981).

D.D. Morrison (1968), "Optimization by least squares", *SIAM J. Numer. Anal.*, Vol. 5, pp. 83–88; (4.45), (7.1).

T.S. Motzkin and I.J. Schoenberg (1954), "The relaxation method for linear inequalities", *Canad. J. Math.*, Vol. 6, pp. 393–404; (3.2).

H. Mukai and E. Polak (1976), "On the implementation of reduced gradient methods", in *Optimization Techniques: Modelling and Optimization in the Service of Man, Vol. 2*, ed. J. Cea, Springer-Verlag (Berlin); (4.70).

H. Mukai and E. Polak (1978), "On the use of approximations in algorithms for optimization problems with equality and inequality constraints", *SIAM J. Numer. Anal.*, Vol. 15, pp. 674–693; (4.70).

J.M. Mulvey (ed.) (1981), "Testing and validating mathematical programming software", in preparation; (7.2).

J.M. Mulvey: *see also* Crowder, Dembo and Mulvey (1979); Jackson and Mulvey (1978).

W. Murray (1969), "An algorithm for constrained minimization", in *Optimization*, ed. R. Fletcher, Academic Press (London); (2.4), (4.2).

W. Murray and M.L. Overton (1979), "Steplength algorithms for minimizing a class of nondifferentiable functions", *Computing*, Vol. 23, pp. 309–331; (2.4), (4.4).

W. Murray and M.L. Overton (1980), "A projected Lagrangian algorithm for nonlinear minimax optimization", *SIAM J. Sci. Statist. Comput.*, Vol. 1, pp. 345–370; (2.4), (2.46), (4.4).

W. Murray and M.L. Overton (1981), "A projected Lagrangian algorithm for nonlinear ℓ_1 optimization", *SIAM J. Sci. Statist. Comput.*, Vol. 2, pp. 207–224; (2.4), (4.4).

W. Murray and M.H. Wright (1978), "Projected Lagrangian methods based on the trajectories of penalty and barrier functions", Report SOL 78-23, Department of Operations Research, Stanford University; (4.2), (4.4), (4.38), (6.5).

W. Murray and M.H. Wright (1980), "Computation of the search direction in constrained optimization algorithms", Report SOL 80-2, Department of Operations Research, Stanford University (to be published in *Math. Programming Stud.*); (4.2), (4.17), (6.5).

W. Murray: *see also* Gill, Golub, Murray and Saunders (1974); Gill and Murray (1974a, 1974b, 1977a, 1977b, 1978a, 1978b, 1979);

Gill, Murray, Picken and Wright (1979); Gill, Murray and Saunders (1975); Gill, Murray and Wright (1981).

B.A. Murtagh and R.W.H. Sargent (1969), "A constrained minimization method with quadratic convergence", in *Optimization*, ed. R. Fletcher, Academic Press (London); (3.1).

B.A. Murtagh and M.A. Saunders (1978), "Large scale linearly constrained optimization", *Math. Programming*, Vol. 14, pp. 41—72; (3.1), (3.2), (3.4), (4.4), (4.5), (6.3).

B.A. Murtagh and M.A. Saunders (1980), "A projected Lagrangian algorithm and its implementation for sparse nonlinear constraints", Report SOL 80-1R, Department of Operations Research, Stanford University (to be published in *Math. Programming Stud.*); (4.4), (4.5), (4.17), (4.21), (5.1).

G.E. Myers (1968), "Properties of the conjugate gradient and Davidon methods", *J. Optim. Theory Appl.*, Vol. 2, pp. 209—219; (1.2).

W.C. Mylander (1974), "Processing nonconvex quadratic programming problems", Ph.D. thesis, Department of Operations Research, Stanford University; (3.8).

L. Nazareth (1979), "A relaionship between the BFGS and conjugate gradient algorithms and its implications for new algorithms", *SIAM J. Numer. Anal.*, Vol. 16, pp. 794—800; (1.2).

L. Nazareth (1980), "Some recent approaches to solving large residual nonlinear least squares problems", *SIAM Rev.*, Vol. 22, pp. 1—11; (6.1).

J.S. Newell and D.M. Himmelblau (1975), "A new method for nonlinearly constrained optimization", *AIChE J.*, Vol. 21, pp. 479—486; (6.4).

J. Nocedal (1980), "One and two step methods based on conic optimization models", presented at the SIAM 1980 Fall Meeting (Houston); (3.61).

J. Nocedal: *see also* Bjorstad and Nocedal (1979); Bourgeon and Nocedal (1981).

E.A. Nurminskii: *see* Lemaréchal and Nurminskii (1980).

D.P. O'Leary (1980), "A generalized conjugate gradient algorithm for solving a class of quadratic programming problems", *Linear Algebra Appl.*, Vol. 34, pp. 371—399; (3.22).

R.P. O'Neill (1981), "A comparison of real-world linear programs and their randomly-generated analogs", presented at the COAL Meeting on Testing and Validating Algorithms and Software (Boulder, Colorado); (7.30).

S.S. Oren (1974), "On the selection of parameters in self-scaling variable metric algorithms", *Math. Programming*, Vol. 7, pp. 351–367; (1.1).

J.M. Ortega and W.C. Rheinboldt (1970), *Iterative Solution of Nonlinear Equations in Several Variables*, Academic Press (New York); (3.3).

M.R. Osborne and G.A. Watson (1969), "An algorithm for minimax approximation in the nonlinear case", *Comput. J.*, Vol. 12, pp. 63–68; (2.4).

M.R. Osborne and G.A. Watson (1971), "On an algorithm for discrete nonlinear ℓ_1 approximation", *Comput. J.*, Vol. 14, pp. 184–188; (2.4).

M.R. Osborne: *see also* Anderson and Osborne (1977).

A.K. Osman-Hashim (1981), "An investigation of Yamashita's method for constrained optimization", Mathematics Degree Project Report, The Hatfield Polytechnic; (4.71).

L.J. Osterweil (1981), "TOOLPACK: an integrated system of tools for mathematical software development", presented at the COAL Meeting on Testing and Validating Algorithms and Software (Boulder, Colorado); (7.2).

L.J. Osterweil: *see also* Feiber, Taylor and Osterweil (1980).

L.M. Ostresh (1978), "On the convergence of a class of iterative methods for solving the Weber location problem", *Oper. Res.*, Vol. 26, pp. 597–609; (2.2).

M.L. Overton (1981), "A quadratically convergent method for minimizing a sum of Euclidean norms", Technical Report 030, Computer Science Department, Courant Institute, New York University; (4.4).

M.L. Overton: *see also* Murray and Overton (1979, 1980, 1981).

F. Palacios-Gomez, L.S. Lasdon and M. Engquist (1981), "Nonlinear optimization by successive linear programming" (to be published in *Management Sci.*); (5.1), (6.2).

C.Y. Pang: *see* Deo, Pang and Lord (1980).

V. Parisi: *see* Incerti, Parisi and Zirilli (1979, 1981).

B.N. Parlett: *see* Bunch and Parlett (1971).

H.C. Pedersen: *see* Chamberlain, Lemaréchal, Pedersen and Powell (1980).

A.F. Perold and G.B. Dantzig (1978), "A basis factorization method for block triangular linear programs", Technical Report SOL 78-7, Department of Operations Research, Stanford University; (5.9).

A. Perry (1978), "A modified conjugate gradient algorithm", *Oper. Res.*, Vol. 26, pp. 1073—1078; (1.2).

B.C. Peters and H.F. Walker (1978), "An iterative procedure for obtaining maximum likelihood estimates of the parameters for a mixture of normal distributions", *SIAM J. Appl. Math.*, Vol. 35, pp. 362—378; (2.1).

E.L. Peterson: *see* Duffin, Peterson and Zener (1967).

K.H. Phua: *see* Shanno and Phua (1976, 1978a, 1978b).

S.M. Picken: *see* Gill, Murray, Picken and Wright (1979).

D.A. Pierre and M.J. Lowe (1975), *Mathematical Programming via Augmented Lagrangians. An Introduction with Computer Programs*, Addison-Wesley Publishing Co. (Reading, Mass.); (6.4).

T. Pietrzykowski (1969), "An exact potential method for constrained maxima", *SIAM J. Numer. Anal.*, Vol. 6, pp. 294—304; (4.1), (4.4).

E. Polak (1971), *Computational Methods in Optimization — a Unified Approach*, Academic Press (New York); (4.70), (6.38).

E. Polak (1976), "On the global stabilization of locally convergent algorithms", *Automatica - J. IFAC*, Vol. 12, pp. 337—342; (4.40), (4.53).

E. Polak and G. Ribière (1969), "Note sur la convergence de methodes de directions conjuguées", *Rev. Française Informat. Recherche Opérationelle*, 3e Année, No. 16, pp. 35—43; (1.2).

E. Polak, R. Trahan and D.Q. Mayne (1979), "Combined phase I — phase II methods of feasible directions", *Math. Programming*, Vol. 17, pp. 61—73; (4.40).

E. Polak: *see also* Klessig and Polak (1973); Mayne and Polak

(1978); Mukai and Polak (1976, 1978).

B.T. Poljak (1964), "Some methods of speeding up the convergence of iteration methods", *U.S.S.R. Computational Math. and Math. Phys.*, Vol. 4, No. 5, pp. 1–17; (1.5).

B.T. Poljak (1967), "A general method of solving extremum problems", *Soviet Math. Dokl.*, Vol. 8, pp. 593–597; (2.3), (4.55).

B.T. Poljak (1969), "The conjugate gradient method in extremal problems", *U.S.S.R. Computational Math. and Math. Phys.*, Vol. 9, No. 4, pp. 94–112; (3.22).

R.A. Poljak: *see* Zuhovickii, Poljak and Primak (1963).

M.J.D. Powell (1969), "A method for nonlinear constraints in minimization problems", in *Optimization*, ed. R. Fletcher, Academic Press (London); (4.1), (4.3), (7.1).

M.J.D. Powell (1970a), "A new algorithm for unconstrained optimization", in *Nonlinear Programming*, eds. J.B. Rosen, O.L. Mangasarian and K. Ritter, Academic Press (New York); (1.1), (6.1).

M.J.D. Powell (1970b), "A hybrid method for nonlinear equations", in *Numerical Methods for Nonlinear Algebraic Equations*, ed. P. Rabinowitz, Gordon and Breach (London); (1.1), (6.1).

M.J.D. Powell (1971), "On the convergence of the variable metric algorithm", *J. Inst. Math. Appl.*, Vol. 7, pp. 21–36; (5.3).

M.J.D. Powell (1975), "Convergence properties of a class of minimization algorithms", in *Nonlinear Programming 2*, eds. O.L. Mangasarian, R.R. Meyer and S.M. Robinson, Academic Press (New York); (1.4), (5.3), (6.1).

M.J.D. Powell (1976), "Some global convergence properties of a variable metric algorithm for minimization without exact line searches", in *Nonlinear Programming, SIAM-AMS Proceedings Vol. IX*, eds. R.W. Cottle and C.E. Lemke, SIAM Publications (Philadelphia); (1.1).

M.J.D. Powell (1977), "Constrained optimization by a variable metric method", Report DAMTP 77/NA6, University of Cambridge; (4.2).

M.J.D. Powell (1978a), "A fast algorithm for nonlinearly constrained optimization calculations", in *Numerical Analysis, Dundee 1977, Lecture Notes in Mathematics 630*, ed. G.A. Watson, Springer-Verlag (Berlin); (2.4), (4.1), (4.2), (6.4), (6.9), (7.1).

M.J.D. Powell (1978b), "Algorithms for nonlinear constraints that use Lagrangian functions", *Math. Programming*, Vol. 14, pp. 224—248; (4.3), (4.4), (4.5).

M.J.D. Powell and Ph.L. Toint (1979), "On the estimation of sparse Hessian matrices", *SIAM J. Numer. Anal.*, Vol. 16, pp. 1060—1074; (1.1), (6.1).

M.J.D. Powell and Ph.L. Toint (1981), "The Shanno—Toint procedure for updating sparse symmetric matrices", Report 81/1, Department of Mathematics, University of Namur (to be published in *IMA J. Numer. Anal.*); (1.1).

M.J.D. Powell: *see also* Chamberlain, Lemaréchal, Pedersen and Powell (1980); Curtis, Powell and Reid (1974); Fletcher and Powell (1963).

M.E. Primak: *see* Zuhovickii, Poljak and Primak (1963).

B.N. Pschenichny (1970), "Algorithms for a general mathematical programming problem", *Kibernetika (Kiev)*, No. 5, pp. 120—125; (4.3).

B.N. Pschenichny and Y.M. Danilin (1975), *Méthodes Numériques dans les Problèmes d'Extremum* (French translation, 1977), *Numerical Methods in Extremal Problems* (English translation, 1978), M.I.R. Publishers (Moscow); (2.3), (4.3).

D.A. Pyne: *see* Byrd and Pyne (1979).

R.E. Quandt and J.B. Ramsey (1978), "Estimating mixtures of normal distributions and switching regressions", *J. Amer. Statist. Assoc.*, Vol. 73, pp. 730—752; (2.1).

R.E. Quandt: *see also* Goldfeld, Quandt and Trotter (1966).

K.M. Ragsdell (1981), "The evaluation of optimization software for engineering design", presented at the COAL Meeting on Testing and Validating Algorithms and Software (Boulder, Colorado); (7.4).

K.M. Ragsdell: *see also* Fattler, Sin, Root, Ragsdell and Reklaitis (1980); Gabriele and Ragsdell (1976, 1977); Sandgren and Ragsdell (1979).

J.B. Ramsey: *see* Quandt and Ramsey (1978).

C.R. Rao (1955), "Estimation and tests of significance in factor analysis", *Psychometrika*, Vol. 20, pp. 93—111; (2.12).

M.W. Ratner: *see* Lasdon, Waren, Jain and Ratner (1978); Lasdon,

Waren, Ratner and Jain (1975a, 1975b).

R.A. Redner (1980), "Maximum likelihood estimation for mixture models", Report SR-JO-04007-JSC-16832, NASA Space Center, Houston; (2.1).

R.A. Redner and H.F. Walker (1981), "Mixture densities, maximum likelihood, and the EM algorithm", in preparation; (2.1), (2.2).

C.M. Reeves: *see* Fletcher and Reeves (1964).

J.K. Reid: *see* Curtis, Powell and Reid (1974); Duff and Reid (1976, 1979).

C.H. Reinsch (1967), "Smoothing by spline functions", *Numer. Math.*, Vol. 10, pp. 177–183; (1.4).

C.H. Reinsch (1971), "Smoothing by spline functions II", *Numer. Math.*, Vol. 16, pp. 451–454; (1.4).

G.V. Reklaitis: *see* Fattler, Sin, Root, Ragsdell and Reklaitis (1980); Sarma, Martens, Reklaitis and Rijckaert (1978); Sarma and Reklaitis (1979).

W.C. Rheinboldt: *see* Ortega and Rheinboldt (1970).

G. Ribière: *see* Polak and Ribière (1969).

M.J. Rijckaert and X.M. Martens (1978), "Comparison of generalized geometric programming algorithms", *J. Optim. Theory Appl.*, Vol. 26, pp. 205–242; (7.1), (7.12).

M.J. Rijckaert: *see also* Avriel, Rijckaert and Wilde (1973); Sarma, Martens, Reklaitis and Rijckaert (1978).

K. Ritter (1979), "Local and superlinear convergence of a class of variable metric methods", *Computing*, Vol. 23, pp. 287–297; (1.1).

K. Ritter (1980), "On the rate of superlinear convergence of a class of variable metric methods", *Numer. Math.*, Vol. 35, pp. 293–313; (1.1).

K. Ritter (1981), "Global and superlinear convergence of a class of variable metric methods", *Math. Programming Stud.*, Vol. 14 (Mathematical Programming at Oberwolfach), pp. 178–205; (1.1).

K. Ritter: *see also* Best, Bräuninger, Ritter and Robinson (1981); Best and Ritter (1976).

F.D.K. Roberts: *see* Barrodale and Roberts (1973).

S.M. Robinson (1972), "A quadratically convergent algorithm for general nonlinear programming problems", *Math. Programming*, Vol. 3, pp. 145–156; (4.1), (4.4), (4.17), (4.21), (7.1).

S.M. Robinson (1974), "Perturbed Kuhn-Tucker points and rates of convergence for a class of nonlinear programming algorithms", *Math. Programming*, Vol. 7, pp. 1–16; (2.4), (3.2), (4.17), (7.1).

S.M. Robinson: *see also* Best, Bräuninger, Ritter and Robinson (1981); Jennrich and Robinson (1969).

R.T. Rockafellar (1970), *Convex Analysis*, Princeton University Press (Princeton, N.J.); (2.3).

R.T. Rockafellar (1973), "A dual approach to solving nonlinear programming problems by unconstrained optimization", *Math. Programming*, Vol. 5, pp. 354–373; (7.1).

R.T. Rockafellar (1976), "Solving a nonlinear programming problem by way of a dual problem", *Symposia Mathematica*, Vol. 19, pp. 135–160; (4.3).

R.R. Root (1977), "An investigation of the method of multipliers for engineering design", Ph.D. dissertation, Purdue University; (6.4).

R.R. Root: *see also* Fattler, Sin, Root, Ragsdell and Reklaitis (1980).

J.B. Rosen (1960), "The gradient projection method for nonlinear programming. Part I. Linear constraints", *J. Soc. Indust. Appl. Math.*, Vol. 8, pp. 181–217; (4.4).

J.B. Rosen (1961), "The gradient projection method for nonlinear programming. Part II. Nonlinear constraints", *J. Soc. Indust. Appl. Math.*, Vol. 9, pp. 514–532; (4.1).

J.B. Rosen (1978), "Two-phase algorithm for nonlinear constraint problems", in *Nonlinear Programming 3*, eds. O.L. Mangasarian, R.R. Meyer and S.M. Robinson, Academic Press (New York); (4.17), (4.21).

J.B. Rosen and J.L. Kreuser (1972), "A gradient projection algorithm for nonlinear constraints", in *Numerical Methods for Nonlinear Optimization*, ed. F.A. Lootsma, Academic Press (London); (4.17), (4.21).

D.B. Rubin: *see* Dempster, Laird and Rubin (1977).

H. Rubin: *see* Anderson and Rubin (1956).

D. Rufer (1978), "Users' guide for NLP — a subroutine package to solve nonlinear optimization problems", Report No. 78-07, ETH, Zürich; (6.4).

B.G. Ryder (1973), "The FORTRAN verifier: user's guide", Computing Science Report 12, Bell Telephone Laboratories, Murray Hill; (7.2).

B.G. Ryder (1974), "The PFORT verifier", *Software Practice and Experience*, Vol. 4, pp. 359—377; (6.1).

B.G. Ryder and A.D. Hall (1979), "The PFORT verifier", Computing Science Report 12 (revised April, 1979), Bell Telephone Laboratories, Murray Hill; (6.1).

Th.L. Saaty (1977), "A scaling method for priorities in hierarchical structures", *J. Math. Psych.*, Vol. 15, pp. 234—281; (7.1).

Th.L. Saaty (1980), *The Analytic Hierarchy Process*, McGraw-Hill Book Co. (New York); (7.1).

R. Saigal (1977), "On the convergence rate of algorithms for solving equations that are based on methods of complementary pivoting", *Math. Oper. Res.*, Vol. 2, pp. 108—124; (3.3).

R. Saigal (1979), "The fixed point approach to nonlinear programming", *Math. Programming Stud.*, Vol. 10 (Point-to-Set Maps and Mathematical Programming), pp. 142—157; (3.3).

E. Sandgren (1977), "The utility of nonlinear programming algorithms", Ph.D. thesis, Department of Mechanical Engineering, Purdue University; (7.1), (7.4).

E. Sandgren (1981), "A statistical review of the Sandren—Ragsdell comparative study", presented at the COAL Meeting on Testing and Validating Algorithms and Software (Boulder, Colorado); (7.4).

E. Sandgren and K.M. Ragsdell (1979), "On some experiments which delimit the utility of nonlinear programming methods for engineering design", presented at the Tenth International Symposium on Mathematical Programming (Montreal); (7.1).

R.W.H. Sargent (1974), "Reduced gradient and projection methods for nonlinear programming", in *Numerical Methods for Constrained Optimization*, eds. P.E. Gill and W. Murray, Academic Press (London); (3.7).

R.W.H. Sargent (1978), "An efficient implementation of the Lemke algorithm and its extension to deal with upper and lower bounds", *Math. Programming Stud.*, Vol. 7 (Complementarity and Fixed Point Problems), pp. 36—54; (3.7).

R.W.H. Sargent (1981), "Recursive quadratic programming algorithms and their convergence properties", presented at the Third IIMAS Workshop on Numerical Analysis (Cocoyoc, Mexico); (4.65).

R.W.H. Sargent: *see also* Murtagh and Sargent (1969).

P.V. Sarma, X.M. Martens, G.V. Reklaitis and M.J. Rijckaert (1978), "A comparison of computational strategies for geometric programs", *J. Optim. Theory Appl.*, Vol. 26, pp. 185—203; (7.1).

P.V. Sarma and G.V. Reklaitis (1979), "Optimization of a complex chemical process using an equation oriented model", presented at the Tenth International Symposium on Mathematical Programming (Montreal); (4.5).

M.A. Saunders: *see* Gill, Golub, Murray and Saunders (1974); Gill, Murray and Saunders (1975); Murtagh and Saunders (1978, 1980).

P.B. Saunders: *see* Crowder and Saunders (1980).

H.E. Scarf and T. Hansen (1973), *The Computation of Economic Equilibria*, Yale University Press (New Haven); (3.3).

H. Scheffé (1959), *The Analysis of Variance*, John Wiley & Sons (New York); (7.1).

K. Schittkowski (1980a), "A numerical comparison of optimization software using randomly generated test problems", in *Numerical Optimization of Dynamic Systems*, eds L.C.W. Dixon and G.P. Szegö, North-Holland Publishing Co. (Amsterdam); (4.2).

K. Schittkowski (1980b), *Nonlinear Programming Codes, Lecture Notes in Economics and Mathematical Systems 183*, Springer-Verlag (Berlin); (6.4), (7.1), (7.4).

K. Schittkowski (1981a), "Nonlinear programming methods with linear least squares subproblems", presented at the COAL Meeting on Testing and Validating Algorithms and Software (Boulder, Colorado); (4.2), (6.4), (7.4).

K. Schittkowski (1981b), "A model for the performance evaluation in comparative studies", presented at the COAL Meeting on Testing and Validating Algorithms and Software (Boulder, Colorado); (7.4).

K. Schittkowski: *see also* Hock and Schittkowski (1981).

R.B. Schnabel (1977), "Analysing and improving quasi-Newton methods for unconstrained optimization", Ph.D. thesis, Report

TR-77-320, Department of Computer Science, Cornell University; (5.3).

R.B. Schnabel (1981a), "Unconstrained minimization using conic models and second derivatives", in preparation; (1.1).

R.B. Schnabel (1981b), "Determining feasibility of a set of nonlinear inequality constraints" (to be published in *Math. Programming Stud.*); (3.45).

R.B. Schnabel: *see also* Dennis and Schnabel (1979, 1981, 1982); Frank and Schnabel (1981).

I.J. Schoenberg: *see* Motzkin and Schoenberg (1954).

L.K. Schubert (1970), "Modification of a quasi-Newton method for nonlinear equations with a sparse Jacobian", *Math. Comp.*, Vol. 24, pp. 27–30; (6.1).

L.F. Shampine and M.K. Gordon (1975), *Computer Solution of Ordinary Differential Equations*, W.H. Freeman (San Francisco); (6.5).

D.F. Shanno (1970), "Conditioning of quasi-Newton methods for function minimization", *Math. Comp.*, Vol. 24, pp. 647–656; (1.1).

D.F. Shanno (1978), "Conjugate gradient methods with inexact searches", *Math. Oper. Res.*, Vol. 3, pp. 244–256; (1.1), (1.2), (3.4).

D.F. Shanno (1980), "On variable metric methods for sparse Hessians", *Math. Comp.*, Vol. 34, pp. 499–514; (1.1), (5.3).

D.F. Shanno (1981), "On preconditioned conjugate gradient methods", in *Nonlinear Programming 4*, eds. O.L. Mangasarian, R.R. Meyer and S.M. Robinson, Academic Press (New York); (1.2).

D.F. Shanno and R.E. Marsten (1979), "Conjugate gradient methods for linearly constrained nonlinear programming", Technical Report 79-13, Department of Management Information Systems, University of Arizona at Tucson; (3.4), (4.5), (4.62).

D.F. Shanno and K.H. Phua (1976), "Minimization of unconstrained multivariate functions", *ACM Trans. Math. Software*, Vol. 2, pp. 87–94; (5.26).

D.F. Shanno and K.H. Phua (1978a), "Matrix conditioning and nonlinear optimization", *Math. Programming*, Vol. 14, pp. 149–160; (1.1).

D.F. Shanno and K.H. Phua (1978b), "Numerical comparison of several variable metric algorithms", *J. Optim. Theory Appl.*, Vol. 25, pp. 507–518; (1.1).

F. Shepardson: *see* Marsten and Shepardson (1978).

N.Z. Shor (1970), "Utilization of the operation of space dilation in the minimization of convex functions", *Kybernetika (Kiev)*, Vol. 6, pp. 6—12 (translated in *Cybernetics*, Vol. 6, pp. 7—15); (3.2).

N.Z. Shor (1977), "New development trends in nondifferentiable optimization", *Kibernetika (Kiev)*, Vol. 13, pp. 87—91 (translated in *Cybernetics*, Vol. 13, pp. 881—886); (2.27).

Y.T. Sin: *see* Fattler, Sin, Root, Ragsdell and Reklaitis (1980).

J.W. Sinclair: *see* Bartels, Conn and Sinclair (1978); Conn and Sinclair (1975).

S. Smale (1976), "A convergent process of price adjustment and global Newton methods", *J. Math. Econom.*, Vol. 3, pp. 107—120; (1.5).

S. Smale: *see also* Hirsch and Smale (1979).

R.E. Small: *see* Beale, Hughes and Small (1965).

T. Smedsaas: *see* Engquist and Smedsaas (1980).

B.T. Smith (1977), "Fortran poisoning and antidotes", in *Portability of Numerical Software, Lecture Notes in Computer Science 57*, ed. W.R. Cowell, Springer-Verlag (Berlin); (6.1).

B.T. Smith, J.M. Boyle and W.J. Cody (1974), The NATS approach to quality software", in *Software for Numerical Mathematics*, ed. D.J. Evans, Academic Press (London); (6.5).

D. Sokolowsky: *see* Hoffman, Mannos, Sokolowsky and Wiegmann (1953).

D. Solow (1980), "Comparative computer results of a new complementary pivot algorithm for solving equality and inequality constrained optimization problems", *Math. Programming*, Vol. 18, pp. 169—185; (3.3).

D. Solow: *see also* Gochet, Loute and Solow (1974).

D.C. Sorensen (1977), "Updating the symmetric indefinite factorization with applications in a modified Newton's method", Report ANL-77-49, Argonne National Laboratory, Illinois; (3.2).

D.C. Sorensen (1980a), "The Q-superlinear convergence of a collinear scaling algorithm for unconstrained optimization", *SIAM J. Numer. Anal.*, Vol. 17, pp. 84—114; (1.1), (1.3), (1.17).

D.C. Sorensen (1980b), "Newton's method with a model trust-region modification", Report ANL-80-106, Argonne National Laboratory, Illinois (to be published in *SIAM J. Numer. Anal.*); (1.1), (1.4), (6.1).

D.C. Sorensen (1981), "An example concerning quasi-Newton estimation of a sparse Hessian", *SIGNUM Newsletter*, Vol. 16, No. 2, pp. 8–10; (6.1).

D.C. Sorensen: *see also* Moré and Sorensen (1979, 1981).

B. Speelpenning (1980), "Computing fast derivatives of functions given by algorithms", Report R-80-1002, Department of Computer Science, University of Illinois at Urbana; (6.1).

R.L. Staha and D.M. Himmelblau (1973), "Evaluation of constrained nonlinear programming techniques", Technical Report No. 11, Department of Chemical Engineering, University of Texas; (7.1).

T. Steihaug (1980), "Quasi-Newton methods for large scale nonlinear problems", Ph.D. thesis, Report 49, School of Organization and Management, Yale University; (6.1).

T. Steihaug (1981), "The conjugate gradient method and trust regions in large scale optimization", Technical Report 81-1, Department of Mathematical Sciences, Rice University; (1.1).

T. Steihaug: *see also* Dembo, Eisenstat and Steihaug (1980); Dembo and Steihaug (1980).

G.W. Stewart: *see* Dongarra, Moler, Bunch and Stewart (1979).

R.A. Stewart: *see* Griffith and Stewart (1961).

E. Stiefel (1960), "Note on Jordan elimination, linear programming and Tchebycheff approximation", *Numer. Math.*, Vol. 2, pp. 1–17; (2.4).

E. Stiefel: *see also* Hestenes and Stiefel (1952).

R.B. Stillman: *see* Lyon and Stillman (1974).

K.A. Stordahl (1980), "Unconstrained minimization using conic models and exact second derivatives", M.Sc. thesis, Department of Computer Science, University of Colorado at Boulder; (1.1).

T.A. Straeter (1973), "A parallel variable metric optimization algorithm", NASA Technical Note D-7329, Langley Research Center; (7.3), (7.33).

T.A. Straeter and A.T. Markos (1975) "A parallel Jacobson–

Oksman optimization algorithm", NASA Technical Note D-8020, Langley Research Center; (7.3), (7.33).

J.J. Strodiot: *see* Lemaréchal, Strodiot and Bihain (1981).

G.P. Szegö: *see* Dixon and Szegö (1975, 1978).

R.A. Tapia (1977), "Diagonalized multiplier methods and quasi-Newton methods for constrained optimization", *J. Optim. Theory Appl.*, Vol. 22, pp. 135—194; (4.2), (4.4).

R.A. Tapia and J.R. Thompson (1978), *Nonparametric Probability Density Estimation,* Johns Hopkins University Press (Baltimore); (2.1).

R.N. Taylor: *see* Feiber, Taylor and Osterweil (1980).

L.S. Thakur (1978), "Error analysis for convex separable programs: the piecewise linear approximation and the bounds on the optimal objective value", *SIAM J. Appl. Math.*, Vol. 34, pp. 704—714; (5.45).

M.N. Thapa (1980), "Optimization of unconstrained functions with sparse Hessian matrices", Ph.D. thesis, Department of Operations Research, Stanford University; (1.1).

J.R. Thompson: *see* Tapia and Thompson (1978).

M.J. Todd (1976), *The Computation of Fixed Points and Applications*, Springer-Verlag (Berlin); (3.3).

M.J. Todd (1980), "A quadratically convergent fixed point algorithm for economic equilibria and linearly constrained optimization", *Math. Programming*, Vol. 18, pp. 111—126; (3.3).

M.J. Todd: *see also* Awoniyi and Todd (1981).

Ph.L. Toint (1977), "On sparse and symmetric matrix updating subject to a linear equation", *Math. Comp.*, Vol. 31, pp. 954—961; (1.1), (5.3), (6.1).

Ph.L. Toint (1978), "Some numerical results using a sparse matrix updating formula in unconstrained optimization", *Math. Comp.*, Vol. 32, pp. 839—851; (6.1).

Ph.L. Toint (1979), "On the superlinear convergence of an algorithm for solving a sparse minimization problem", *SIAM J. Numer. Anal.*, Vol. 16, pp. 1036—1045; (1.4).

Ph.L. Toint (1980), "A sparse quasi-Newton update derived variationally with a non-diagonally weighted Frobenius norm", Report 80/1, Department of Mathematics, University of Namur (to be published in *Math. Comp.*); (1.1).

Ph.L. Toint (1981), "Towards an efficient sparsity exploiting Newton method for minimization", in *Sparse Matrices and their Uses*, ed. I. Duff, Academic Press (London); (1.2).

Ph.L. Toint: *see also* Griewank and Toint (1981a, 1981b); Powell and Toint (1979, 1981).

J.W. Tolle: *see* Boggs and Tolle (1980, 1981); Boggs, Tolle and Wang (1979).

J.A. Tomlin: *see* Beale and Tomlin (1970).

R. Trahan: *see* Polak, Trahan and Mayne (1979).

H.F. Trotter: *see* Goldfeld, Quandt and Trotter (1966).

H. Uzawa: *see* Arrow, Hurwicz and Uzawa (1958).

G. Van der Hoek (1980), *Reduction Methods in Nonlinear Programming*, Mathematisch Centrum (Amsterdam); (4.2), (7.4).

K.S. Vastola: *see* Bertsekas, Gafni and Vastola (1979).

M. Vidyasagar: *see* El-Attar, Vidyasagar and Dutta (1979).

H.F. Walker: *see* Dennis and Walker (1980); Peters and Walker (1978); Redner and Walker (1981).

P. Wang: *see* Boggs, Tolle and Wang (1979).

A.D. Waren and L.S. Lasdon (1979), "The status of nonlinear programming software", *Oper. Res.*, Vol. 27, pp. 431–456; (6.2), (7.1).

A.D. Waren: *see also* Lasdon and Waren (1978, 1980); Lasdon, Waren, Jain and Ratner (1978); Lasdon, Waren, Ratner and Jain (1975a, 1975b).

H. Watanabe: *see* Ishizaki and Watanabe (1968).

G.A. Watson (1979), "The minimax solution of an overdetermined system of nonlinear equations", *J. Inst. Math. Appl.*, Vol. 23, pp. 167–180; (2.4).

G.A. Watson (1981), "An algorithm for linear ℓ_1 approximation of continuous functions", *IMA J. Numer. Anal.*, Vol. 1, pp. 157–167; (4.4).

G.A. Watson: *see also* Fletcher and Watson (1980); McLean and Watson (1980); Osborne and Watson (1969, 1971).

L.T. Watson (1979), "An algorithm that is globally convergent with probability one for a class of nonlinear two-point boundary value problems", *SIAM J. Numer. Anal.*, Vol. 16, pp. 394–401; (3.39).

L.T. Watson (1980), "Solving finite difference approximations to nonlinear two-point boundary value problems by a homotopy method", *SIAM J. Sci. Statist. Comput.*, Vol. 1, pp. 467–480; (1.5), (6.1).

D.G. Watts: *see* Bates and Watts (1980).

B.E. Weiss (1980), "A modular software package for solving unconstrained nonlinear optimization problems", M.Sc. thesis, Department of Computer Science, University of Colorado at Boulder; (1.1), (1.7).

R.E. Welsch: *see* Dennis, Gay and Welsch (1981).

N. Wiegmann: *see* Hoffman, Mannos, Sokolowsky and Wiegmann (1953).

A.P. Wierzbicki (1978), "Lagrangian functions and nondifferentiable optimization", Report WP-78-63, IIASA, Laxenburg, Austria; (2.4).

D.J. Wilde: *see* Avriel, Rijckaert and Wilde (1973).

R. Wilson: *see* Manne, Chao and Wilson (1980).

R.B. Wilson (1963), "A simplicial algorithm for concave programming", Ph.D. thesis, Graduate School of Business Administration, Harvard University; (2.4), (4.1), (4.2), (6.4).

P. Wolfe (1963), "Methods of nonlinear programming", in *Recent Advances in Mathematical Programming*, eds. R.L. Graves and P. Wolfe, McGraw-Hill Book Co. (New York); (3.6).

P. Wolfe (1967), "Methods of nonlinear programming", in *Nonlinear Programming*, ed. J. Abadie, North-Holland Publishing Co. (Amsterdam); (4.5).

P. Wolfe (1969), "Convergence conditions for ascent methods", *SIAM Rev.*, Vol. 11, pp. 226–235; (1.1).

P. Wolfe (1971), "Convergence conditions for ascent methods II: some corrections", *SIAM Rev.*, Vol. 13, pp. 185–188; (1.1).

P. Wolfe (1975), "A method of conjugate subgradients for minimizing nondifferentiable functions", *Math. Programming Stud.*, Vol. 3 (Nondifferentiable Optimization), pp. 145–173; (2.3).

P. Wolfe: *see also* Held, Wolfe and Crowder (1974).

R.S. Womersley (1981), "Numerical methods for structured problems in nonsmooth optimization", Ph.D. thesis, Department of Mathematics, University of Dundee; (4.1).

M.H. Wright (1976), "Numerical methods for nonlinearly constrained optimization", Ph.D. thesis, Report CS-76-566, Department of Computer Science, Stanford University; (2.4), (4.38), (6.5).

M.H. Wright (1978), "A survey of available software for nonlinearly constrained optimization", Technical Report SOL 78-4, Department of Operations Research, Stanford University; (6.2).

M.H. Wright: *see also* Gill, Murray, Picken and Wright (1979); Gill, Murray and Wright (1981); Murray and Wright (1978, 1980).

H. Yamashita (1979), "A globally convergent quasi-Newton method for equality constrained optimization that does not use a penalty function", Technical Report, ONO Systems Ltd, Tokyo; (4.71).

J.A. Yorke: *see* Kellogg, Li and Yorke (1977).

W.I. Zangwill (1967a), "Nonlinear programming via penalty functions", *Management Sci.*, Vol. 13, pp. 344–358; (4.4).

W.I. Zangwill (1967b), "An algorithm for the Chebyshev problem with an application to concave programming", *Management Sci.*, Vol. 14, pp. 58–78; (2.4).

W.I. Zangwill: *see also* Garcia and Zangwill (1980).

C. Zener: *see* Duffin, Peterson and Zener (1967).

F. Zirilli (1981), "The solution of nonlinear systems of equations by second order systems of o.d.e. and linearly implicit A-stable techniques" (to be published in *SIAM J. Numer. Anal.*); (1.5).

F. Zirilli: *see also* Aluffi, Incerti and Zirilli (1980a, 1980b); Incerti, Parisi and Zirilli (1979, 1981).

S.I. Zuhovickii, R.A. Poljak and M.E. Primak (1963), "An algorithm for the solution of the problem of Cebysev approximation", *Soviet Math. Dokl.*, Vol. 4, pp. 901–904; (2.4).

Index